D0934453

# THE STORY OF ASTRONOMY

# THE STORY OF ASTRONOMY

Lloyd Motz
and
Jefferson Hane Weaver

PLENUM PRESS • NEW YORK AND LONDON

Library of Congress Cataloging-in-Publication Data

On file

ISBN 0-306-45090-9


Plenum Press is a Division of Plenum Publishing Corporation
233 Spring Street, New York, N.Y. 10013-1578

10 9 8 7 6 5 4 3 2 1



Printed in the United States of America

# Preface

Of all the scientific disciplines Astronomy stands out as the all-encompassing one in that all the other sciences grew out of astronomy and are still influenced by it. A history of astronomy, to be complete, should trace these interscience relationships to some extent, but the story of astronomy is not subject to these constraints. By the very designation as a "story," this book was designed, and so written, to delineate the high points of astronomy and to trace the evolution of the great astronomical ideas from their birth as pure speculations in the minds of the great astronomers of the past to their present fully developed and fully accepted state.

This, of course, entails a fuller discussion of the astronomers themselves than one might find in a history of astronomy or in a straightforward treatise in astronomy. We have emphasized this phase of the story of astronomy in this book. This has been very fruitful and revealed the intimate relationship among the sciences: particularly astronomy, physics, and mathematics.

One can hardly speak of Kepler, Galileo, or Newton without describing their holistic approaches to astronomy. Thus Galileo considered himself a mathematician who speculated all phases of nature and Newton was the great polymath who contributed to all phases of mathematics, astronomy, and physics, seeking in each of these disciplines the theoretical bridges to the others, and the great nineteenth century mathematician Carl Friedrich Gauss greatly enriched astronomy by applying his great mathematical skill to the solution of complex astronomical problems. And so it went, culminating in Einstein's great theoretical discoveries of the photon and the theory of relativity that ushered in our current rational approach to cosmology.

Our concern in writing this book was to see how close we could come to this ideal and to present to the reader one of the most exciting stories in the history of civilization. We hope that we have succeeded.

Lloyd Motz
Jefferson Hane Weaver

# Contents

CHAPTER 1 The Origins of Astronomy *1*

CHAPTER 2 The Ancient Cosmologies *13*

CHAPTER 3 The Greek Philosophers and the Early Greek Astronomers *21*

CHAPTER 4 From Aristarchus to Ptolemy: The Birth of Accurate Observational Astronomy *39*

CHAPTER 5 The Revival of European Astronomy *57*

CHAPTER 6 Tycho Brahe and Johannes Kepler *69*

CHAPTER 7 Galileo, the Astronomical Telescope, and the Beginning of Modern Astronomy *89*

CHAPTER 8 The Newtonian Era *107*

CHAPTER 9 The Rise of Modern Astronomy *129*

CHAPTER 10 Post-Newtonian Astronomy *143*

CHAPTER 11 The Beginning of the New Age of Astronomy: Beyond the Solar System *159*

CHAPTER 12 Astronomy as a Branch of Physics *185*

CHAPTER 13 The New Physics and Its Impact on Astronomy *205*

CHAPTER 14 Relativity and Astronomy *233*

CHAPTER 15 The Origin and Development of Astrophysics *257*

CHAPTER 16 Stellar Evolution and the Beginning of Galactic Astronomy *279*

CHAPTER 17 Beyond the Stars: The Galaxies *309*

CHAPTER 18 Cosmology *331*

Epilogue *353*

Bibliography *361*

Index *363*

# THE STORY OF ASTRONOMY

CHAPTER 1

# The Origins of Astronomy

*If I had been present at the creation, I would have given some useful hints for the better arrangement of the universe.*

—ALFONSO THE WISE, KING OF CASTILE

Astronomy, as an orderly pursuit of knowledge about the heavenly bodies and the universe, did not begin in one moment at some particular epoch in a single society. Every ancient society had its own concept of the universe (cosmology) and of humanity's relationship to the universe. In most cases, these concepts were certainly molded by three forces: theology (religion), nature (climate, floods, winds, natural disasters), and the assumed influence of the stars and planets on the fortunes and fate of people and their societies (astrology).

Because theology deals with the creation of the universe and everything in it, the various religions were quite naturally the precursors of the ancient astronomies. To the ancients the apparent division of their universe into water, land, and sky pointed to a creator or creators who could dwell primarily in the sky. The study of the sky became an important phase of religion around the world. The astronomy that stemmed from these studies was, of course, extremely primitive. Gods were believed by many ancient societies to inhabit not only the heavens but also the highest mountains and the deepest oceans.

We can see the profound influence of religion on the development of astronomy most clearly when we consider that from the time of the Babylonians to the Roman era, astronomical knowledge and the management of the calendar were confined to the priesthood in most cultures. The Babylonians believed the heavens, the dwellings of the gods, was a "solid vault" with its foundations supported by the oceans and that the earth was a huge island at the very center. These ideas are also found in the Old Testament, with one God envisioned as playing the dominant role as the creator and the regulator of earth, sky, and water. In many of these primitive cosmologies, the earth was distinct from the oceans, which were described as having been formed first, with the "earth arising later from the waters."

One of the most famous theological cosmogonies is contained in the book of Genesis, with time, space, and matter beginning with "God's creation of the heavens, and the earth." Interestingly, the oceans are not specifically stated as having been created; rather the earth is pictured as spread out over the waters, which were "divided" to accommodate the creation of the earth. The stars were described as the "host of heaven" which probably meant the soldiers or angels of God.

The acceptance or assumption of the concept of the creation of the universe implies the acceptance of the existence of a void before that creation. The book of Genesis implies such a void when it states that God then placed the sun and moon "in the firmament." Ancient Egyptian theology, by contrast, did not accept the concept of a void but pictured the universe as a boundless manifold of water. The common elements in most of these theological concepts of the universe, however, were the initial void, a creator, and the act of creation. This led inevitably to the concept of the immutability of all elements of the universe. If God (or several gods) had created the universe, it would have had to remain exactly as God had created it. This concept of immutability was carried over to the Christian theology which dominated Western astronomical thinking until the time of Galileo; this theology allowed no room for any astronomical phenomena that differed even in a minute way from the original creation and ultimately retarded progress in astronomy. We note that this astronomically restrictive feature was, and to a great extent, still is, characteristic of all theologies. Fortunately, it no longer has the power to curtail astronomical progress as it did in ancient and medieval times.

The influence of nature (natural forces) on the development of astronomy became more important as societies grew and the contributions of individuals to the well-being of society became increasingly specialized. Specific examples of this phenomenon come immediately to mind. Although most of the people in any particular society may not have been overly concerned about climate, the food-producing population (the farming and animal husbandry section) was deeply concerned about it and ultimately turned to and depended upon the astronomer for information about the changing seasons. In the earliest years, much of this information was probably guesswork, but in time it became apparent that careful records needed to be kept; these records were the precursor of the modern precise calendar. Although, strictly speaking, the construction of a calendar is not a branch of astronomy, it owes much to astronomy and astronomers.

The importance of the understanding of natural forces to the development or understanding of astronomy, and the importance of astronomy to the understanding of natural forces, are dramatically illustrated by the annual flooding of the Nile, which is of utmost importance to Egyptian agriculture. The ancient Egyptians knew that the flooding of the Nile occurs when the star Sirius is first visible on the eastern horizon shortly after the sun sets. Arriving at this conclusion required careful study of the rising and setting of the stars and the Egyptians eventually learned to track and record the positional astronomy of the stars. Had this study been pursued faithfully by the early Egyptian stargazers, they might have discovered that this heliacal rising of Sirius occurred at slightly different times as the years advanced and might thus have discovered the precession of the equinoxes, which was credited to the Greek astronomer, Hipparchus, hundreds of years later.

Because the ebb and flow of the tides were of great importance to the marine activities of the ancient coastal civilizations, their skywatchers must have become aware of the relationship of the high and low tides to the positions of the sun and moon in the sky. This knowledge could have prompted these ancient observers to pursue astronomy more assiduously than they might otherwise have done. Whether or not this effort influenced the development of astronomy to any great extent we cannot say, but it certainly had some influence on it.

The flow of time and its measurement probably influenced the development of astronomy more than any other natural phenomenon. Before the first clock and watches were introduced, various crude devices such as the hourglass and water clocks were used to measure short periods of time; these periods were measured by the sun and stars moving across the sky. The length of the day measured by these apparent motions was divided into basic equal intervals which later became the hour. The observation that the length of the day is related to the rising and setting of the sun led to the concept of the noon hour, and the division of the day into morning and afternoon as determined by the sundial, which was the earliest timekeeper, and was probably introduced by the Chaldean astronomer Berossus in the third century BC. The evening hours were measured by the rising and setting of the stars. Thus the accurate measurement of short time intervals, before the introduction of clocks, demanded accurate monitoring of the rising and setting of the sun and stars. As an example of how this led to new astronomical concepts we note that solar noon was defined as the moment the sun was at its highest point above the observer's horizon (on a great celestial circle called the observer's meridian), which differs from observer to observer. Before the sun reached the observer's meridian—the morning period—the hours were referred to as antemeridian (AM); the period after the sun passed the meridian was called postmeridian or PM (afternoon). These designations are still used, though most people do not know why AM refers to morning hours and PM to afternoon hours.

The need to keep accurate records of daily events and commitments (appointments, work times, etc.) ultimately led to the invention of clocks and watches. Before such inventions, however, all daily activities were governed by the apparent positions of the sun, which led to the concept of "solar time" and the "solar day." The solar day, which is still used, is defined as the time interval between two successive passages of the sun across the observer's meridian. The solar day was then divided into exactly 24 equal intervals, called solar hours. Each hour was divided into 60 equal parts called minutes, and each minute was in turn divided into 60 equal intervals called seconds. Solar time is still used today to guide and order our daily activities.

Even the earliest stargazers, the forerunners of the astronomers, knew that solar time cannot be used to follow the stars, and so they introduced star time or "sidereal time," which is used by astronomers today. Just as solar time is based on a solar day, sidereal time is based on a sidereal day which is defined as the time interval between successive passages of a given star, such as Sirius, across the observer's meridian. This is called the "sidereal day" which is 3 minutes, 56 seconds shorter than the solar day. The difference between the length of the solar day and the sidereal day stems from the apparent eastward motion of the sun in the sky. The skywatchers noted that if the sun rose in the east on any given day just when some particular star was setting, then the next day the sun rose about 4 minutes later than that same star set on the same day. These early stargazers described this phenomenon in a general way by saying that the stars rise 4 minutes earlier than the sun. Here we see how timekeeping became an astronomical activity so that astronomy acquired practicality and, in a sense, became an "applied science" before it became a true science.

The early skywatchers and astronomers explained the difference between solar and sidereal time by assigning a real eastward slow motion to the sun on the sky. They then noted that the sun moves about 1 degree per day eastward on the sky, thus traversing 12 different constellations along its apparent "path." These constellations, called the "signs of the zodiac," have names which the ancient Greeks took from the Hindu astronomers after Alexander the Great acquainted the Greeks with Hindu astronomy. Discovering the apparent motion of the sun among the constellations was the beginning of the development of calendars and contributed considerably to the origin of astronomy.

The demands of accuracy in keeping daily time required very careful monitoring of the sun's apparent eastward motion, which showed that the rate of this motion varies from season to season; it is slowest during the summer months and fastest during the winter months. The ancient observers had no idea as to the cause of this variable apparent motion of the sun, which, as previously noted, they interpreted as a real motion, and which they described in terms of a solar circular orbit around the earth. This was the beginning of the geocentric model of the solar system, which was given its most sophisticated exposition by the Egyptian Ptolemy at the beginning of the Christian era.

As civilizations developed from primitive tribal organizations into complex interrelated societies, the need for methods and devices for keeping records of societal activities over long periods of time grew, ultimately becoming societal imperatives. These needs were met by the invention of the calendar, which appeared in various forms in different societies. Because all calendars divide the year into smaller time intervals, knowledge of the length of the year in days was absolutely essential for constructing a calendar. Astronomy thus became indispensable to calendar makers. Because the same phases of the moon reappear periodically, this periodicity—the month—became the basis of all calendars going back as far as the Babylonians.

Such calendars could be constructed only if the length of the year and the length of the month were fairly accurately known. Determining these periods seems to be a fairly simple procedure to the casual student as it simply requires counting the number of days between two successive appearances of the sun in the same stellar constellation in one case, and the number of days between two successive appearances of the same phase of the moon, in the other case. This would, indeed, be fairly simple if the number of days in the year were an exact integer and if that were also true for the length of the month. But neither of these is true, so the length of the year and that of the month must be expressed in integers plus fractions of a day. This, of course, requires very accurate observations of the apparent motions of the sun and moon. Because no fractional days appear in a calendar, a question that naturally arises is how a calendar keeps in step with very long passages of time.

The very earliest calendars were based on "year lengths" ranging from 354 to 365 days. The accurate length of the sidereal year in solar time units as measured today, with modern astronomical methods, is 365 solar days, 5 hours, 48 minutes, and 45.5 seconds.

Here, however, we distinguish between the sidereal year and the tropical or seasonal year, which is the time interval between two successive beginnings of spring (two successive passages of the sun across the vernal equinox). The tropical year is slightly shorter than the sidereal year—the length of the sidereal year is 365.256 days and the length of the seasonal or tropical year is 365.242 days.

Because the changes of phase of the moon made it easy to divide the year into lunar months, the earliest calendars were lunisolar. As the moon revolves around the earth once every 27.32 days (the sidereal period of the moon, or the sidereal month) one might think that the calendar year should be divided into about 13 such months. This unit would then be called a lunar year. But the months in present-day calendars are not sidereal months but synodic months, each of which—the time between two identical phases of the moon—is 29 days, 8 hours.

The Babylonians introduced a lunisolar calendar based on 12 lunar months, each of 30 days, which added up to 360 days. To keep the calendar in step with the season, they added months whenever necessary. The ancient Egyptians were the first to construct and use a solar calendar with a 365-day year, which did not refer to the moon. This year was divided into 12 months, each 30 days long; five extra days were added to this calendar in Egyptian chronology. The Egyptians may have known that the length of the year is 365 days, 6 hours because King Ptolemy III, in 238 BC, ordered that one full day be added to the Egyptian calendar every fourth year. The Egyptians charged their priests with the task of keeping their calendar in step with the true length of the year. The priesthood, therefore, acquired considerable power for they could thus direct the populace when to celebrate holy days, when to plant, and when to harvest. They themselves could not get too far out of line because the flooding of the Nile trumpeted the beginning of planting time. By comparing the time of the flooding of the Nile from year to year with the height of the sun above the horizon at noontime, they probably discovered that the time of Sirius' rising shifts slightly from year to year with respect to the flooding of the Nile. This phenomenon was not explained until some centuries later when the Greek astronomer Hipparchus discovered the slow shift (precession) of the earth's axis of rotation.

As the celebration of special feast days and religious holidays was very important in all early cultures, it was natural that priests became the keepers of the calendar. Thus priests themselves were astronomers of a sort, principally concerned with the study of the apparent motion of the sun and the appearance and disappearance of the well-known constellations from the evening sky. In time, as astronomy began to evolve into a precise science, the professional astronomer became the calendar authority, laying down the guideline for calendar improvements. Although astronomy did not originate with the construction of the early calendars, the need for accurate calendars certainly encouraged and contributed to the study of the apparent motion of the sun and thus to astronomy.

Two technologies, map making and navigation, that are extremely important to trade and commerce owe much to astronomy and in turn contributed to its study. Map making dates back to the Babylonians who constructed maps on clay tiles about 40 centuries ago. Because maps introduce directions relative to an observer on the surface of the earth, it was natural for the map makers to base their definitions of directions on the position of the sun. The east was defined as the direction of the rising sun and the west as the direction of the setting sun. The north–south direction was then defined as the direction of the sun half-way between its rising and setting—at noon. All of this terminology was infused with precision by introducing lines of longitude and latitude. One could then locate any position on the map of the earth by giving the latitude and longitude lines that intersect at that position.

The concepts of latitude and longitude were carried over to navigation, which is the technique of finding the position of a ship on the sea. In the early days of navigation, one did this by finding the altitude of the sun (called "shooting the sun") which is the height in degrees of the sun above the horizon. This datum, together with the reading of a chronometer, enabled the navigator to determine the north–south direction, from which he could then determine the direction of the ship's motion. Initially, this was done by eye but, in time, a special optical device, called a sextant, was invented to perform this task very accurately.

Because no sun is available to the navigator at night time, he had to use the stars to navigate the oceans. This practice led to a sophisticated navigational technique called "celestial navigation," which required introducing a system of great circles on the sky equivalent to the circles of longitude on the earth and a system of parallel small circles on the sky similar to circles of latitude on the earth. Thus the celestial coordinate system that is still used in astronomy was first introduced by the early navigators.

These early navigators who plied the Mediterranean, Thracian, and Aegean seas as well as the shallow waters along the European and African shores, were well acquainted with the stars of the circumpolar constellations. These are constellations, such as Ursa Major (the Big Dipper is part of this constellation) which circle the north celestial pole without setting. The further north one moves, the more of these constellations one sees. As one goes south, however, the number of circumpolar constellations visible to the observer decreases. These very early navigators could have interpreted these observations in only one way: the waters of the earth lie on a spherical surface. That the earth is round must therefore have occurred to the early navigators even though such speculations cannot be verified because there are no written records of their astronomical or cosmological concepts.

Astronomy was also stimulated by the spiritual needs of people, their religious beliefs, and their belief that the stars and planets influence their affairs. This belief led to the birth of astrology which greatly stimulated the study of astronomy. Indeed, the original study of the positions of the heavenly bodies was called astrology. Though later recognized as a pseudoscience and dropped from the true science labeled "astronomy," it did much up until the time of Newton to contribute to knowledge of the stars and their apparent motions in the sky. (This was particularly true of the period of time from Copernicus to Galileo.) Astrology was probably practiced by the ancient Egyptians, Hindus, Chinese, Etruscans, and Chaldeans of Babylonia. Knowing that the sun greatly influenced their lives, they believed the moon, the planets, and the stars did so as well.

A complicated system of predicting the "influences" on human destiny of the positions of the planets in the various constellations along the ecliptic (the signs of the zodiac) was thus developed. This method of "predicting" human events—astrology—is still popular but, as noted, has no influence at all on astronomy.

Astrology benefitted astronomy in a rather indirect way in the sixteenth century during the period when Tycho Brahe and Johannes Kepler lived and did their important astronomical work. Although astronomy by that time was recognized as an important branch of science, worthy of devoted study in its own right, astronomers and mathematicians were also expected to pursue astrology. Thus Kepler published an annual calendar of astrological forecasts during the 4 years he was teaching mathematics in Graz, Austria. Although he recognized astrology as a "dreadful superstition," he also saw it as the door to accurate astronomical observations of the planets and the stars.

To complete our discussion of the origins of astronomy we consider briefly the contributions of the Chinese, the Hindus, and the early Hebrews. Early Chinese astronomy was driven primarily by the need to construct an accurate calendar. By the year 2000 BC, the Chinese astronomers had determined that the length of the year is slightly more than 365 days, but there is no evidence that they had refined their estimates of the length of the year beyond 365 days, 6 hours by the beginning of the common era. Being deeply concerned about the occurrence of lunar and solar eclipses, the Chinese populace held astronomers in great esteem, maintaining them as officials of the imperial court. The main activity of these astronomers was the prediction of such eclipses.

Although astronomy was actively pursued by Hindu astronomers, they observed the heavens primarily in conjunction with their deep interest in numerology. Thus they did not treat astronomy as requiring exact observations, contenting themselves with a year length of 366 days. The notion that the length of the year lies between 365 and 366 days was repugnant to their sense of propriety, which was governed by numerology. Because astrology was important to the early Hindus, they worked with year lengths that ranged from 324 to 378 days, according to their fancy.

It was easy enough for astrologers to acquire prestige and achieve a lofty status in society by using coincidences between terrestrial phenomena and celestial events. Thus the ancient Sumerians referred to the seven stars in Pleiades as "wicked demons" because they noted that torrential storms occurred when the Pleiades arose early in the evening in spring at the time of the new moon. They argued then that the Pleiades caused these storms, which naturally occur in spring owing to climatic changes. Here we see that astronomy was used to bolster astrology.

Venus, owing to its brilliance, played a large role in the astrology of the Babylonians. The Chinese astrologers, observing the apparent retrograde motion of Jupiter, labeled Jupiter "an omen of famine" instead of "a harbinger of good fortune" as they had previously considered it. Here we see how early astronomy had played into the hands of astrologers. Thus astronomy and astrology affected each other in a very asymmetrical way. Astronomy became the handmaiden of astrology, presenting astrologers with celestial data that astrologers used in any way that suited them. Astrology could not have grown without astronomy, but early astronomy gained nothing from astrology, except justification in the eyes of the public, most of whom then were devout believers in astrology. Astrologers today, as in the past, freely use whatever astronomical data they believe can be twisted to serve their purposes.

In this chapter we have limited our discussion of the beginning of astronomy to those early civilizations that influenced and contributed to the evolution of astronomy from its initial amorphous state to its status as a rigorous science. Outstanding among these early societies were the Babylonians, the Greeks, and the Egyptians. This does not mean that astronomical studies and speculations were not important in other great early civilizations such as the Chinese, the Hindu, the pre-Roman British, and the Mayan. As we have noted, the Chinese and the Hindus had developed sophisticated astronomical concepts and astronomical techniques but their self-imposed isolation greatly limited their influence on the burgeoning of Greek and Egyptian—particularly Alexandrian—astronomy which became the basis and forerunner of medieval astronomy. That the ancient Britons had a flourishing ceremonial-based astronomy as an adjunct to their religious practices is indicated by the famous Stonehenge ritual monuments. Careful studies of these massive stone structures indicate that they date from the Late Stone and Early Bronze Ages. A computer analysis of these megaliths shows that they were used as recently as 1500 BC to predict the summer and winter solstices and the vernal and autumnal equinoxes, as well as solar and lunar eclipses. These predictions must certainly have led them to the construction of a calendar. But astronomy as a science, did not arise from these astronomical studies. The astronomical pursuits of the early Mayan Indians in South America and their relationship to astronomy as a science were similar to those of the Stonehenge Britons. Although the Mayans developed a very elaborate calendar, the most accurate until the Gregorian calendar was introduced in Europe in the sixteenth century, it did not lead them to a science of astronomy.

Because megaliths are found in all parts of the world, from Carnac in France to Easter Island west of the coast of Chile, we may conclude that celestial observations were widely pursued by most of the ancient civilizations. But only the ancient Mediterranean civilizations produced the *science* of astronomy. The contributions of much of the rest of the ancient world were to amount to little more than unsystematic observations and the creation of elaborate myths to coincide with and to explain those observations.

## CHAPTER 2

# The Ancient Cosmologies

*The highest wisdom has but one science—the science of the whole—*
*the science explaining the whole creation and man's place in it.*

—LEO TOLSTOY

In our first chapter, which dealt primarily with the origins of astronomy, we touched on the ancient cosmologies that were the seeds of modern astronomy. We now consider a few of these cosmologies in more detail, emphasizing the difference between what we call cosmology today and what we mean by ancient cosmologies. This difference can be best understood by comparing how the cosmologists operate today and how the ancient cosmologists studied the universe. Modern cosmologists operate from the same fundamental scientific base as that from which all current scientific enterprises stem. Cosmology differs from atomic physics, nuclear physics, and chemistry only in that its domain of activity, subsuming all of these fields, is the entire universe, and its principle aim is to construct a model of the universe that is in agreement with all the laws of nature. Primitive cosmologists, having no such laws to guide them or to restrain their wild speculations, introduced and promulgated many fanciful ideas about the universe.

Insofar as the intellectual pursuits in any society have been associated with the cultural development of that society, we may argue that the ancient primitive cosmologies were most highly developed in the culturally advanced Babylonian and Sumerian civilizations. Their cosmologies probably date back to around 4000 BC. Although the Babylonians knew the periods of the sun, moon, and planets, their cosmology dealt primarily with the earth's place in the cosmos, and, in particular, the space of their small territory lying between the Tigris and the Euphrates rivers on the shores of the Persian Gulf.

Governed by their polytheism, the Babylonians assigned a special god to everything with which they dealt. Because these gods "acted" in accordance with very definite laws that governed their actions, we may argue that the Babylonians believed in a well-ordered universe. Each god controlled a different part of the heavens, the earth, the air, and the seas. The Babylonian cosmology may also be considered an orderly cosmology because the Babylonians believed that no god could act in any way to conflict with any other god. How the rules that governed the gods were invented and assigned to the gods is not clear but the gods could not alter the order of natural phenomena. Presumably the gods were the cosmological "watchdogs" who could only preside over the natural flow of phenomena but not alter that flow; their gods were thus guardians whom the Babylonians identified with the moon, the planets, and the prominent constellations.

The Babylonian cosmology thus naturally led to astrology because with all-powerful gods present, the constantly changing configurations of the sun, moon, and planets against the background of the evening sky meant that the gods were sending some kind of message to the Babylonians about their fate and fortune. Because these constantly changing messages were understood by relating the observed changing celestial configuration to their own experiences, the Babylonians became astute observers of the celestial bodies and, thus, developed a study of astronomy. Nothing much in the nature of astronomical knowledge emerged from this Babylonian astronomy because it was pursued not to further astronomy but to develop a "meaningful" astrology. We must therefore conclude that the Babylonians did not develop any meaningful cosmology, even though astronomy as a systematic study of the motion of the celestial bodies originated in Babylonia. One can, indeed, trace the evolution of primitive astronomy from the Babylonians to the Greeks, but, as we shall see, cosmology, as a nonmythological study of the cosmos, as distinct from mythological cosmogony, began with the Greek stargazers.

Ancient Egyptian cosmology was as primitive as the early Babylonian cosmology but not burdened with as many gods. Unlike the Babylonians, the Egyptians believed that the earth and heavens had not existed forever but were created from a vast reservoir of primeval water which contained the germs of the earth, heavens, and stars. Just how these germs began to germinate and became the observable universe is not described. The Egyptians introduced a "water spirit" which did the trick, changing the primeval sea into a rich source of life. The Egyptians went beyond the Babylonians in introducing an actual "physical model" of the universe. Governed by their reverence for the Nile which flows from south to north, they pictured the whole universe as a rectangular box parallel to the Nile. The bottom of the box with Egypt at the center constituted the earth, and the top of the box the heavens at a vast distance. The Nile was described as a branch of a vast river that flowed all around the earth. This river carried a boat in which a "living god" named Ra carried a dish of fire. Ra died every night and was reborn every morning. Except for a drastic reduction in the number of gods involved, this Egyptian cosmology represented no advance over the Babylonian cosmology.

The first real advance in cosmology in this area of the world was produced by the ancient Greeks, who conceived of a primeval sea from which the universe sprang in its full-blown form. Like the oriental and Egyptian cosmologies, the Greek cosmology emerged and was closely related to Greek mythology. New gods and supernatural beings were invented who controlled every aspect of the observable universe.

The early Greeks differed in their thinking from other ancient civilizations in one very important respect: they tried to find the underlying order in the universe that was not subject to the whims or wishes of their gods. This theme was later expounded by the great mathematician Pythagoras as the "harmony of the spheres." This search for order was probably the forerunner of the Greek mathematics of Pythagoras, Euclid, and Archimedes. Whether the early Greek cosmology led the Greeks to mathematics or the mathematics to cosmology does not matter; in time the Greeks began to describe the kinematics of the celestial bodies mathematically and to seek physical explanations for the observed celestial phenomena.

That the ancient Greeks knew a great deal about the geometry and geography of the earth is evidenced by the Homeric poems which date back to about 1100 BC. In these poems the earth is described as a flat disk surrounded by a mighty river called Okeanos. The heavens were pictured as a huge bell hovering over the entire earth which was partly covered by the Mediterranean Sea, extending westward to a vast ocean. This geography indicates that Homer and his contemporaries knew about the north–south and east–west expanses of the seas and the ocean. Homer's detailed description of Odysseus' journey through the treacherous waters of the Greek islands indicates that he knew a good deal about celestial navigation. In particular, he was aware of the circumpolar stars. Why the ancient Greeks embellished their accurate knowledge of the apparent motions and positions of the celestial bodies with mythological tales is not clear. But it may be that the authors of these tales looked upon the ordinary Greek citizens as children who had to be entertained by myths if they were to be taught the truths about the sun, moon, planets, and stars.

In these mythological stories the Greek writers presented certain basic philosophical ideas mixed with pure fancy. Cosmogony and cosmology were presented as poetic revelations. Thus the Milky Way (galaxy) is pictured as the path of the disastrous celestial fire that Phaeton, the son of Helios, unleashed when he lost control of the solar chariot he was driving across the sky. This may be compared with the Egyptian story that the Milky Way is the celestial counterpart of the Nile river. In their mixture of mythology, religion, and observed celestial phenomena, the ancients, particularly the Greeks, tried to develop a philosophy of man, nature, God, and the universe. It is difficult for us to believe that such clever and observant people as the Greeks of the pre-Christian era would accept their own myths as explanations of celestial phenomena. Why then, did they promulgate these myths as true descriptions of cosmogony and cosmology? The reason may well be that the early Greek philosophers saw the myths as an easy way to interest people in the study of astronomy. Just as we use fantasy today to bring sophisticated ideas to our children, the Greek philosophers and poets may have taught their people by employing well-known myths and stories as a way of introducing their new ideas.

It may well be that the Greek stargazers who lived before the era of the great classical Greek philosophers and mathematicians accepted the celestial mythologies as truth, but such brilliant philosophers as Aristarchus, Eratosthenes, and Archimedes, who flourished before the common era, considered these mythologies to be nothing more than fanciful tales with which the story-tellers entertained their audiences.

The stellar constellations which astronomers still use as mnemonic stellar designs for memorizing the celestial positions of prominent stars are the last remnants of this ancient mythology that appear in current astronomical literature. To those of us who are acquainted with Greek mythology and the relationship of the names of the constellations to the Greek myths, the names of the constellations are not too far-fetched. The relationship of the figure to a great hunter produced by the stars in Orion is easy to accept, as is the figure of a bull—the constellation Taurus—which can be pictured as charging the Hunter Orion who is defending himself against the charging Taurus with a great club of stars. For thousands of years the various constellations, appearing at different seasons as the earth revolves around the sun, were the only calendar available to people. The Big Dipper, or Ursa Major, was the evening clock, just as the sun was the daytime clock for most early people.

Though we think of the constellations in terms of the Greek images assigned to them, every early civilization had its own set of images for the constellations. This is best illustrated by the Big Dipper part of Ursa Major (The Great Bear), the complete pattern of which is somewhat difficult to discern. In Britain the Big Dipper is called the "Plough" whereas in certain European countries it was called the "Wagon." The ancient Egyptians referred to the seven bright stars of the Big Dipper as the "Bulls' Thigh," whereas the Chinese called these stars the "Government" and also "The Northern Measure." In a similar vein, the early Hindus called these stars the "Seven Authorities." The Scandinavians named these stars "The Wagon of the Great God Thor," and the Germans named them "Chariot of Heaven."

Just as the ancients assigned various images to the dominant constellations with special names attached to these images, they attached names to the brightest stars in these constellations. Thus the Arabs called the middle star in the handle of the Big Dipper "Mizar." At about the same time the Arabs discovered a companion star of Mizar which they called Alcor. We now believe that Mizar and Alcor form a binary or visual double star which consists of two stars gravitationally bound to each other. The Arabs are said to have used Alcor as a test of good eyesight. The Arabs also called Alcor the "Forgotten One," whereas the Chinese called it the "Supporting Star."

Of all the celestial constellations, those that lie along the zodiac, which we discussed previously, played the most important part in the lives of the ancients and in the rapid growth of astrology. From the names of these stellar figures, sociologists believe that several early totem tribal animals played a big role in naming the constellations. We believe that in the early Euphration astronomy the zodiac consisted of only six symbols: bull, crab, maiden, scorpion, sea goat, and fish. The early Egyptians, the Israelites of biblical times, and the Chinese introduced their own zodiacal figures, in all of which the names of various animals were prominent. It may well be that totemism was the basis of the constellation ideology of ancient civilizations. This was a natural way for early tribes to express their relationship to and dependence on the animals that they lived with. Moreover, it was an easy way to record totemic loyalties and the heroic qualities of various tribes.

We have called this chapter "Ancient Cosmologies," but in a very fundamental sense, this is a misnomer if, by cosmology, we mean a rational model of the universe which encompasses its origin, its evolution, and its end. Since the ancients had no basic science to guide them in constructing such a model, they really had no cosmology. What we call "Ancient Cosmology" in this chapter is a melange of fairy tales, superstitions, primitive religions, and tribal beliefs. Nevertheless, these primitive concepts fit into a book such as this which emphasizes the narrative aspect of astronomy. But astronomy as a science, which deals with all the phenomena that occur in celestial bodies, could not have evolved until the basic natural laws had been developed. This development began with Kepler, Galileo, and Newton. Before we lead the reader to these mathematically oriented, revolutionary scientists, however, we devote some time to the Greek philosophers and mathematicians and to the Greek and Alexandrian stargazers.

CHAPTER 3

# The Greek Philosophers and the Early Greek Astronomers

*The civilization of one epoch becomes the manure of the next.*

—CYRIL CONNOLLY

Because the ancient Greeks drew no distinction between philosophers and astronomers, we cannot discuss one without the other. Indeed, the Greek philosophers based their philosophies on their interpretations of celestial events so that Greek philosophy and astronomy advanced together until the beginning of the Christian era, when Christianity, particularly Catholicism, began to replace Greek philosophy as the correspondent of astronomy and continued to do so for the first 1550 years of the Christian era.

Of the early Greek philosopher-astronomers, Thales of Miletus (640–560 BC) and his younger disciple Anaximander (611–645 BC) were the first to propose celestial models that are based, at least to some degree, on the movements of heavenly bodies and not merely the manifestations of mythological beasts and superstitions. These ideas, so elementary to us, represented tremendous progress in our understanding of the universe as an orderly system.

Anaximander did not progress much beyond Thales but carried on and propagated Thales' ideas and teachings. He discovered the obliquity of the ecliptic, which is expressed as the tilt of the plane of the earth's equator to the plane in which the earth revolves around the sun (the plane of the ecliptic). Anaximander did not know anything about the motion of the earth around the sun, but he observed that the height of the sun above the horizon at noon changes from season to season. From this observation, he deduced that the circle along which the sun appears to move from day to day is not parallel to (does not lie in the same plane as) the circle in which the sky appears to rotate during the day. Anaximander is also credited with introducing the sundial and inventing cartography. His philosophy is perhaps best remembered for his "concentric cylinder" model of the universe, with the outermost cylinder containing the sun, the middle one the moon, and the innermost one the stars, with the earth at the common center of these cylinders. By imparting rotational motions of different speeds to these cylinders he tried to describe the observed motions of the sun, moon, and the heavens, on which the stars were "fixed." This model, however primitive, represented a great departure from the pre-Grecian anthropomorphic cosmic mythology. It was, indeed, the beginning of the epicycle concept developed hundreds of years later by Hipparchus and Ptolemy.

Like all early Greek philosophers Anaximander had his own theory of the origin of all things. He postulated that the universe originated from the separation of opposites. Hot "naturally" separated from the cold, followed by the separation of the dry from the wet. He completed these ideas by postulating that all things eventually return to their original elements.

Anaxagoras, a younger contemporary, and probably a student of Ananimander's (they were both of the Ionian School of Greek philosophy), taught that all matter had existed originally as "atoms or molecules," infinitely numerous and infinitesimally small. Anaxagoras declared that these atoms had existed "from all eternity" and that all forms of matter are different aggregates of these basic atoms. In this statement, he laid the basis for the Greek philosopher Democritus, who is credited today with having created atomic theory. Leucippus probably greatly influenced Democritus, who was his student, in his formulation of the atomic theory. But what little we do know about Leucippus stems from Aristotle's commentaries—not Leucippus' surviving works. Aristotle himself thought very highly of Leucippus, crediting him with the invention of atomic theory. Democritus and Leucippus had similar ideas about the relationship of the sun and moon to the earth and the stars. They both presented theories about solar and lunar eclipses which were probably not developed in isolation from one another.

Other Greek philosophers such as Metrodeus of Chios and Empedocles of Sicily, who lived in the fifth century BC, offered cosmic theories that contributed to the intellectual dominance of Athens. This proud city-state became the undisputed home of philosophers, who flocked there from all corners of the known earth. The importance of these philosophers for the growth of astronomy was not in the correctness of their primitive speculations, but in their insistence on careful observations of the celestial bodies and their motions. Mythology was thus beginning to give way to rationalism as the proper way to understand the universe and to explain natural phenomena.

To complete our discussion of this remarkable pre-Pythagorean era we note that Anaximenes of Miletus, Xenophanes of Kalophon, Parmenides of Elea, and Heracleitus of Ephesus, all contemporaries or disciples of Thales, speculated about the nature of the sun, moon, planets, and the stars and developed primitive cosmologies, which have one thing in common; they were all based on the atomic theory of Democritus of Abdera. This was the beginning of unity in astronomy, but still a far cry from the Greek astronomy that finally began to emerge from these ancient cosmologies in the fourth century BC.

These philosophers were concerned primarily with discovering "first principles" from which all the properties of the universe could be deduced by pure philosophical reasoning. Though these speculations did little to further astronomy, they were very useful in that they forced those who followed these philosophers to compare earlier philosophical deductions about the universe with observations of the celestial bodies. This led to the beginning of observational astronomy which culminated in the work of the Greek Alexandrian astronomers at the beginning of the Christian era.

Of all the early Greek philosopher–astronomers, Pythagoras was probably the most influential in turning the attentions of astronomers to the importance and usefulness of mathematics in constructing cosmological models that could be compared more or less accurately with the observed motions of the celestial bodies. Pythagoras founded a school of philosophy whose main concern was to interpret all natural phenomena in terms of numbers. Born in Samos and spending most of the 50 years of his life in Croton, he laid down the basic principle of his philosophy: number is everything. Numbers not only express the relationships among various natural phenomena but, in Pythagoras' opinion, cause all of these phenomena. This school of thought became so popular and dominant that it survived for 200 years—not because of its mathematics and astronomy but because of the religious mysteries surrounding it.

With their emphasis on number as the basis of natural phenomena, the Pythagoreans quite naturally sought in nature phenomena that would provide a basis for their numerology, and they were quick to find this justification in the great regularity they observed in the motions of the celestial bodies. The sun rose without fail every day as did the stars. Everything seemed to repeat itself in a precise period to which a definite integer could be assigned. This regularity led the Pythagoreans to the concept of cosmic harmony; indeed, Pythagoreans introduced the word "cosmos" to designate the universe. They were convinced that numbers would lead them to a complete understanding of cosmology and reveal the basic unifying cosmic principle.

The Pythagoreans were encouraged in their search for harmony in the universe by Pythgoras' discovery that the most harmonious sounds, pleasing to the ears, are those whose vibrations are related to each other in simple numerical ways. The Pythagoreans extended this harmony to the heavens and called it the "harmony of the spheres." In this cosmology, the universe is described as a "cosmic union" with one celestial skin inside another, each revolving at a different rate. The Pythagoreans hoped to prove that each such rotation produces musical tones which are harmoniously related. More than 2000 years later, Kepler tried unsuccessfully to use this Pythagorean concept of cosmic harmony to describe the motions of the planets around the sun. When Kepler discovered the third law of planetary motion, he called it the "harmonic law."

Pythagoras' devotion to integers led him to the concept that everything in the universe can be explained in terms of just four basic elements: earth, water, air, and fire, and that all figures can be reduced to spheres and regular polygons. The famous Pythagorean theorem that the square of the hypotenuse of a right triangle equals the sum of the squares of the other two sides, led the Pythagoreans to the belief that all geometry can be reduced to arithmetic and that this is also true of space, and therefore, of the universe itself.

Although historians differ in their evaluation of the specific contribution of Pythagoras to astronomy they all agree that his contribution to the way one must think about nature and about astronomy, in particular, was a revolutionary departure from everything that had occurred before. Specifically, however, it is believed that Pythagoras promulgated the doctrine of a spherical earth in conformity with the spherical appearance of the sky which does not change as one moves from point to point on the surface of the earth. There is also some historical evidence that Pythagoras believed that the earth rotates, thus producing day and night. Further, the Pythagorean philosophers who followed Pythagoras stated that he was the first to discover that Phosphorus and Hesperus, the "morning and evening" stars, are the same celestial body, later called the early planet Venus by the Romans. He was also said to have been the first to discover that the planets move in separate orbits tilted at different angles to the plane of the ecliptic.

That the Pythagoreans knew of the concept of the motion of the earth "around a central fire" (the heliocentric concept) is indicated from the Greek biographer and essayist Plutarch (46–120 AD) who stated that "the Pythagorean Philolaus believed that the earth moves in a closed inclined orbit around a central fire." Because Philolaus was a disciple of Pythagoras and was not known for independent thought, we can interpret Plutarch's statement about Philolaus as evidence that Pythagoras himself proposed the concept of a heliocentric solar system.

An interesting but incorrect conclusion about the number of planets or nonstellar objects associated with the earth stemmed from Philolaus' devotion to numbers as the rules of celestial harmonies. The nine visible bodies: earth, moon, sun, the five planets, and the sphere of the fixed stars, left a gap in what Philolaus considered the perfect numerical harmony that 10 bodies would constitute. He therefore proposed a tenth body which he assumed to be "another earth" on the opposite side of the sky which he called the "antichthon" or "counterearth." This error persisted for some time, but we can clearly see the significance of the fact that it was predicted by the Pythagoreans as a consequence of their requirement that numerical harmony apply and was not based on any mythological or religious reasons. This perhaps was the first instance in which a theory, however faulty, led to a prediction. Indeed, this is the way science progresses today, with scientists using basic physical laws to guide them in their predictions rather than fanciful notions about the predictive powers of numbers.

The Pythagoreans made remarkably accurate observations, discovering that the phases of the moon complete a cycle every 29 days, 12 hours. The moon itself was pictured as revolving around "the central fire" during that period. They contrasted this prediction with the sun's motion around the same "central fire," finding that the sun completed its journey once every 364 days, 12 hours—a period of time which they identified with the length of the year. At no time did they identify the sun with the "central fire"; instead they placed the earth at the center, without worrying about how the earth could exist at such a center. They used this rather curious unphysical model of the solar system, with its rotating sky, to account for the observed (apparent) motions of the celestial bodies. This model suggested a calendar based on the lunar period, with the lunar period defining the month. This calendar, perhaps the earliest lunar–solar calendar, did not keep step with the seasons and so one month had to be inserted into the calendar every 3 years.

Pythagoras is best known today for his famous theorem about the sides of a right triangle, but his greatest contribution was his invention of a consistent philosophy and procedure for explaining natural phenomena which did not call upon the gods, fairies, or spirits. In spite of the great attraction that Pythagorean philosophy and numerology held for people, the Pythagorean influence began to fade and practically vanished by the beginning of the Christian era. But from the many references to Pythagoras present in the writings of Plato and Aristotle, it is clear that their natural philosophies were affected by the Pythagoreans.

Although Plato himself did not develop any original cosmological models, he carried on the Pythagorean principle that only the application of pure thought can fathom the "true harmony" of the universe. Plato also argued that symmetry is the basis of all natural phenomena, which led him to the belief that the sphere is the only admissible shape for celestial bodies and the circle is the only admissible orbit of a celestial body. Plato's concept of symmetry also led him to a numerical model of the distances between the sun, moon, and planets which is completely Pythagorean. Using the sequences 1, 2, 4, 8 and 1, 3, 9, 27, which are the successive powers of 2 and 3, he assigned the distances 1, 2, 3, 4, 8, 9, 27 to the moon, sun, Venus, Mercury, Mars, Jupiter, and Saturn, each moving in its own circle with a radius given by the number in the sequence assigned to it. Plato did not believe that his deductions about the motions of the planets required any observational verification for he was firmly convinced that pure reason could lead to no other numerical model. This is as far as Plato pursued his astronomical speculations which seemed to him to describe the motions of the sun, moon, planets, and stars precisely. Plato was the essential mystic who viewed the universe as the Kosmos, the manifestation of the divine principle or, alternatively, as a divine work of art which needed no explanation. This divine work carried within it the "breath of life" which required no "scientific explanation" or verification. In this synthesis of ideas he considered pure philosophy as superior to mathematics, which had, in Plato's opinion, a vulgar element in it because mathematics was associated with commerce and trade.

Although Plato's influence, as a philosopher, dominated the intellectual lives of the Greek philosophers of that period, his younger disciples and students included Eudoxus, Aristotle, Herakleides, and Aristarchus, all of whom advanced science (astronomy, in particular) far more than had Plato. Eudoxus attended Plato's Academy in Athens for some months but then went to Egypt where he studied planetary motion with the priests of Heliopolis. Eudoxus was an excellent mathematician. Indeed, some historians believe that he wrote Euclid's fifth book of geometry. Plutarch states that Plato considered Eudoxus to be the foremost mathematician in Athens. He knew the length of the year quite accurately and suggested that a solar year of 365 days be accepted for 4 years and that every fifth year consist of 366 days. This suggestion later became the basis of the Julian calendar, which was the first of our modern calendars. Plato held Eudoxus in such great esteem that he suggested Eudoxus tackle the problem of planetary motions (i.e., why they sometimes moved [apparent motion] with and at other times opposite to the rotation of the sky).

Eudoxus produced a very ingenious model of the apparent planetary motions which depicted them in accordance with the observational data known at the time. The model introduced by Eudoxus is known as the homocentric sphere model because Eudoxus assigned a sphere to each planet and assumed that these spheres nest in each other and are concentric with the earth which he assumed to be at the common center of these spheres. Because a single sphere for each planet, rotating at its own speed, could not explain the retrograde as well as the direct observed motions of the individual planets and the observed motions of the sun and moon, Eudoxus introduced subsidiary spheres arranged to rotate in such a way as to reproduce the observed motions, which were all assumed to be circular. Unfortunately the book *On Velocities* written by Eudoxus, which describes this theory in detail, is lost, but Aristotle knew about this book, and had probably read it, because he gave a detailed account of Eudoxus' homocentric spheres. Though this model is incorrect, Eudoxus' work set an important standard for all astronomers who followed him: to let the observations guide the astronomers in their search for models of the motions of celestial bodies.

Eudoxus was most accurate in his model for the lunar motion, which he asserted was controlled by three different spheres spinning at different rates. These spheres correctly had the moon rise 52 minutes later each day and move around the earth in its circle once every 27 days, 8 hours. He also knew about the 19-year period of the regression of the lunar nodes (the two opposite points on the celestial sphere through which the moon's line of nodes passes). It is not clear from the available historical literature whether or not Eudoxus did any observing at all. Whatever evidence we have indicates that his observational data were very meager, for it is known that his students tried to improve his homocentric spheres' model to be in better accord with the observations.

Aristotle, Plato's most famous and most productive student, differed dramatically from Plato in his insistence that one's observations of nature be the guiding principle in the study of nature. In spite of this practical philosophy, Aristotle applied metaphysical reasoning and guidance in developing a model of the universe. He discussed his astronomical speculations in his books *On the Heavens*, and *Meteorologica*. His main concern was to explain the apparent motions of the celestial bodies. Because he had no laws of motion nor any real understanding of the nature of motion—particularly the change of the state of motion of a body—he had to introduce some very primitive concepts about how the celestial bodies acquired their motion. In this approach he differentiated between perfect motion, which he pictured as the "quickest," from the so-called lesser motions. He assigned the "most perfect" motion to the celestial sphere of the stars and lesser motions to the moon, sun, and planets. He assumed that these were primordial motions that were assigned to the various celestial spheres by a "divine power." He further assumed that this divine power acted to keep the celestial bodies moving continuously. He missed entirely the important difference between unchanging motion and accelerated (changing) motion and therefore missed the opportunity to discover the basic laws of motion. In particular, he did not understand the concept of inertia which remained for Galileo to discover some 20 centuries later.

But Aristotle had a true scientific attitude and approach to the problems and puzzles that confronted him. Thus he wondered why stars (as opposed to planets) twinkle, and explained this difference in terms of the eye, which he suggested tears and shakes when viewing distant objects such as stars but remains vigorous and steady when viewing closer objects such as the planets. He explained the "circular motions" of the stars and planets by arguing that the material of the stars and planets is "circular motion" material, as opposed to earth material which can travel in straight lines, like flowing water and rising fire. Aristotle also asserted that the sun's heat was generated by the friction caused by its movement through the ether (space).

Aristotle accepted the spherical shape for the celestial bodies because, he argued, the sphere is a perfect shape and therefore the only one fitting for celestial bodies. He arrived at the spherical shape for the earth by reasoning that the earth was formed from particles that all moved toward a common center. These particles then formed a series of concentric spheres, the result of which formed the surface of the earth. He estimated the diameter of the earth to be 400,000 stadia or 12,361 miles, almost one and one-half times its true value, which was a remarkably accurate estimate for that time. He strengthened his argument that the earth is spherical by pointing out that different stars appear overhead as one moves north or south and that the horizon also changes with these movements.

The importance of Aristotle's speculations about astronomy is that he established a new standard for pursuing and investigating natural phenomena. Nothing was to be accepted without presenting a reasonable explanation. Reasonable in this context meant either mathematically acceptable, consistent with experience, or both, but without recourse to gods or myths.

Of the four disciples of Plato listed previously in this chapter, Herakleides and Aristarchus were closer to modern astronomers in their thinking than Eudoxus and Aristotle. Both Herakleides and Aristarchus discarded the concept of a rotating sun, replacing that concept with that of a rotating earth. Although Herakleides is said to have suggested that the earth rotates on its own axis and that it probably revolves around a central fire, historians do not believe that he proposed a heliocentric model of the solar system. However, he is credited, as we shall see, with having introduced the concept of the epicycle to explain the observed motions of the planets.

The early Greek astronomers were greatly puzzled by the apparent periodic retrograde motions of the planets (from east to west). Herakleides argued that this motion can be explained by assigning to each planet a loop in its observed orbit. He discovered that the observed motion of Jupiter around the earth can be explained by assigning 12 loops to its complete circular orbit; Saturn's observed motion, by contrast, requires 29 loops. These loops were the forerunners of Ptolemy's epicycles introduced about 400 years later. In his discussion of the apparent motions of Mercury and Venus, however, Herakleides proposed the novel idea that these planets revolve around the sun, with Mercury revolving in a smaller circle than Venus. This model was accepted by Tycho Brahe some 1800 years later, and he made it the basis of what we now know as the Tychonic model of the solar system. Later authors wrote that Herakleides had adopted this model from the Egyptians, but there is no clear evidence to support this theory.

Because all but a few fragments of Herakleides' own writings are lost, we find it difficult to separate what Herakleides actually discovered and proposed from what commentators wrote about following his death. But the many references made to Herakleides by Aristotle and other contemporary philosophers suggest that Herakleides was one of the most influential of the early Greek philosophers.

Of all these ancient Greek astronomers, Aristarchus of Samos, who lived in the third century BC, was the closest to modern astronomers in spirit and approach to the solution of astronomical problems. Indeed, he was the first to propose a self-consistent heliocentric model of the solar system, with the planets, starting with Mercury, arranged in the order from the sun which we accept today. He was the first to attempt to determine the dimensions and distances from the earth of the sun and moon. His approach in these exercises was mathematically and physically impeccable, which he described in a treatise, *The Dimensions and Distances of the Sun and Moon*. This is the only one of his original treatises that has survived. In this treatise Aristarchus presented a brilliant analysis of the relationship between the phases of the moon and the geometrical arrangements of the earth, moon, and sun for the various phases. From this analysis he concluded correctly that the sun is at a much greater distance from the earth than is the moon, and that the sun must therefore be many times larger than the moon and the earth.

We now describe Aristarchus' reasoning briefly, which deals with the appearance of the moon when it is in its first quarter. Here the "lunar phase month" of 28 days is divided into quarters, with the first quarter defined as the phase beginning 7 days after the new moon. Because the new moon is the phase when the moon is between the earth and sun, the phase of first quarter, beginning 7 days later, is often called the "half moon" because only half of the face of the moon that is toward the earth is lit up by the sun's rays.

If the moon is then viewed when it is due south (on the observer's meridian), the rays of light from the setting sun that illuminate the moon are coming from the right, that is, from due west. These rays are therefore at right angles to the line from the observer to the moon. Aristarchus noted then that the line from the earth to the sun makes a very small angle with these rays—smaller than 3 degrees. This was correctly interpreted by Aristarchus to mean that the sun is at least 30 times as large as the moon.

Although this calculation still greatly underestimated the size of the sun, it led Aristarchus to the conclusion that the sun cannot be revolving around the earth because it was unreasonable to have so large a body as the sun revolve around so small a body as the earth. The Greek mathematician Archimedes, a younger contemporary of Aristarchus, commented on this solar system model of Aristarchus, noting that this model implies that "the world [cosmos] is many times larger than had previously been thought." The Greek biographer and historian Plutarch (46–120 AD) in his book *On the Face in the Disk of the Moon* remarked that Aristarchus proposed the hypothesis that the "heavens stand still and the earth moves in an elliptic circle at the same time as it turns round its axis." This is essentially the Copernican heliocentric model of the solar system, so that Copernicus may be called the "Aristarchus of the modern era." Interestingly enough, Copernicus, in seeking ancient authoritative support for his heliocentric solar system, referred to Aristarchus.

With the decline of Athens after the rule of Pericles (the Golden Age of Athens) and its defeat by the Macedonians in 338 BC, Plato's Athenian Academy began to lose its preeminence as an intellectual center. This preeminence slowly shifted to the Egyptian city of Alexandria, which Alexander the Great founded in 322 BC. Athens still continued to attract foreign students when it came under Roman rule after 146 BC, but the best of the Athenian intellectuals migrated to Alexandria and to the Greek city of Rhodes which, for a century and a half before the Christian era, rivaled Alexandria as the center of Greek literary and intellectual life. However, Alexandria retained its prominence owing to the beneficence of the Macedonian king of Egypt, Ptolemy, surnamed Soter, and his progeny, the famous Ptolemies, ending with the death of Queen Cleopatra VII of Egypt in 30 BC.

The famous Alexandrian Library, established by Ptolemy I Soter, and expanded by his son Ptolemy II Philadelphus, greatly enhanced the intellectual attractiveness of Alexandria. By the end of the third century BC, this library had the largest collection of books in the ancient world. In the time of Ptolemy II, its book collection increased to nearly 500,000 volumes or rolls, and its annex in the temple of Serapes contained an additional 43,000 volumes. The library was the source of manuscripts for libraries throughout the ancient civilized world, for it provided an unbroken stream of copies of its original manuscripts to all libraries that wanted them. This remarkable dissemination of knowledge about the intellectual activity of the ancient civilizations made it possible for many of the advances of these early societies to survive to the present day. In the incessant warfare between Rome and Alexandria during the Ptolemaic era, the Alexandrian Library burned in whole or in part several times: It was destroyed in 47 BC when Pompey the Great besieged Julius Caesar in Alexandria, in AD 27 by the order of the Roman Emperor Lucius Domitius Aeoreliones, in AD 391 under the Roman Emperor Theodosius I, and in AD 640 by Muslims commanded by the Caliph Dinar I.

The Alexandrian Library was famous for the great Greek literary men of letters such as Zenodotus of Ephesus, Callinachus, and Aristophanes, who were its librarians. It is most famous among astronomers and geographers for having had the Greek mathematician, astronomer, geographer, and poet Eratosthenes for its librarian and director from 240 BC to 194 BC. Although Eratosthenes measured the obliquity of the ecliptic with an incredibly small error of 7 minutes of arc and drew up a catalogue of about 700 fixed stars, he is most famous for having measured the circumference of the earth, also with great accuracy.

As Eratosthenes' procedure for measuring the circumference of the earth is similar to what geographers and surveyors do today to determine the shape, as well as the size of the earth, we describe this procedure in some detail. Eratosthenes did what we now call measuring the length of a degree on the earth's surface. Because the earth is a sphere, the direction of the vertical at any point on the earth's surface changes with respect to the direction of the earth's axis of rotation as the latitude of the point changes. If the earth were a perfect sphere, this change would always be the same if the distance on the earth between two points, measured along a great circle, were the same. Specifically, if the earth were a perfect sphere and we changed our latitude by one degree, going either north or south, we would have to walk (north or south) by the same number of miles. If this number were $n$, then the circumference of the earth would be $n$ multiplied by 360 or, alternatively stated, $360n$, because a complete rotation equals 360 degrees. The number $n$ would then be called the length of the degree on the surface of the earth and this would be the same no matter where we measured this distance on the earth.

Careful measurements carried out with our best geodetic instruments today show that the length of the degree is not the same at all latitudes but varies from a minimum value of 110.572 kilometers at the equator to a maximum value of 111.33 kilometers at the north pole. This difference tells us that the earth is not a perfect sphere but is flattened at the poles. The average length of a degree on the earth's surface is about 69 miles, varying from 68.7 miles at the equator to 69.4 miles at the poles.

Returning now to Eratosthenes, we note that he measured the length of the degree by comparing the length of the shadow cast by a vertical shaft (gnomon) in the ground at noontime at two different cities at the same time of the year. He stated that at Syene this vertical shaft cast no shadow at noon on the day of the summer solstice. In other words, he noted that the sun was directly overhead at Syene when it was highest in the sky (noon) on the day of the summer solstice. At Alexandria, however, on the same day, the gnomon cast a shadow whose length showed that the direction of the vertical (gnomon) at Alexandria differs from its direction at

Syene by slightly more than 7 degrees. Eratosthenes then noted that Syene is 5000 "stadia" in distance from Alexandria as measured along a north–south meridian. This means that the length of the earth's circumference is as many times larger than 5000 stadia as 360 degrees is larger than 7 degrees. Because 7 degrees is contained in 360 degrees slightly more than 50 times, Eratosthenes concluded that the earth's circumference equals about 250,000 stadia. We do not know with any certainty the length of the "stadium" adopted by Eratosthenes, but all the evidence available in the written records that have survived indicates that one of Eratosthenes's stadia is 516.7 feet. Eratosthenes's circumference calculation is thus 24,662 miles, which is remarkably close to the accepted value today.

This was the beginning of attempts by the early Greek astronomers to find (i.e., measure or estimate) the "dimensions of the world." We have already mentioned that Anaximander proposed that the sun's distance is 27 times that of the earth's radius and that the moon's distance is 19 times that of the earth's radius. Plato had speculated about the planetary distances, trying to arrange them according to the integers 1, 2, 3, 4, 8, 9, 27. But Aristarchus was the first to outline a fairly precise mathematical procedure for obtaining the relative sizes of the earth, moon, and sun. The result was a solar distance vastly larger than the lunar distance.

Archimedes (287 BC–212 BC), Greece's greatest mathematician other than Euclid, and a younger contemporary of Aristarchus, did not accept Aristarchus' model of the earth revolving around the sun because it meant to Archimedes that the earth would have to be revolving around the stellar sphere as well. This was unacceptable to Archimedes because, he argued, the earth could not revolve around a sphere of which it was the center. Here we see how correct reasoning can lead one to wrong conclusions if one's initial assumptions are incorrect. Archimedes placed the earth at the center of the celestial sphere because of the apparent motions of the stars around the earth.

One might have thought that the Greeks, with their very clever and accomplished mathematicians, would have developed a much more advanced and rigorous astronomy than they did, but most of the early Greek mathematicians did not see astronomy as a mathematical challenge. In their belief system, everything in the heavens moved according to the preordained order, with concentric circles as the true celestial orbits. That being so, the Greek mathematicians had nothing to add to this "perfect model." Indeed, the intrusions of mathematicians could only detract from this preordained perfection.

But the Greek observers of the apparent planetary motions, in time, began to detect "flaws" in this perfection; this was particularly true of the apparent motion of the planet Mars, which traversed its apparent "circular" orbit around the earth in a noticeably irregular way. The Babylonians had already noted that Mars appeared to cover a greater distance along its orbit, month for month, during one half of its apparent orbit than during the other half. To the orderly minded Greek philosophers, this irregularity could only be an illusion and not a real phenomenon. But very careful observations over a period of years finally convinced the Greek astronomers and mathematicians that the apparent motion of Mars and some of the other planets could not be accommodated in perfectly circular orbits. We have already seen how Herakleides tried to solve the overall problem by introducing circular loops in the "apparent orbits" of Jupiter and Saturn but this geometrical artifice could not be applied to the apparent orbit of Mars which remained to plague Greek astronomers until one of the greatest of the early Greek mathematicians, Apollonius of Perge, introduced an entirely new and philosophically disturbing geometric concept to explain the orbit of Mars.

Apollonius of Perge, who spent most of his productive life at Alexandria, became famous among his Greek contemporaries for his development of the mathematics of "conic sections" which deals with the various curves cut along the surface of a cone by planes tilted at various angles to and cutting through the cone such as the circle, the ellipse, the parabola, and the hyperbola. He introduced a simple symmetry to account for the asymmetry in the apparent motion of Mars. He shifted the center of the apparent circular orbit of Mars from the center of the earth to a point at some distance from the earth. This meant, according to Apollonius' model, that Mars did not revolve in a circle around the earth's center but around an eccentric point. Some 300 years later, this conclusion led Ptolemy to the concept of the epicycle.

Apollonius wrote a treatise of eight books on conic sections, seven of which have survived and which deal only peripherally with the motions of the planets but which influenced both Hipparchus, the greatest of the early Greek celestial observers, and Ptolemy, who dominated astronomical thinking for some 1500 years. That Apollonius greatly influenced early Greek astronomy is indicated by Ptolemy's reference to him in *The Almagest*. From Ptolemy's reference to Apollonius it is clear that mathematics was about to become an important feature of astronomy. In a sense, Apollonius may be called the father of theoretical astronomy, an enterprise which was carried on at the beginning of the Christian Era by Ptolemy, the father of the theory of epicycles, which, probably more than anything else, retarded astronomy for the first 1500 years of the Christian era.

CHAPTER 4

# From Aristarchus to Ptolemy: The Birth of Accurate Observational Astronomy

*Except the blind forces of Nature, nothing moves in this world which is not Greek in its origin.*

—SIR HENRY JAMES SUMNER MAINE

As we saw in the last chapter, Aristarchus and Apollonius were the first two theoretical astronomers who based their astronomical conclusions on the careful mathematical analysis of observational data. Aristarchus was more of a theoretician than Apollonius, who was the greater mathematician of the two. Unfortunately, only a few scattered remnants of Aristarchus' great treatise *On the Dimensions and Distances of the Sun and Moon* remained to influence his followers who were opposed by those who considered his views to be heretical. According to Plutarch, Aristarchus was charged with impiety for daring to propose that the earth is not at the center of the universe. Because Apollonius did not speculate about the motion of the earth but simply presented mathematical models (conic sections) for the orbits of the planets, his theories did not provoke the resentment of the populace. However, his suggestions that the planetary orbit might well be elliptical instead of circular aroused some criticism.

It is difficult to trace the developments of astronomy from Aristarchus and Apollonius to Hipparchus, who was the greatest of the early Greek astronomers, and best known for having introduced precision in the study of astronomy. We know that both Aristarchus and Apollonius influenced Hipparchus, but just how is not clear because no historians of astronomy flourished in the years between

Aristarchus and Hipparchus. Only one early Greek historian of astronomy is known—Eudemus of Smyrna. Eudemus, a disciple of Aristotle, wrote a history of astronomy from Thales to Aristotle, but no records of the work of Aristarchus other than those found in the writings of Hipparchus and Ptolemy are extant. That both Hipparchus and Ptolemy praised Aristarchus and Apollonius is evidence enough of the great influence of these two theoretical astronomers on the development of astronomy.

In reviewing the evolution of astronomy from Thales to Hipparchus, a period of some 500 years, we are struck by the lack of direction in the growth of astronomy. Part of this lack is certainly due to the absence of reliable communications and record keeping, part to the intrusion of religion and mythology, and part to the absence of a plan of systematic observations of the heavens that were carried on from generation to generation. But more important than these factors as a deterrent to the development of Greek astronomy from Thales to Ptolemy, was the absence of a body of natural laws that could guide the early Greek astronomers in their pursuit of an understanding of the motions of the planets. In particular, the Greeks had no laws of motion, that is, no understanding of the relationship between the motions of bodies and the forces acting on these bodies. Aristotle, for example, had no concept of inertia and believed that bodies had to be pushed or pulled all the time just to keep them moving at a constant speed in the same direction. With the introduction of the concept of inertia by Galileo and the discovery of the basic law of motion by Newton, a rigorous theoretical astronomy was born, which threw off the Ptolemaic shackles of its more primitive precursor.

We can see the limits that the absence of theory imposes on pure observational astronomy, however precise such observations may have been, when we consider the observations of Hipparchus who was, by far, the greatest of the Greek astronomers of the pre-Christian era. Born in Nicaea, Bithynea, in about 190 BC, Hipparchus spent most of his life in Rhodes, one of the most prominent states of Greece. Like Alexandria, Rhodes was a center of intellectual life with great activity in literature, astronomy, and mathematics. The only writing of Hipparchus still extant is his book written in 140 BC in which he expounded upon the importance of a continuous pursuit of accuracy in tracing the apparent motions of the sun and the planets, particularly the apparent motion of the sun. Most of what we know about Hipparchus comes from the writings of Ptolemy in his famous work, *The Almagest*.

Hipparchus' emphasis on observational precision led him to perhaps his greatest discovery, the precession of the equinoxes. The equinoxes are two imaginary points on opposite sides of the visible sky (the celestial sphere) where the two imaginary great celestial circles—the celestial equator and the ecliptic—intersect. These circles are defined by the diurnal (daily) rotation of the earth and its annual revolution around the sun. The rotation of the earth defines a plane and an axis of rotation which is perpendicular (normal) to the plane which cuts the earth in a great circle called the earth's equator. If the plane of the earth's rotation is extended infinitely in all directions, it cuts the celestial sphere in a great celestial circle called the "celestial equator." The axis of the earth's rotation, an imaginary line through the earth's center at right angles to the equatorial plane, cuts the sky (celestial sphere) at two points: the north celestial pole and the south celestial pole. We can easily find the north celestial pole by finding the Big Dipper in the northern sky and tracing a line along the pointers in the Big Dipper (the two stars in the bowl) to a bright star called the North Star. The north celestial pole is about one degree away from this point. The altitude of the north celestial pole (the number of degrees above the observer's horizon) equals exactly the observer's latitude.

As the earth revolves around the sun, it defines another plane, called the plane of the ecliptic, which cuts the celestial sphere (the sky) in a great imaginary circle called the ecliptic. The two imaginary celestial circles, the celestial equator and the ecliptic, intersect each other in two diametrical imaginary celestial points called the vernal and autumnal equinoxes. The earth's equatorial plane and the plane of the ecliptic are tilted 23.5° with respect to each other. Astronomers call this angle the "obliquity of the ecliptic" which was known to the ancient Greeks, particularly to Eratosthenes who gave its value as 23°51′20″. Schoolchildren are taught this "obliquity" as the "tilt of the earth's axis," which has no meaning unless

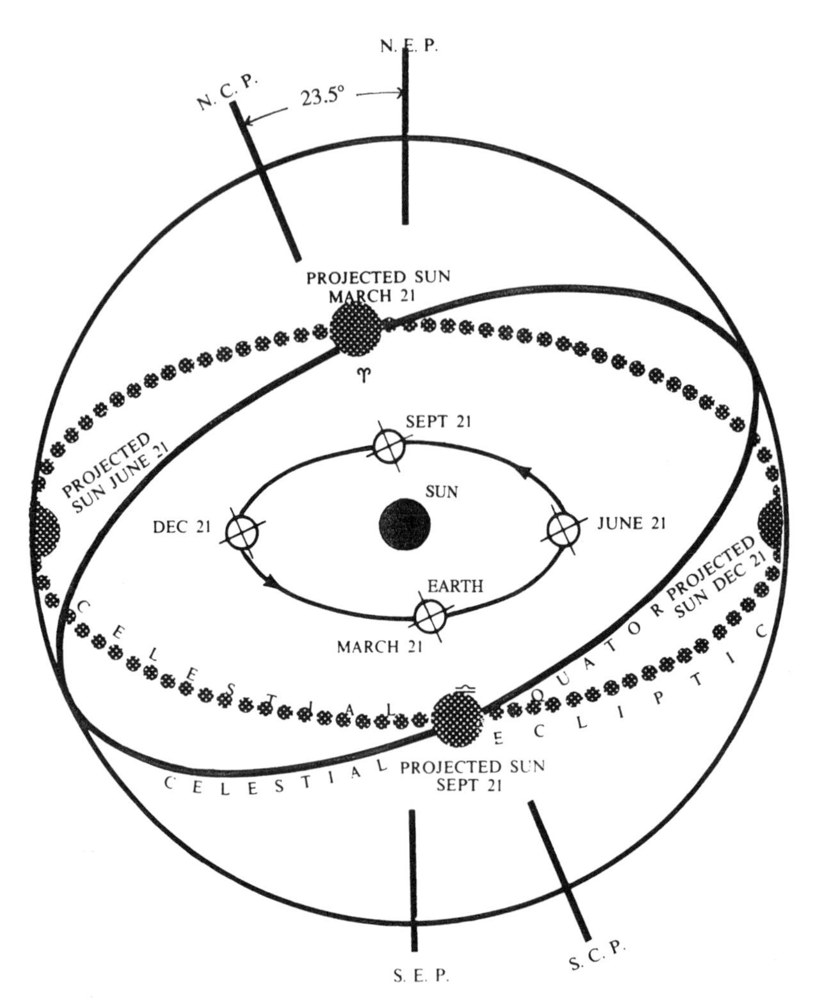


Figure 4.1. Schematic diagram showing the sun projected onto the celestial sphere at different times of the year—March 21 (vernal equinox), December 21 (winter solstice), September 21 (autumnal equinox), and June 21 (summer solstice)—as seen from the earth. The earth's axis and its tilt (23°) are also shown. NCP = north celestial pole; SCP = south celestial pole.

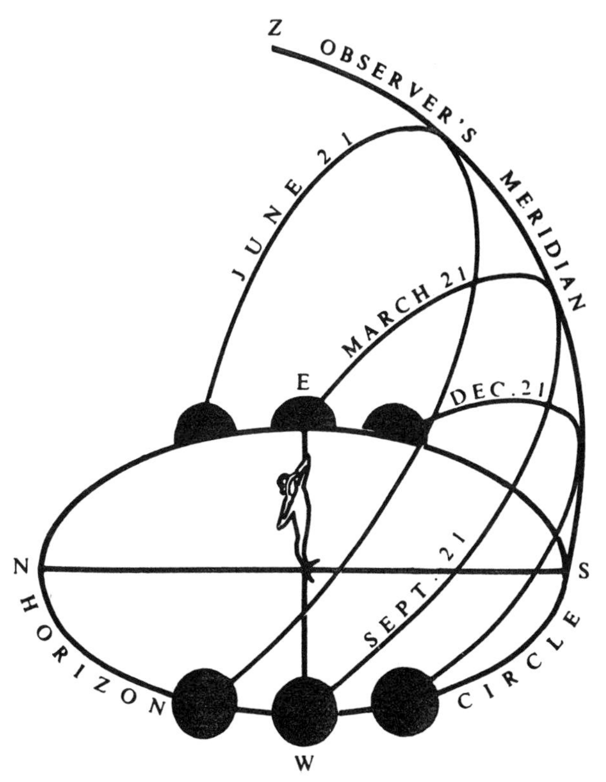


Figure 4.2. The noontime altitude of the sun changes from month to month as seen by a fixed observer. (Here a northern-hemisphere observer is shown.)

the "tilt" is defined with respect to some fixed direction. From this point of view the obliquity of the ecliptic is the angle between the earth's polar axis and the polar axis of the ecliptic (the imaginary line passing through the center of the earth and perpendicular to the plane of the ecliptic).

To follow the celestial observations that led Hipparchus to his discovery of the precession of the equinoxes, we imagine ourselves studying the apparent motion of the sun in the sky, believing this to be a true solar motion instead of an apparent motion imparted to the sun by the motion of the earth around the sun in the plane of the ecliptic. As the earth moves eastward in the plane of the ecliptic, the sun is projected onto the sky and appears to move eastward along the ecliptic circle by about one degree per day, so that the stars appear to rise about 4 minutes earlier each night. As the sun "moves" eastward each day, the length of daylight (the time between the rising and setting of the sun) changes, becoming longer and then shorter during each year, year after year. At the same time the sun changes its position with respect to the celestial equator. Because the ecliptic and the equator cross each other at two points (twice a year), the sun is exactly on the celestial equator at the two moments when the sun is at these two points. These two moments are called the equinoxes because the duration of daylight exactly equals the duration of darkness everywhere on earth on the 2 days of the year when this happens. On those 2 days, the sun rises exactly at 6:00 AM and sets exactly at 6:00 PM.

One of these equinoxes is called the vernal equinox (about March 21st) because it marks the beginning of spring, and the other is called the autumnal equinox (about September 21st) because it marks the beginning of autumn (fall). The interval of time between two successive passages of the sun across the vernal equinox (or between two successive solar passages across the autumnal equinox) is defined as the tropical year, which is the year of our calendar, since it is in step with the changing seasons. The tropical year is about 20 minutes shorter than the sidereal year which is the time between two successive passages of the sun across any particular star. Put differently, and expressed in terms of the earth's revolution around the sun (which Hipparchus did not believe), the interval between two successive beginnings of spring (the vernal equinox) is about 20 minutes shorter than the time it takes the earth to revolve once around the sun (the length of the sidereal year). This means that, in its journey around the sun, the earth, starting from the vernal equinox, meets the vernal equinox again 20 minutes before it completes its journey around the sun. This leads to two different year lengths: the sidereal year, the period of a complete trip around the sun, namely 365.2596 solar days and the tropical year (from vernal equinox to vernal equinox), namely 365.2422 days. Our calendar follows the tropical year because the calendar must keep in step with the seasons.

The explanation of the difference in length between the sidereal and tropical year, which Hipparchus discovered, is that the equinoxes are not fixed points on the ecliptic but move westward along the ecliptic by a small amount each year, moving all the way around the ecliptic once every 26,000 years. This is called the precession of the equinoxes, which may have been known to the ancient Egyptians and, probably, to the ancient Babylonians. Hipparchus discovered this precession by comparing the positions of certain stars relative to the vernal equinox with these positions as given by Greek sky observers 150 years earlier. From these observations he calculated that the vernal equinox, and therefore the autumnal equinox, were shifting westward along the ecliptic by about 1.4 degrees per hundred years, which gives the change in the direction of the vernal equinox every hundred years as seen from the earth.

Although Hipparchus is most famous for his discovery of the precession of the equinoxes, his most important contribution to the development of astronomy was his introduction of precision and systematic recording in observational astronomy. Up to the time of Hipparchus, all observational astronomy was more or less of a random occupation. The need for systematic observation of the celestial bodies was not recognized because the early Greeks believed that everything was precisely ordered and, hence, unaltered from eon to eon. Thus nothing would be gained by a night to night observation of the sky.

Hipparchus pursued astronomy in the manner of modern observers, which meant keeping careful records of the stars and the planets. The only book by Hipparchus which still exists, was written in 140 BC, before he discovered the precession of the equinoxes. This book deals primarily with the rising and setting of the stars and lists the times of culmination (passing across his meridian) of stars from hour to hour. The accuracy of these observations tells us that the clock that Hipparchus used was probably an accurate water clock. We may judge the accuracy of Hipparchus' observations from his determination of the lengths of the tropical and sidereal years, both of which he determined to within 6.5 minutes of modern measurements.

Hipparchus' stellar astronomy was also highly accurate and it greatly influenced the astronomers who followed him, particularly Ptolemy, who, in his *The Almagest* (also known as *The Great Syntax*) gave a detailed record of Hipparchus and his achievements. Ptolemy records that Hipparchus accurately determined the positions relative to the vernal equinox and the ecliptic (celestial latitudes and longitudes) of 1080 stars and classified their brightness. This is the first example in astronomical history of a star catalogue. Hipparchus probably discovered the precession of the equinoxes by comparing the positions of the stars in his catalogue with the positions of those stars that were recorded by the earlier Greek astronomers. His catalogue is remarkable not only because it listed the precise positions of the stars but because it also gave the apparent brightness of each star. For the first time in the history of astronomy the concept of the magnitude (as a measure of brightness) of a star—now universally used—was introduced. From Ptolemy's description of Hipparchus' stellar observations we know that Hipparchus recorded the first observed nova in 125 BC Hipparchus' discovery was also recorded by the Roman historian Pliny the Elder, in AD 50 in the second volume of his *Natural History*. The story in the *Natural History* is that Hipparchus was so amazed at seeing what he considered to be the birth of a new star that he decided to devote himself to measuring the brightness of stars.

He introduced the classification of stellar brightness by arranging the stars on an importance scale which he associated with brightness. Thus to Hipparchus, the brightest stars were the most important "stars and therefore, stars of the first magnitude." "Magnitude" in this context has nothing to do with the size of a star, which could not be measured until 2000 years later (the second decade of the twentieth century). In his stellar catalogue Hipparchus divided the visible stars into six magnitude classes, with those in the first class the brightest to the naked eye, and those in the sixth class barely visible to the naked eye. This magnitude concept was extended a few hundred years later by Ptolemy, but it did not become a precise astronomical unit of brightness until 1850 when it was placed on a sound arithmetic basis by the British astronomer Norman Pogson (it is now known as the Pogson magnitude scale). The magnitude scale is a reverse order arithmetic scale, with large magnitudes assigned to faint stars and small magnitudes assigned to bright stars.

Hipparchus is renowned not only for his star catalogue and his discovery of the precession of the equinoxes, but also for his investigations into the orbits of the planets, his attempt to determine the sizes of the sun and moon, and his contributions to mathematics. In his study of planetary motions he was most concerned about the irregularity in their motions and their retrograde motions. He tried to explain both of these by introducing different centers for their circular orbits. Because this did not account for both their irregular motions and their retrograde (from east to west) motions, he proposed the concept that two different circles of motion must be assigned to each planet. This concept did not explain all the observations, but it was the beginning of what we now call the theory of epicycles, which was taken over and greatly extended by Ptolemy.

All Hipparchus could measure in his attempt to determine the sizes of the sun and moon was their apparent sizes. The apparent size of an object at a given distance from the eye is the angle the object subtends at the eye (the angle formed by two lines to the eye, one from each side of the object). In the case of the sun or the moon it is the angle that the diameter of the disk of either one of these bodies subtends at the eye of the observer. If one knew the distance of the object one could then calculate the true diameter of the object, or vice versa. Alternatively, we may picture the full moon and imagine looking first at the left edge of the moon and then turning our eye until we are looking exactly at its right edge. The angle through which we rotate (turn) our eye from left to right is the angle subtended at the eye by the moon. This is called the apparent size of the moon, which is about half a degree.

Hipparchus knew this and saw that the apparent sizes of the sun and moon are almost equal. To determine the true sizes of the sun and moon from these observations, Hipparchus needed to know the distances of these bodies. Though the distance of the moon was probably fairly well known, the same could not be said of the sun's distance. Aristarchus had already estimated, from his observations of the lunar phases, that the sun is at least about 20 times more distant than the moon and therefore at least some 20 times larger than the moon. Hipparchus was critical of this method and actually tried to find the relative sizes of the disks of the sun and moon by measuring the time the moon takes to pass through the earth's shadow during a lunar eclipse. This required a bit of trigonometry, which Hipparchus knew thoroughly. From these measurements Hipparchus found that the moon's distance from the earth is about 59.1 earth radii (the accepted value today is very nearly 60 earth radii) and that the sun's distance is about 2550 earth radii (a flagrant underestimate as the sun's true distance is 93 million miles). Still, Hipparchus's remarkable work was a drastic departure from the work of earlier Greek astronomers (Aristarchus excepted). It was a bold attempt to treat the sun as a physical body with measurable dimensions. The historian Pliny mentions that the Greek astronomer Poseidonius had attempted to measure the size of the sun, with very little success.

Hipparchus was an accomplished mathematician and contributed some important theorems to trigonometry. In particular, he prepared a table of the chords of a circle which is equivalent to a table of the sines of angles. He showed that the sine of half the angle, subtended at the center of a circle of unit radius by an arc of a circle connecting two points on the circumference, equals half the length of chord (straight line) connecting the two points. Thus the lengths of various chords in a unit circle give sines of angles.

Hipparchus' importance to the story of astronomy cannot be overemphasized because he changed the quality of the study of astronomy from an undisciplined, disorganized random collection of observations of the celestial bodies to a precise pursuit of knowledge. Precision was all-important to Hipparchus but attaining that observational precision required infinite patience and oft-repeated measurements of the same phenomena, month after month and year after year. This was most forcefully demonstrated by his discovery of the precession of the equinoxes. Thus Hipparchus was an innovator in every sense of the word insofar as Greek astronomy went. Indeed, as the first "modern" astronomer among the early Greeks he represented the beginning of a new era and, of course, the end of an old one. But two hundred years elapsed before Hipparchus's great contributions were recognized by Ptolemy whose *The Almagest* is devoted in great part to the work of Hipparchus. Without Ptolemy's remarkable description of Hipparchus' astronomical discoveries, we would have very little information about Hipparchus because he left hardly any record of his own work.

Claudius Ptolemaus (Ptolemy) (AD 100–170) was probably born in Greece but his Latin named indicated that he possessed Roman citizenship. His earliest celestial observations are dated AD 127, the eleventh year of the Roman emperor Hadrian, and his stellar catalogue is dated AD 137. Because he flourished in Alexandria, he is often called Ptolemy the Egyptian. He deserves his title of Ptolemy the Great for his summary of the work of the outstanding Greek astronomers such as Aristarchus and Hipparchus. In his *The Almagest,* Ptolemy examined and commented on every problem in astronomy that had challenged his predecessors. But the principal problem that concerned him was the explanation of the motions of the planets. In this area of his work he accepted the concept of epicycles as proposed by Hipparchus, but improved on Hipparchus' work by developing a very elegant epicycle model of the motions of Venus and Mercury. Ptolemy had to explain why the apparent motions of Mercury and Venus differ drastically from the apparent motions of Mars, Jupiter and Saturn.

It is sufficient to consider the apparent motion of Venus alone to understand the problem that confronted both Hipparchus and Ptolemy and how Ptolemy solved it using special epicycles. When we speak of the "apparent motion" of a celestial body, we mean motion relative to our fixed earth; the earth is our fixed frame of reference. To Hipparchus and Ptolemy, the concept of a frame of reference had no meaning because it never occurred to them that the earth could be movable. But we see quite clearly today that if we take into account the earth's motion around the sun and correct for it in our analysis of planetary motions, we obtain what we may call the "true motions of the planets" or, more accurately, the motions of the planets relative to the sun.

Returning now to the planet Venus, we note that it always appears to be near the sun, either to its left or right, but never straying more than 48 degrees on one side or the other, shifting its position from one side to the other periodically as though it were attached to the sun like a pet attached by a rope to a post in a yard. When Venus is to the right (to the west) of the sun, it rises early in the morning, before the sun rises. Venus is then called the Morning Star. The early Greeks called it Phosphorus. It then sets before the sun sets. When Venus is to the left of the sun (east of the sun), it is visible in the early evening and sets after the sun does. Called Hesperus by the early Greeks, we now call it the Evening Star. Pythagoras was probably the first to discover that Hesperus and Phosphorus are the same planet, which was named Venus by the Romans.

Unlike the planets Mars, Jupiter, and Saturn which rise and set at all times of day without regard to the sun, the rising and setting of Venus and Mercury closely coincide with the risings and settings of the sun. To explain this "strange" behavior of the planets, Ptolemy accepted the loops proposed by Herakleides for the motions of Mars, Jupiter, and Saturn, but such loops would not work for Venus and Mercury. He solved the problem very neatly by picturing Venus as moving around a small circle (the epicycle) whose center is on a large circle at the center of the earth. Ptolemy then had the line from the center of the earth to the center of the epicycle pass through the center of the sun. By choosing the radius of Venus' epicycle just right, Ptolemy kept Venus in its observed relationship with respect to the sun as the center of the epicycle revolved around the earth. This was so clever a model (which also applied to Mercury but with a smaller epicycle) that the epicycle concept for planetary motions dominated astronomers' thinking for some 1500 years and was swept away only with the theories later proposed by Copernicus in the sixteenth century.

As an ever increasing number of epicycles had to be introduced in time to account for the changes in the observed motions of the planets that the increasing accuracy of observations revealed (e.g., the additional observed motions produced by the precession of the equinoxes), one may consider why a patently incorrect model of planetary motions can agree with the observations. The epicycle model is not "incorrect" insofar as the observations go if we take the earth as our frame of reference. The observations can then be understood in terms of a famous mathematical theorem in trigonometry discovered in the eighteenth century by the French mathematician and physicist Joseph Fourier. This theorem states that any motion, however complex, can be represented as a sum of smaller and smaller circular motions (the Fourier series). By the time Copernicus came on the scene, astronomers had found it necessary to introduce more than 100 epicycles to account for the observed motions of the planets produced by the earth's detailed motions.

One may wonder why the astronomy of Hipparchus and Ptolemy did not mark the beginning of a new astronomy but only the end of the ancient Greek astronomy. Both were excellent observers and theoreticians, but, of course, they lacked the laws of motion and the law of gravity which were developed by Galileo and Newton. Copernicus also lacked the knowledge of those laws, and yet he took the tremendous leap which forever banished the Ptolemaic cosmos from astronomy. The only way we can account for this 1500-year gap is to argue that philosophy and religion were to blame for the stagnation of astronomy. Ptolemy had found favor in the eyes of the Christian priesthood, and Aristotle had produced the philosophy and metaphysics to support the Ptolemaic geocentric cosmology which thus left nothing to be questioned or to be desired.

In addition to these two brakes on the development of science was the deterrence produced by the destruction of the Roman Empire and the rise of the "barbarians." The last Roman emperor was deposed in AD 476, but, in time, the domination of the barbarians gave way to the Holy Roman Empire as Christianity was gradually adopted throughout Europe. This development had its good and bad points. The good points were that Christianity, to some extent, encouraged a devotion to contemplation and study and that the convents and monasteries became the repositories of manuscripts of all kinds including the writings of the Greek astronomers which might otherwise have been lost. The bad points were that the Christian hierarchy insisted on a narrow-minded literal interpretation of every word in the Scriptures. Every departure from these strictures and any attempt to question Christian authority were rejected, scorned, and severely punished. It is no wonder, then, that hardly any significant astronomy was produced during the medieval period in Europe from AD 500 to 1500.

A further deterrent to the growth of independent creative thinking and, therefore, of astronomy, was that such thinking in general was limited to those who had time to devote themselves to thinking, as the Greek and Alexandrian philosophers did. Instead of such free thinkers who flourished in the early pre-Christian era and exchanged their ideas in such free institutions as Plato's Academy and the great library in Alexandria, the thinkers in the European medieval era were associated with monasteries and convents and therefore constrained by the religious rules of the institutions to which they were bound.

As evidence of the low level of thought that dominated monasteries during the medieval period in Europe, we point out that extensive tracts were written by religious leaders warning that the acceptance of any thesis based on a moving earth or a denial of the central position of the earth was heretical.

When the mathematician Gerbert ascended the papal throne in AD 999 as Sylvester III, the restrictions on accepting the earth as a sphere were removed so that some freedom of thought about astronomy was allowed. Geographers thus became the leaders of astronomical thinking. This was a far cry from the intellectual freedom of the pre-Christian Greeks and Alexandrians. But the writers of this period were free to read the early Greek and Roman histories of the Greeks. Most important among these was the work of Pliny whose history of the early Greek astronomers stimulated many of the medievalists to stray into forbidden astronomical territories. But the combined efforts of all of these brave intellectual explorers did very little to ease the grip of Christian theology on astronomical thinking or to offer any alternative to Ptolemy's epicycles and his geocentric model of the solar system. One need only read the writings of the Church fathers to see what an uphill struggle it was to free astronomy from doctrinaire religion, let alone to advance it beyond Ptolemy. Certainly most of what was written in the Middle Ages in Europe actually deterred astronomical study instead of advancing it. Even those medievalist writers who dared to wander into astronomy did not go beyond Aristotle, who was accepted as the authority on all things scientific. This adherence to Aristotelian thinking evolved into what we now call the scholastic school led by Thomas Aquinas (1227–1274) and Albertus Magnus (1193–1280). If anything, scholasticism was more of a drag on astronomy than was theology since it bore the imprint of a great Greek philosopher.

One independent thinker who went against the theology and scholasticism of the Middle Ages was the British scholar Roger Bacon (1214–1294), who knew the Greek philosophers very well and had mastered the Ptolemaic model of the solar system. Bacon was even acquainted with the Arabian astronomy of that period. In all his writings, Bacon argued passionately for the need to separate the Scriptures from the study of the stars and planets. Indeed, he was courageous enough to point out difficulties and contradictions in various passages of the Old Testament. He ridiculed the writings of the Church leaders on astronomy, arguing that they were without merit because they were based not on observations and measurements but on speculation and superstition. Bacon would certainly have projected astronomy into its modern mode if he had not been ordered by his supervisors in the Franciscan monastery to which he belonged to give up all scientific work, particularly astronomy. Indeed, his books were banned and he was imprisoned for some ten years.

The extent of astronomical knowledge in the Middle Ages is best exemplified by Dante's *Divine Comedy* in which Dante attempted to incorporate Ptolemaic astronomy into a complete cosmological treatise in agreement with the accepted Catholic theology of the day. However, he was guided more by Aristotle's philosophy than Ptolemy's astronomy. Dante seems to have accepted the idea of a spherical earth because in his description of his descent into Hell, he remarks on passing through the "center of Hell" and looking back so that they (Dante and his guide Virgil) saw Lucifer "upside-down" which meant that they had "commenced their ascent to the other side of the earth."

Summing up the contributions of the Middle Ages to astronomy, we may say that they amounted to very little of substance. Dante died in 1321, about 1000 years after Christianity had been proclaimed the state religion of the Roman Empire. Christian theology frowned upon the study of "paganism" as the Greek philosophy and spirit of antiquity were called. Even the acceptance of Aristotle as the "official" philosopher of Christianity did little to encourage the free flow of ideas. Indeed, the acceptance of Aristotelianism was used as a device for discouraging any ideas that threatened the Ptolemaic cosmology.

That very little astronomy was pursued in medieval Europe does not mean that astronomy was dead everywhere. Indeed, it flourished in countries such as India, the Arab countries, and China, but this activity influenced the development of Western astronomy only very slightly, primarily because the contact between medieval Europe and India and the Arab countries was very superficial. The Indian astronomers taught that the earth is a sphere, floating freely in space, with all the other planets revolving around the earth. Some Hindu astronomers promulgated the concept of a rotating earth to explain the diurnal rising and setting of the stars, but the idea of a spinning earth was rejected by most Indian astronomers as contrary to the observations that objects on the earth are not flying about helter-skelter as they "would be if the earth were spinning."

The conquest of Persia by the Arabs in the seventh century AD brought the Arabs in contact with the Hindus whose philosophies and science had penetrated into Persia. The Caliphs of that period had become interested in the motions of the stars and planets and had ordered various of the Hindu books on the stars and planets to be translated into Arabic. These books greatly encouraged not only the study of astronomy but also of mathematics. The Caliph Mamun, the son of Harun Al Rashid (813–833), was a patron of science and, in fact, had enlarged the observatory near Damascus that had been constructed by the Omayyad Caliphs. All of this, however, had little impact on medieval astronomy in Europe because the Europeans were unaware of Arabian astronomy.

The contrast between Arabian astronomy and medieval European astronomy is that no theological restrictions were placed on Arabian astronomers. In any case, no real progress was made by the Arabs or the Europeans since neither advanced beyond Ptolemy. Indeed, as long as astronomers accepted the wrong notion that the earth is fixed in the solar system they could not advance beyond Ptolemy.

We have said nothing about the contributions of the Chinese to early astronomy primarily because China was such a closed society that the Europeans received very little information about Chinese astronomical research. Even after Marco Polo had returned to Italy in 1292 from his stay in China, with exciting stories about the wealth and technical skills of the Chinese, the Europeans had no contact with the Chinese, certainly not with their astronomical work. We now know, however, that the Chinese were brilliant astronomers and superb observers. The best example of their astronomical achievements was their observations of the Crab Nebula in Taurus, which the Chinese described in 1054 as a "guest star" (now known as the remnants of a supernova) which exploded on July 4, 1054. The Chinese followed this "guest star" for many days until it faded from sight. It was discovered in the twentieth century with the aid of modern telescopes. That the medieval European astronomers have left no records of this amazing event shows how far ahead the Chinese astronomers were of their European counterparts at that time. One may wonder why the sudden appearance of this celestial object escaped the attention of the European astronomers since it became as bright as Venus and remained so for many days. It may well be that European astronomers did not report it for fear of being labeled as heretics since the heavens and celestial bodies had been ordained by the Scriptures to remain unchanged forever.

We have seen, in our discussion of medieval European astronomy, that Ptolemy's work, instead of being an advance in the development of astronomy, was really an obstacle to it. How, then, did the truth finally break through the barrier of religious terror, bigotry, superstition, scholasticism, and Aristotelian error and misconception? It broke through only with the acceptance of the motions of the earth as a fact. But even then, the transition from Ptolemaic astronomy to modern astronomy took hundreds of years and a revolution in man's concept of nature and its laws. In the next chapter we discuss the first step in this remarkable transition.

CHAPTER 5

# The Revival of European Astronomy

*Here and elsewhere we shall not obtain the best insights into things until we actually see them growing from the beginning.*

—ARISTOTLE

The fifteenth century in Europe was remarkable in that two great intellectual movements began to sweep across the continent almost simultaneously: the Renaissance and free scientific inquiry—particularly astronomy. Scholasticism was a mixture of Greek reasoning and Christian revelation as epitomized in the philosophical writings of Saint Thomas Aquinas in Italy, Duns Scotus in Scotland, and Saint Albertus Magnus in Germany. Accepting Socrates not only as the father of logic, but also as the infallible authority in physics, astronomy, and biology, the scholastics considered the Scriptures and Christian theology in general as the guiding truth to be followed in studying and understanding nature at all levels. Acting as a pall on the freedom of thought, particularly on the pursuit of science, for almost two centuries, it began to fade by the end of the thirteenth century. It had not been without its merits, however, because it had awakened an enormous interest in the Greek philosophers so the monasteries were besieged by requests for ancient Greek manuscripts. Fortunately, the printing press with movable type appeared at about this time so that knowledge and ideas were widely disseminated.

The Renaissance contributed to the growth of new ideas in that it encouraged art, literature, architecture, and exploration. This led to technological developments and the invention of such things as the mariner's compass (the magnetic compass). Ultimately these technological developments led to the astronomical telescope. At some point in the decline of scholasticism and the rise of the Renaissance, the pursuit of forbidden ideas had to receive the imprimatur and encouragement of an outstanding scholastic himself, who was beyond censure. This occurred when Nicholas of Cusa, a German prelate, philosopher, and mathematician, anticipated the heliocentric theory of Copernicus. Born in 1401 at Cues on the Moselle, he studied at the universities of Heidelberg, Bologna, and Padua and learned astronomy from the geographer Paolo Toscanelli. A friend and firm supporter of Pope Pius II, who appointed him a Cardinal, Cusa wrote extensively on mathematics; he was perhaps the first of the dominant intellectual figures of that time to suggest that the universe is infinite and therefore has no center, which is in line with modern cosmological concepts. He argued that the earth is moving and that its inhabitants do not know it is moving; they are moving along with it and cannot detect its motion in a universe that looks the same in all directions and from all points. This led him to reject the concept of a center of the universe or, put differently, he believed that every point can equally well serve as a cosmic center. He also argued that motion is a natural state of all bodies in the universe and that, therefore, the earth being one such body must also be moving. In spite of Cusa's revolutionary astronomical concepts, his work had relatively little influence on the growth of astronomy. This was probably owing to the very speculative nature of his concepts which had little observational material to support them. He confined himself to generalities and always considered himself a disciple of Ptolemy. All in all we cannot say that Cusa anticipated Copernicus, for in his later writings he rejected the notion of a moving earth and accepted Ptolemy completely. But Cusa's speculations were straws in the wind that presaged the flow of new ideas that were to culminate in the Copernican revolution.

Between the time of Cusa and Copernicus, a few writers on astronomy are worth mentioning, not because they contributed to the Copernican revolution that was to come but because they kept alive the flow of new ideas. Among them were Celio Calcagnini (1479–1541), Johannes de Monte Regio (1436–1476), George Reurbach (1423–1461), Girolamo Gracastoro (1483–1553), and Giovanni

Battista Amiri (1500–1538). These astronomers, if we may call them that, were all concerned with trying to reconcile the Ptolemaic geocentric cosmology with the planetary and stellar observations, but as the accuracy of these observations increased, an ever growing number of epicycles had to be introduced to account for the observations. As an example, we note that the precession of the equinoxes, as first observed by Hipparchus, required more epicycles than those introduced by Ptolemy. So the number of epicycles grew until more than 140 had been introduced by the time Copernicus came on the scene to support the Ptolemaic geocentric model of the solar system. The most stubborn observations that required extensive epicycles were the variable speeds of the individual planets and the eccentric motions as observed from the earth. To Copernicus these additional circles were intolerable and indicated fatal flaws in Ptolemy's astronomy.

Nicolaus Copernicus was born on February 19, 1473 in Thorn on the Vistula. He was educated as an Aristotelian and, in his early years, accepted the Ptolemaic model of the solar system, as taught at the University of Cracow by the outstanding authority on astronomy, Albert Brudzew (Brudzewski), who had written the first of a series of commentaries on the outstanding book on planetary motions by Reurbach.

Whether Brudzewski had imbued Copernicus with doubt about Ptolemy's theory is not clear, but Copernicus' thinking about the motions of the earth and the planets was influenced much more by his travels in Italy than by his studies at Cracow. In 1494, when Copernicus returned home from Cracow, his maternal uncle, Lucas Watzelrode, Bishop of Ermland since 1489, made him a canon of the Church in the Cathedral of Frauenburg. Watzelrode insisted, however, that his nephew spend time at various Italian universities before assuming the canonry. Copernicus left for Italy in 1496 and returned briefly to Frauenburg to accept the canonry, but was immediately granted another leave of absence to complete his studies in Italy, where he remained until 1506.

Copernicus had started his Italian schooling at the University of Bologna, where he studied with the astronomer Domeniro Maria de Novara, who was a very careful observer. But the excitement

Nicolaus Copernicus (1473–1543) (Courtesy AIP Neils Bohr Library)

of observational astronomy never rubbed off onto Copernicus, who was far more comfortable working with data supplied by observers than with doing any practical observing himself. All told, he seems to have made no more than 30 observations of his own. This does not detract in any way from Copernicus' great contribution to astronomy and to the advancement of science as a whole.

Before leaving Italy for the first time Copernicus spent a year (1500) at the University of Rome, where he gave a course on mathematics. When Copernicus returned to Italy in 1502 he went to Padua where he continued his studies of law, mathematics, and medicine, receiving a degree of Doctor of Canon Law. By the time he left Italy in 1506 he was a master of theology and knew the classics thoroughly; he had also mastered all the mathematics and the astronomy known at that time.

When Copernicus returned to his canonry duties in Ermland, he was delighted by the leisure he had to pursue his astronomical studies and by the small demands on his time made by his clerical commitments. Copernicus, untrained an observer as he was, discovered many celestial phenomena that are greatly simplified by a heliocentric solar system, and so he decided to devote the rest of his life to formulating a heliocentric system that would be acceptable to everyone, even to the Pope. He was driven by the conviction that if one accepted, as a fact, that the earth and all the other planets revolve around the sun, the astronomy of the solar system would be greatly simplified. His reasoning was very simple and basic: he understood that the apparent motions of the celestial bodies—the moon, sun, planets, and the stars—mirror to some extent the motion of the earth. The true motions of the celestial bodies can then be obtained or deduced from the observed, that is, apparent celestial motions by subtracting out the earth's motion. We know that if we are on a train that starts moving gently we are startled by seeing what appear to be the motions of fixed objects gliding past us. This startled feeling lasts but a moment as we become aware of our mistake and note that our train is moving. Copernicus knew this but he could not bring himself to the point of announcing his heliocentric theory without first finding some justification in the opinions and theses of the great and noble philosophers and mathematicians who had come before him and had questioned the truth of the geocentric solar system.

In his dedication to Pope Paul III at the beginning of his great book, Copernicus remarks that he was prompted to develop a new model of the solar system by the inadequacies and inconsistencies he found in the Ptolemaic solar system. This convinced him that some essential feature had been left out by Ptolemy and all his disciples, which, if included, would have simplified everything and led to a correct and consistent model of the solar system. Believing that the motion of the earth is this "essential feature," he decided to read the essays of philosophers of the past which might say something about the motion of the earth.

Beginning with the Roman histories he discovered that Cicero had stated that Nicetus had believed the earth to be moving and that Plutarch, the Greek biographer, had stated the same thing, noting that Philolaus, Herakleides, and Ekphantes had also played around with the idea of a moving earth. Plutarch's discussion and description of the work of Aristarchus of Samos was probably the argument that was most convincing as far as Copernicus was concerned. According to Plutarch, "Aristarchus of Samos was charged with impiety. It was said that he would not believe the Earth to be the center of the world; he averred that it runs in an inclined plane and simultaneously turns around its axis—and that he would have this so in order to be able to calculate more exactly the phenomena of the heavens." Emboldened by these statements by the ancient Greeks, Copernicus rejected Ptolemy and placed the sun at the center of the solar system, noting at once that this correctly accounts for the apparent back and forth motions of Mercury and Venus without having to introduce epicycles. In his dedication to Pope Paul III, as recounted by J.L.E. Dreyer, Copernicus includes a passionate and poetic panegyric to the sun: "How could the light [the sun] be given a better place to illuminate the whole temple of God? The Greeks called the Sun the guide and soul of the world; Sophocles spoke of it as the All-seeing One; Trismegistus held it to be the visible embodiment of God. Now let us place it upon a royal throne, let it in truth guide the circling family of planets, including the Earth. What a picture—so simple, so clear, so beautiful."

Copernicus ends his tribute by appealing to the Pope to understand why he, Copernicus, had finally decided to propose the motion of the earth as the only solution to the problems raised by the Ptolemaic system. He explained that he had thought and anguished about his theory for some 36 years before proposing it. Noting the disagreements and conflicting ideas among the Greeks about the earth's motion, he said

> Occasioned by this [the disagreements] I also began to think of the motion of the earth, and although the idea seemed absurd, still as others before me had been permitted to assume certain circles [the epicycles and loops] in order to explain the motions of the stars, I believed it would readily be permitted me to try whether on the assumption of some motion of the earth better explanations of the revolutions of the heavenly spheres might be found. . . . When the motions of the other planets are referred to circulation of the earth and are computed for the revolution of each star, not only do the phenomena necessarily follow therefrom, but the order and magnitude of the stars and all their orbs and the heavens itself are so connected that in no part can everything be transposed without confusion to the rest and to the whole universe.

Copernicus was particularly pleased that the motions of Venus and Mercury and the motions of the outer planets fell neatly into place as did the asymmetry in the brightness of Mars when rising in the evening and rising in the morning. One can easily detect, as Copernicus did, that the brightness of Mars when it rises in the morning is distinctly different from its brightness when its rises in the evening. He reasoned that such a marked difference in its brightness in those two instances cannot be produced with epicycles because epicycles cannot produce the necessary difference in Mars' distance from the earth. An orbit around the sun does it because the change in brightness is due to the changes in the distance of Mars from the earth by as much as the diameter of the earth's orbit around the sun, which is 186 million miles. Copernicus did not know this distance precisely but he knew it to be large enough to produce the observed brightness difference.

Copernicus, deeply concerned about his commitments to his church and his devotion to the Pope, was very reluctant to publish his heliocentric theory, even though he was strongly urged to do so by some church dignitaries themselves. Thus one of the cardinals in Rome gave a lecture to the Pope on Copernicus' heliocentric theory, and sent his secretary to Frauenburg to obtain copies of Copernicus' proof that the earth revolves around the sun. In a letter to Copernicus he stated "If you fulfill this wish of mine you will learn how deeply concerned I am of your fame, and how I endeavor to win recognition of your deed. I have closed you in my heart." This letter was equivalent to an official sanction from the Church because the cardinal was its chief censor. This was followed by the encouragement from the Bishop in Copernicus' diocese that Copernicus' book be published as a duty to the world.

As rumors began to spread that Copernicus had developed a heliocentric theory that eliminated most of the difficulties that plagued the Ptolemaic system, his fame grew and he began to receive requests for a written exposition of his work. To satisfy his friends and circle of admirers and disciples, he prepared a short resume of his work (*Commentariolus*) which was widely distributed. Thus Tycho Brahe received a copy in 1575 from the physician of the Emperor of Denmark.

The *Commentariolus,* far from satisfying the ever-increasing number of Copernicans who were becoming ever more vociferous and penetrating into the very highest circles of society, made them more insistent than ever that Copernicus publish his complete works. Copernicus, in his own words, indicated that he was afraid of two things if he published his book: that he would appear ridiculous ("hissed off the stage of history," as he put it) in the eyes of his peers, and that his daring theory would generate a storm of protest and even violence against him. He was finally persuaded to publish by Georg Joachim Rheticus, a young professor in Wittenberg. He visited Copernicus in Frauenburg in 1539 and spent two years there studying Copernicus' great work, which was published in Danzig in 1540. This immediately made Copernicus famous and probably induced him to yield to pleas of his friends to publish his grand work *De Revolutionibus*. When Copernicus finally consented to the publication of his book, it was first edited by Rheticus, and later by Andreas Osiander, a Lutheran theologian of Nürnberg.

In the final version of the book Copernicus defined his philosophy about the role of the scientist, particularly that of the astronomer. He argued that the astronomer should not be blamed for celestial phenomena, for all that astronomers can do is to report these phenomena, confessing that the astronomer may have no understanding of why they occur. All the astronomer must do is record his observations as accurately as possible so that from them calculations can be made as to the true nature of the phenomena. He went on to state that any hypothesis the astronomer makes need not be true or even probable as long as it is heuristic and ultimately leads to a correct understanding of events. If the hypothesis leads to a model that correctly describes the events, then the hypothesis may be assumed to be correct.

Critics have taken this statement as an apology or even a denial by Copernicus of his own theory, but further reading of the original manuscript shows that Copernicus wanted to be as honest as he possibly could be about his own doubts, ending his statement by saying that since science cannot explain or even understand the cause of the observed irregularities of the motions of the planets, the astronomer is free to, and should, choose the hypothesis that gives the simplest and most admirable explanation. In his letter to Osiander about publishing *De Revolutionibus*, Copernicus stated his belief, with great firmness, that he had to proclaim "his convictions before the world, even though science should be condemned." In spite of Copernicus's discussion of hypothesis versus reality, all who spoke or listened to him knew he firmly believed that the motion of the earth is real and not a mere hypothesis.

Copernicus received a copy of his book on his deathbed in 1543 and died before he could read it. It is divided into six sections or "books" as they were then called. The first of these books contains a general discussion and description of his heliocentric system. Using general arguments about the perfection and beauty of a sphere, he argues that the universe and all bodies in it, including the earth, are spheres, pointing out that drops of water form into spheres quite naturally. From this simple argument he goes on to accepting the motions of all celestial bodies as uniform and circular. He then proposes the idea that bodies moving in smaller circles must travel faster than those moving in larger circles. He thus accounts for the rapid changes of Mercury and Venus with respect to the sun as compared to the changes of Mars, Jupiter, and Saturn. The epicycles of Venus and Mercury proposed by Ptolemy were thus replaced by circular orbits (a small one for Mercury and one about $2^1/_2$ times as large for Venus). To enable readers to follow his arguments, Copernicus ended this book with two chapters on plane and spherical trigonometry.

In the books that followed, Copernicus carefully examined all the objections that had been raised to a moving earth, going back to Ptolemy, Hipparchus, and Aristotle and showed that all of these objections can be answered in a simple way. Thus to the objection that "if the earth were moving, the atmosphere would be left behind," he answered that the atmosphere is attached to the earth and moves along with the earth as the earth revolves around the sun and rotates with the earth's rotation so that each of us sees an unchanging atmosphere.

In the five books that follow the first and complete his great work, Copernicus considers in turn, spherical astronomy, the precession of the equinoxes, the motion of the moon, the motions of the planets in longitude (east–west) and their motion in latitude (north–south). Each of these subjects was considered with great care and concern.

Using his hypothesis that the planets closest to the sun move more rapidly than the more distant planets, Copernicus calculated the mean distances of the planets from the sun as measured in terms of the earth's distance which he took as 1. On that scale he obtained the distances 0.376, 0.719, 1.520, 5.219, and 9.1743 for the planets Mercury, Venus, Mars, Jupiter, and Saturn respectively. Copernicus's great error, in all of this brilliance, was in assigning circular orbits to the planets, which forced him to introduce his own epicycles. This flaw in his remarkable theory detracts only slightly from his great achievement, which brought astronomy from the scholasticism of the medievalists to the rigorous science of Newton, which was promulgated more than a century later.

After its appearance in 1543, *De Revolutionibus* became the astronomy textbook of choice, quickly displacing Ptolemy's *Almagest*. The most remarkable thing about this tale was that the acceptance of the motion of the earth, as Copernicus wove it into his heliocentric theory over a period of some 30 years, overthrew a 1600-year-old tract that had enthralled the greatest minds for centuries. But the best was yet to come, for Copernicus had opened the floodgates to a vast flow of new discoveries and concepts that revealed an incredibly pent up desire among intellectual leaders to throw off the theological and scholastic chains that had bound humanity for centuries. Even though Copernicus had constructed a complete, self-consistent astronomy on the basis of his assumption of a moving earth and a fixed central sun, almost 90 years elapsed before the last remnants of Ptolemaic astronomy were banished forever by a precise mathematical formulation of Copernican astronomy. This remarkable synthesis, the bridge that led from Copernicus to Newton and modern science, was built by Tycho Brahe, Johannes Kepler, and Galileo Galilei. In the following chapter, we describe the work of Brahe and Kepler, whose works are intricately interwoven.

CHAPTER 6

# Tycho Brahe and Johannes Kepler

*The die is cast; I have written my book; it will be read either in the present age or by posterity, it matters not which; it may well await a reader, since God has waited six thousand years for an interpreter of his words.*

—JOHANNES KEPLER

Taking all of the pertinent things into account we must place Copernicus' achievement on a very high level even though his importance to astronomy is often downgraded because his detractors argue that he made very few observations. This in itself is a testament to the greatness of his achievement. It is one thing to be forced to a particular conclusion by incontestable observations and quite another (and vastly superior) one to construct a correct theoretical model on the basis of pure thought, as Copernicus had done in concluding that the heliocentric model of the solar system is superior intellectually and aesthetically to the Ptolemaic system in every way.

But Copernicus had little choice but to work outside the realm of observational astronomy. With no observational equipment available to him, and lacking the temperament that would have permitted or driven him to make the necessary hundreds of observations, night after night, that would have convinced him of the truth of his model, he would have had to rely on observations made in earlier years and these were highly inaccurate. Copernicus initially accepted the observations of the early Greeks and, particularly, those of Hipparchus and Ptolemy as highly accurate and irreproachable, but in his later years he had to confess that these observations were replete with errors. These were particularly egregious in the observed positions of the planets which were particularly important to astrologers who were revered far more than astronomers were at the time of Copernicus.

The most important planetary tables used at that time, known as the Alfonsine Tables, were some 300 years old and had been prepared at the time of King Alfonso X of Castille. They were issued in 1252, they had been prepared under the direction of Ishak ben Said and the physician Jehuda ben Mose Cohen. When Copernicus began his great synthesis, the Alfonsine tables were in error to the extent of 20 minutes of arc. To explain and describe the intolerable magnitude of this error, we note that observational astronomers expressed, or measured, the positions of stars in terms of angles. We further note that angle is a measure of rotation or turning. If we therefore look directly at a distant object such as a star and turn completely around until we are looking at that same star again, we describe that as a rotation of exactly 360 degrees. This example introduces the degree as the unit of rotation (the 360th part of a complete turn). The minute (represented by ′ ) is the 60th part of one degree and the second (represented by ″ ) is the 60th part of one minute. The errors in the Alfonsine planetary tables were a great source of frustration to Copernicus.

To recognize and understand the vast gap between the accuracy of observational astronomy in the days of Copernicus and today, we point out that 1 minute of arc is the resolving power of the normal human eye. This means that if two points are so far away that the observer has to turn his eye less than 1 minute of arc in shifting its attention from one point to the other, the two points appear as one point. They appear as two points only if the eye has to turn by at least 1 minute of arc. The second of arc is 60 times smaller. We may, perhaps, understand this concept more dramatically if we picture viewing a dime or a penny from a distance of $2^1/2$ miles and following a bug creeping across the coin from one edge to the opposite edge. Such an observation would require the eye to turn 1 second of arc. The modern observational astronomer, using the best modern telescopes and recording devices, not only can do this but also note accurately when the bug has moved one hundredth of the way across the coin. We shall see that Tycho Brahe improved observational astronomy by increasing the observational naked-eye accuracy by a factor of 20 from 20 minutes of arc to 1 minute of arc. This is the degree of accuracy, as we shall see, that Kepler needed to discover the laws of planetary motion, one of the greatest of all scientific discoveries.

Tycho Brahe, the last and greatest of all naked-eye observers, was born in 1546 in Knudstrup, then a province of Denmark. His father had been governor of Helsingborg and his uncle was a country squire and a vice admiral. Tycho was adopted and raised by his childless uncle who planned to steer his nephew into a career of law, statesmanship, or diplomacy, to uphold the honor and noble name of the Brahe family. But Tycho would have none of that and early in life became devoted to astronomy. He was, in line with his uncle's wishes, sent to the University of Copenhagen to study philosophy and rhetoric, but he turned completely away from these studies at the end of his first year in Copenhagen when he watched a partial eclipse of the sun, which had been previously announced. To Tycho it was a divine revelation that one could know the motions of the celestial bodies with such accuracy that one could then predict the exact time of a solar eclipse. He decided then that he would devote his life to acquiring that kind of divinity. This was a fortuitous choice for the immediate future of astronomy.

Having made this choice Tycho turned his attention to the astronomical literature available at the time and never swerved from his goal of becoming the greatest astronomer of the day and being recognized as the equal of Hipparchus and Ptolemy. Even while he was studying this literature he was also observing the stars, the planets, and the moon. So careful an observer did he become that he discovered serious errors in the Alfonsine tables of the planets and the Ptolemaic tables that the German astronomer Erasmus Reinhold had prepared and published in 1551 to correct the Alfonsine tables. Tycho had studied the stars so assiduously that he knew by heart the positions of all the visible stars in the sky by the end of his fourteenth year.

Very little of the astronomical literature of the time departed from *The Almagest*, which Tycho had studied with great care and had absorbed thoroughly. Tycho knew of Copernicus and his *De*

*Revolutionibus* but very little literature had been published to support it. Indeed, even in Germany the Copernican doctrine had only a few followers, although in England such writers as Robert Recorde, Thomas Digges, John Field, and the physicist William Gilbert, author of *De Magnete*, supported the Copernican revolution. If Tycho had remained in Copenhagen, it is doubtful that he would have become sufficiently acquainted with Copernicus' work to have been influenced by his thinking. But Tycho's uncle decided, after Tycho had spent three years at Copenhagen, that his nephew should travel to a foreign country and, accordingly, sent him to Leipzig, supervised by a young tutor. Undeterred by the tutor in any way from pursuing his astronomical observations, Tycho continued his studies at the Universities of Wittenberg, Rostock, Basle, and Augsburg, until his twenty-sixth year. By this time he had become famous on the continent as the young, brilliant nonpareil Danish astronomer.

During this period Copernican astronomy was making very little progress on the continent, with arguments pro and con: Even though Copernicus' errors had aroused considerable controversy, the main body of his work had attracted enough adherents to convince thinkers at all levels that astronomy could no longer be pursued as in the past. Tycho had been caught up in this swirl of conflicting ideas. Being primarily an observer and not a "model builder," he felt that there was no need for him to make a choice. But there were those who felt that Tycho's influence could be very important in furthering the Copernican revolution. This was particularly true of a young, brilliant French mathematician Pierre de la Ramee, or Petrus Ramus, a professor of philosophy and rhetoric at the College Royale at Paris.

An opponent to Aristotelian natural philosophy from his youth, Ramus was convinced that only mathematics could extricate astronomy from the welter of hypotheses that impeded it. He admired Copernicus for rejecting all the old hypotheses that lacked a sound mathematical basis, and reliable observations to support them. He saw in the Copernican doctrine the basis for a new liberated astronomy that could stand or fall on the basis of logic, mathematics, and careful observations alone. In 1569, Ramus, traveling in Germany, met the 24-year-old Tycho at Augsburg and tried to convince him of the truth of his views about mathematics and astronomy. He spoke passionately about the need for a young man like Tycho to start from scratch and construct a correct mathematical theory of the orbits of the planets (earth included) around the sun that was based on a large number of the positions of the planets. This, as we shall see, Kepler did later with Tycho's precise observations.

It did not take much to convince Tycho about the need for precise observations because he had pursued the practice of very accurate observing over long periods of time. This principle drove him during his whole life but he did not limit himself to observing alone because he continued to draw significant conclusions from his observations. The Copernican model of the solar system disturbed him for he reacted to it with great ambivalence: on the one hand, he was greatly attracted to it for its simplicity and elegance; on the other hand, he was repelled by the concept of a moving earth. He rejected that part of Copernicus' theory because, he correctly argued, "if the earth is moving in a circle around the sun, I [Tycho] should see the stars shift back and forth once a year and since I do not observe the stars doing that, I reject the motion of the earth." Tycho's argument and reasoning are correct; the stars do indeed appear to shift back and forth, which is how we measure stellar distances today—the so-called method of stellar parallaxes. Tycho's mistake was in his assumption that he could detect and measure this apparent shifting of the stars. Here we see Tycho's great arrogance: his assumption that nothing could escape his great skills as a celestial observer. He had of course completely underestimated the true distances of the stars.

Tycho had another reason for rejecting the motion of the earth. He was basically devout and could not reconcile the motion of the earth with certain passages in the Scriptures. This dichotomy in his approach to the Copernican doctrine bothered him for the rest of his life because he became aware, in time, that Copernicus and his theory were becoming ever more popular among astronomers. When Tycho, in various letters, stated his objections to Rothmann, the German Copernican and chief astronomer to Landgrave Wilhelm IV of Hesse-Cassel, Rothmann pointed out to Tycho that if the Scriptures were interpreted literally much of what had been discovered since the Scriptures were written would have to be discarded. Rothmann and Tycho had carried on a correspondence for years and Tycho greatly respected Rothmann's opinion, so that he did not dismiss Rothmann's arguments lightly. Not to be left behind completely by the rapid advance of the Copernicans, Tycho finally struck upon a new system now known as the Tychonic system which permitted him to keep the earth fixed and yet to accept the Copernican principle that the planets revolve around the sun. In the Tychonic system the earth is still fixed at the center of the solar system, but with the other planets revolving around the sun, which, in turn, is revolving around the earth. We will see later that Tycho's system became a source of trouble between Tycho and Kepler.

We must not suppose that the Copernican doctrine was the concern of astronomers and astrologers only. Indeed, it left its imprint on philosophy, physics, art, literature, and humanism, in general. Scholasticism was dead and the only thing that stood in the way of Copernicanism was theology—Protestantism as well as Catholicism. In spite of this dangerous roadblock to the advance of the new ideas, their growing popularity could not be denied. Thus in Italy Giovanni Battista Benedetti (1530–1590) predated Galileo in rejecting the Aristotelian conceptions of motion and refuting Aristotle's basic errors. He also accepted Aristarchus' theory of the motion of the earth and thus embraced the Copernican theory. Outstanding among all the Copernicans, however, was the Italian philosopher and Dominican monk Giordano Bruno (1548–1600), who proposed the truly revolutionary idea that every star in the sky is a sun and that the universe extends infinitely in all directions so that no center of the universe exists. Bruno paid with his life for these ideas, being burned at the stake by the Inquisition as a heretic. Bruno was thus the founder of modern philosophy, as first developed by Benedict Spinoza, who was greatly influenced by Bruno.

On the other hand, many powerful church leaders in the Protestant and Catholic countries vigorously fought the spread of the Copernican doctrine. Martin Luther in Germany, for example, denounced the Copernicans as "scoundrels" and labeled Copernicus as "the fool [who] will upset the whole science of astronomy" and defy the "Holy Scripture [which] shows that it was the sun and not the earth which Joshua ordered to stand still." The philosopher Philipp Melanchthon, a scholar and religious reformer, strongly supported Luther's condemnation of the Copernicans. Indeed, even before the publication of *De Revolutionibus,* Melanchthon wrote to a correspondent that "wise rulers should suppress such unbridled license of mind."

Despite Tycho's own reluctance to embrace Copernicus' teachings without reservation, his own background and brilliance as an observational astronomer caused him to accept many of the tenets of the Copernican doctrine. Tycho had gained great fame at an early age, becoming so famous by the age of 26 years for his brilliant and accurate observations that the King of Denmark, Frederick II, invited him to come back to Denmark to become the court astrologer. Because astrology in those days was in vogue among the nobility and Tycho himself was passionately devoted to it, he accepted the king's offer. Frederick was so pleased with Tycho that he offered him the island of Hven as the site for an observatory. The "royal instrument" issued by Frederick on May 23, 1576, decreed that Tycho Brahe occupy the island of Hven in perpetuity (as long as he lived) to govern it as he saw fit, to collect all its revenues, and to build an observatory on it.

Tycho accepted this offer and immediately initiated the construction on Hven of the first and one of the most famous observatories in the history of astronomy—Uraniborg, or "City in the Sky," as it came to be known. Financed by Frederick's treasury, Brahe's observatory was constructed like a vast fortress, with underground dungeons, which he held as a threat over workers who failed to share his passion for accurate observing, which was all that mattered to him. Urania was equipped with the best naked-eye instruments Tycho could design, including huge equatorial armillary spheres, quadrants, sextants, and celestial globes. These instruments ultimately ended up in the Jesuit observatory in Peking after Brahe was expelled from his island castle because he refused to obey the commands of the new young King Christian IV, who inherited the Danish crown when his father, Frederick, died.

When Christian deprived Tycho of the observatory and the benefits, Tycho accepted the invitation of the Holy Roman Emperor Rudolph II to come to Bohemia as the royal astronomer. There he was given an estate and castle at Benatky, near Prague. With a stipend of 3000 ducats, Tycho built another, much less pretentious observatory, and laid the basis for his collaboration with Kepler, which began in 1600, a year before Tycho died.

Tycho contributed two important discoveries to the understanding of two astronomical phenomena that were observed at that time. On the evening of November 11, 1572, Tycho, on a walk near his home, suddenly saw a very bright star, near the constellation of Cassiopeia. Far brighter than Venus, this star overwhelmed Tycho for he had never seen such a luminous nonplanetary object. Knowing the positions of all the visible stars, Tycho could only conclude that a new star had suddenly been created. Not believing his own eyes, he called upon all his neighbors to confirm his vision, which they did. This star was so bright that some people could see it during the day. Other astronomers had also seen it. But then as Tycho followed it, night after night, it began to fade. We know now that Tycho had observed a supernova, which is still called the 1572 Tycho nova. In those days the concept of a nova or supernova was not known and so the Tycho nova produced enormous excitement throughout Europe and confirmed the popular belief that Tycho was the supreme astronomer of his day. The wonder this celestial phenomenon produced stemmed from the belief that all changes observed in the sky were produced by the earth's atmosphere in accordance with Aristotelian philosophy and with the Scriptures. Historians, through reading Pliny's *Natural History*, and astronomers knew that Hipparchus had recorded the appearance of a supernova in the year 125 BC but that was before the Christian era and, as a result, did not disturb the theologians of Tycho's time. Tycho considered this apparition as evidence of God's handiwork—the creation of a new star. To commemorate this occurrence, he published his first book *De Stella Nova*, in which he gave a detailed description of his observations of the nova, giving its exact position and an accurate description of the variations in its brightness from night to night. This book assured his everlasting fame.

Tycho's second book discussed the great comet of 1577 and conclusively proved that comets are not atmospheric phenomena but are instead produced in regions beyond the moon. By careful measurements of its change of position during the night and from night to night he proved conclusively that comets come in from distances "at least six times as far as the moon." This, too, was in direct conflict with the Scriptures and Aristotelianism, which, during the Middle Ages, had prevented any serious study of comets. Some attempts, particularly by Regiomontamus in Nürnberg, were made in the fifteenth century to measure the distance of some comets, but without success, because astronomical instruments were very crude in those days. Tycho even attempted to calculate the orbit around the sun of the 1577 comet, but the best he could do was to conclude that the comet moved around the sun in a circle larger than the orbit of Venus. Tycho, of course, emphasized in his second book that the orbit of the comet fit nicely in his (the Tychonic system) model of the solar system. This book, widely distributed in 1588, popularized the Tychonic system.

In his analysis of the 1577 comet he hints for the first time that celestial orbits may not be "circles." In fact, he discovered he could not make a circular orbit for the comet fit his observations and hinted that the orbit "may not be exactly circular, but somewhat elongated like an oval." But he did not pursue this idea further because he did not know what to do with it. In fact, he knew that his mathematics was too shaky to permit him to deduce or construct a meaningful model of the orbit. As another possibility for producing a "correct" (in agreement with observations) model, he suggested introducing another epicycle. But, all in all, Tycho's greatest contribution to the story of astronomy was his body of observations of the motions of the planets, which enabled Kepler to formulate the three laws of planetary motion (Kepler's laws).

The last 18 months of Brahe's life were perhaps his most productive—not because of what he himself did in that highly dramatic year and a half—but because of the work of his assistant, Johannes Kepler. In fact, it would not be too farfetched to suggest that Kepler's labors were anything less than the "redemption of Tycho Brahe." The meeting between these two men, whose personalities were so different from each other, must have been very dramatic. In fact, this meeting was what we may allegorically call a "marriage made in heaven." Each one knew what he wanted, and indeed, needed the other, to complete his life's work. Tycho knew that, without Kepler, the Tychonic system could never be placed on a sound mathematical base, and Kepler knew that all his attempts to establish the truth of the Copernican system would be in vain without Tycho's highly accurate observations of the planets, particularly those of Mars. Kepler believed that if he could fit Tycho's data for Mars into a circular orbit, he could prove the correctness of the Copernican heliocentric model of the solar system.

By all objective standards, and taking into account Kepler's own evaluation of himself and his background, we would hardly have chosen Kepler as the man who revolutionized the study of astronomy and physics. He was born on May 16, 1571 in the township of Weil (called Weil der Stadt), Würtemberg. He described his grandfather Sibald as "remarkably arrogant and proudly dressed . . . short-tempered and obstinate . . . and licentious, eloquent, but ignorant." He predicted that his father, who deserted the family periodically, was "doomed to a bad end," picturing him as "vicious, inflexible, quarrelsome," and cruel to his family. He described himself as constantly ill, suffering from all kinds of diseases. In 1586, he wrote about himself "that man [Kepler] has in every way a doglike nature. In this man there are two opposite tendencies: always to regret any wasted time and always to waste it willingly."

Starting in 1559 he studied theology at the University of Tübingen, where Mastlin, the outstanding scholar and astronomer of the time, introduced him to the Copernican heliocentric theory, with which Kepler was immediately enthralled. He had intended to enter the Protestant church as a Lutheran minister but was repelled by the narrow-minded spirit then prevalent among Lutheran theologians. He was therefore very happy to accept the post of "provincial mathematician" of Styria, which had extensive Protestant estates but was ruled by a Catholic Hapsburg prince. The city of Graz had both a Catholic and a Protestant school. Kepler accepted the chair of mathematics in the Protestant school when its mathematician died. The members of the Tübingen Senate were happy

Johannes Kepler (1571–1630) (Courtesy AIP Neils Bohr Library)

to recommend Kepler, who had become something of a thorn in their sides, because he had adopted Calvinist views and had defended Copernicus in public debates. According to his self-criticism and description of his own character, he was disputatious and a wrangler. Describing himself in the third person, he wrote "He [Kepler] was constantly on the move, ferreting among the sciences, politics, and private affairs. . . ." And again, further on he wrote, "he [Kepler] explored various fields of mathematics, as if he were the first man to do so, which later on he found to have already been discovered. He argued with men of every profession for the profit of his mind."

Kepler was not the usual run-of-the-mill astronomer who dutifully pursued one or two routine tasks, but rather one who entertained all kinds of ideas, however fanciful and unreal they might appear to others. He was particularly driven to find relationships among what, to others, appeared to be disparate, unrelated concepts. He did not believe in coincidences in nature. He was also driven by the need to find and reveal the symmetries that, he was sure, govern the universe. In a word, he operated very much in the way the modern scientist works because he was not afraid to speculate.

A very good example of Kepler's way of thinking and his quickness to correlate concepts that appear to be unrelated but have something in common was his attempt to explain why "only six planets exist." The only planets known at that time were Mercury, Venus, Earth, Mars, Jupiter, and Saturn, and Kepler accepted this number as fulfilling some kind of divine symmetry and as being related to a basic principle or some kind of universal law that governs the sizes of the planetary orbits and the distances between any two neighboring planets. Being an expert mathematician and noting that there are just five such distances, he related these to the five regular solids that are permitted in Euclidean geometry. A regular solid (known as a perfect solid to the Greek geometers) is one with identical faces. Only five such solids can exist: (1) the tetrahedron (pyramid) bounded by four equilateral triangles; (2) the cube with six equal faces; (3) the octahedron (eight equal faces); (4) the dodecahedron (12 equal pentagons); and (5) the icosahedron

(20 equilateral triangles). Kepler was sure that by fitting the five regular solids into the spaces between the six planets he would solve the problem of the sizes of the planetary orbits, and he almost succeeded in doing that, expressing his delight as follows: "The delight that I took in my discovery I shall never be able to describe in words."

Throughout his life Kepler sought the law or laws that bind the solar system together and determine the geometry of the orbits and the dynamics of planets. He was sure he had found the basic geometrical principles in the five regular solids. He was so sure he was right that he published his first great book *Mysterious Cosmographium* (the Cosmic Mystery) explaining his discovery in 1596. Although this geometric explanation is wrong, the book revealed the great power and originality of Kepler's thinking. That Kepler was convinced of the rightness of his views is indicated by this statement: "It is amazing. Although I had no clear idea of the order in which the perfect solids had to be arranged, I nevertheless succeeded in arranging them so happily that later on, when I checked the matter, I had nothing to change." This book, which was distributed widely, particularly among astronomers, established Kepler's fame and reputation as an astronomer and mathematician. This was not because of his regular solids model of the planetary orbits, but instead because of his overwhelming reasons for abandoning the Ptolemaic theory. In this book he also discusses his belief that the planets move along their orbits because they are forced to do so by a force from the sun. Not knowing the nature of this force, nor even its direction, he nonetheless pointed out that the force grows weaker with distance so that the planets further away from the sun move more slowly that those closer to the sun in agreement with the observations. The nature of this force (gravity) was not discovered until Newton promulgated his law of gravity more than 50 years later.

Kepler's book came to Tycho Brahe's attention and convinced Tycho that Kepler was a brilliant astronomer and just the kind of imaginative scientist he needed to prove the validity of the Tychonic model of the solar system. He therefore planned to make Kepler a member of his observatory. There is evidence that Galileo had also read Kepler's book because, following its publication, Kepler and Galileo communicated regularly with each other. With the success of his book, Kepler's greatest desire was to obtain more correct values of the distances of the planets from the sun so that he could establish the absolute truth not only of the Copernican heliocentric model of the solar system but also of his "regular solids" theory of the distances of the planetary orbits. Knowing that Tycho Brahe could give him these data, Kepler was quite enthusiastic about the prospect of working with Tycho. This desire of Kepler's to leave Graz was enhanced by the religious persecution in Styria and so, when the call from Tycho came, at the close of the sixteenth century, Kepler was ready to go. He joined Tycho at the Castle of Benatky as his "official collaborator." A year later in 1601 when Tycho died, Kepler was appointed to succeed him as imperial mathematician. With the Brahe observations of the planetary positions to hand, Kepler began his monumental work on the nature of the planetary orbits. The work lasted for 30 years and was completed with his discovery of his third law of planetary motion, which he called the "harmonic law."

When Kepler began his planetary research his target was Mars because the motion of Mars, as seen from the earth, is more irregular than that of any of the other known planets. Kepler knew that if he could "conquer the recalcitrant Mars" the other planets would be easy. He described Mars as the key to the mystery of the planetary orbits. In his usual flowery language he wrote that "Mars alone enables us to penetrate the secrets of astronomy which otherwise would remain forever hidden from us." Since no one before his time had discovered the true orbit of Mars he considered Mars as the "mighty victor of human inquisitiveness, who mocked all the devices of astronomers." Quoting Pliny, Kepler stated that "Mars is a star who defies observations." To solve the Martian problem Kepler first had to use the method of trigonometric parallaxes to calculate the distance of Mars from the sun for each distance of Mars from the earth as given by Tycho. This is now known as the "Kepler problem." This may be described simply as the method a surveyor uses to find distances by triangulation. With Mars at a given distance from the earth one constructs a triangle by drawing a line from the sun to the earth, a line from the earth to Mars and a line from Mars to the sun. Brahe's observations gave Kepler the lengths of two of the triangle's sides (the line from the sun to the earth and the line from the earth to Mars) and the angle between these two triangles. Using elementary trigonometry Kepler then calculated the length of the third side (the crucial distance of Mars from the sun).

In this way Kepler obtained many points in the orbit of Mars around the sun. This required extremely laborious arithmetic calculations, which Kepler detested but nevertheless continued stoically, fortified by his firm belief that he was fulfilling God's wishes. Fortunately for Kepler, John Napier, with his tables of logarithms, came on the scene when Kepler was hopelessly depressed by the vast task that he knew lay before him. Kepler saw that Napier's tables of logarithms would eliminate the most tedious features of his work, the need to multiply and divide numbers; multiplication could be performed by addition and division could be performed by subtraction. When Kepler wrote to his former teacher Maestlin that "henceforth there is no need to multiply or divide," Maestlin retorted that it was "unworthy of a mathematician to use tricks."

With the worst part of the arithmetic calculations eliminated, Kepler succeeded in deducing the orbit of Mars around the sun, but the result left him greatly discouraged. He did not obtain a circular orbit as he had hoped. Try as he might, he could not make all the points representing the distances of Mars from the sun fall on a circle. The best he could do was to obtain a circle which departed from some of Tycho's observations by about 8 minutes of arc, with which Kepler's contemporaries would have been fully content. But Kepler was not so easily satisfied because he knew that Tycho was too accurate an observer to have made an inadmissible error as large as 8 minutes of arc and so Kepler returned to his mathematical drawing board and saw the answer in a great flash of genius. The orbit had to be an ellipse, with the sun not at the center of the ellipse but at one of its foci. This point would then have to be common to all the planetary orbits. This is the essence of Kepler's first law of planetary motion which he stated as follows: Each planet moves in its own elliptical orbit with the sun at one of the foci of each ellipse, which is, therefore, a common focus of all the ellipses.

That the planets move in elliptical orbits is one of the most remarkable discoveries in the history of science. It was truly revolutionary in its implications because it destroyed, once and for all, the doctrine that all celestial bodies move in uniform circular orbits. Had Kepler done nothing else in his life, the discovery of this first law would have been enough to guarantee his immortality, because all the other properties of the planetary motions can be deduced from this basic principle. Kepler's own reluctance to abandon the circular orbit concept was clear when he wrote that "I [Kepler] searched, until I went nearly mad, for a reason why the planet preferred an elliptical orbit. . . . Ah, what a foolish bird I have been." Here Kepler reveals his unrelenting honesty, rejecting circular orbits which all the pundits of the past had imposed on astronomical thinking.

Kepler saw at once that elliptical orbits can account for the irregular motions of the planets around the sun but he felt that he could not rest on his discovery that the planets move in elliptical orbits. He had to prove the incontrovertible truth of his discovery. He started with the irregular motions, believing that if he could demonstrate why the motions of the planets are irregular, he could prove definitively that the orbits cannot be circles. He thus began his analysis of planetary motions which led him to his second law. Kepler knew that the force that keeps the planets moving around the sun emanates from the sun, and he reasoned that since the planets move in orbits that lie in planes close to each other, this force spreads out along a plane (two-dimensional). This conclusion, if true, means that this force would diminish in strength with increasing distance. The force on a planet twice as far from the sun as the earth would, he reasoned, therefore be half as large as the force on the earth.

Using this hypothesis Kepler reasoned further that the speeds of the planets should decrease in the same proportions that their distances increase. This deduction did not agree with Tycho's observations so Kepler had to discard it. Kepler envisioned the force from the sun as a kind of broom that swept the planets along their orbits as the sun rotated around its axis. To support his hypothesis he borrowed some ideas from William Gilbert, an English physicist and physician, who had published the first book that dealt in a systematic way with magnetism. Using Gilbert's concept of north and south magnetic poles, Kepler pictured the sun as a huge magnet acting magnetically on the smaller planetary magnets. Although wrong, this was the first attempt to explain the motions of the planets in terms of forces. One may speculate here and ask oneself whether Kepler might have discovered the law of gravity if he had pictured the solar force as acting in all directions (three-dimensional) rather than along just the two dimensions of a plane.

Kepler's failure to show that the speeds of the planets decrease in the inverse proportion to their increasing distances did not deter him from looking for a dynamic feature of the motion of a planet that has some universal significance. To this end Kepler noted that the closer any planet gets to the sun, the faster it moves. It then occurred to him that the product of the distance of the planet from the sun at a given point in its orbit and its speed at that point might be a constant, that is, the same for all points of the orbit. Kepler found this to be true; this constancy is the essence of Kepler's second law of planetary motion. It is most easily expressed as the area swept across in a unit of time in the plane of the planet's orbit by a line from the sun to the planet. Considering a planet for a unit time (e.g., a second), we note that the distance it moves is just its velocity $v$. If we multiply that distance by the distance of the planet from the sun, we obtain twice the area swept out by the line from the sun to the planet. As this distance is called the radius vector of the planet from the sun, Kepler stated his second law as follows: The radius vector from the sun to any planet sweeps out equal areas in equal times. We may, as Kepler did, express this law somewhat differently by stating that a planet moving around the sun moves in such a manner that the rate at which its radius vector sweeps out an area is constant. This law was remarkable because it was the first time in the history of science that a scientist (Kepler) had introduced a dynamic constant to describe the motion of a body (the planet) subject to the pull of another body (the sun). Its discovery was all the more remarkable in that Kepler had no basic physical principles to direct him toward this discovery; it was a kind of "hammer and tongs" operation. Without knowing it, Kepler, in his second law, had discovered the first of a series of conservation principles—the principle of the conservation of angular momentum (rotational motion) which applies to dynamic structures ranging from atomic nuclei to electrons in atoms to stars, galaxies and the universe itself. Kepler published his first two laws in his magnum opus with the grand title: *The New Astronomy Based on Causation or a Physics of the Sky Derived from the Investigations of the Motions of Mars Founded on the Observations of the Noble Tycho Brahe.*

Published in 1609, Kepler's book contains a complete description of Kepler's first two laws, stated above, which had been discovered between 1600 and 1606. This book launched Kepler into great public esteem and popularity. As its author, he became the uncontested astronomical authority and the "first astronomer" of Europe, recognized by poets, mathematicians, and other astronomers, including Galileo, as the greatest scientist of the age. Indeed, Galileo turned to Kepler for support when he (Galileo) announced his astronomical telescopic discoveries.

Kepler was still not completely satisfied because the solution of the problem of the relationship between the velocity of a planet in its orbit and its mean distance from the sun still eluded him. Since to Kepler this relationship would, when found, express the "harmony of the world," he began writing his final great book *Harmonice Mundi*" with the hope that he would discover this "harmony" by the time the book was ready for publication. The book itself was completed in 1618 and in it Kepler announced his third law of planetary motion, which gives the correct relationship between the mean (average) distances of the planets from the sun and their periods (the times they take to revolve around the sun, which, of course, depend not only on their velocities but also on their average distances from the sun).

This third, or harmonic law, as Kepler called it, states that if the square of the period of any planet, written as $P^2$ is divided by the cube of its mean distance $a^3$, the number thus obtained is the same for every planet. This is written algebraically as $P^2/a^3 =$

constant. Kepler was so overwhelmed by this discovery that, to him, everything else was trivial compared to it. All that mattered was the harmony in the universe that the law revealed. He particularly castigated the Thirty Year War that had begun in 1618 in these memorable words: "In vain does the God of war growl, snarl, roar. . . . Let us ignore these barbaric neighings . . . and awaken our understanding and longing for the harmonies." Concerning his discovery of the third law he states:

> Having perceived the first glimmer of dawn eighteen months ago, the light of day three months ago, but only a few days ago the plain sun of a most wonderful vision—nothing shall now hold me back. Yes, I give myself to holy raving. I mockingly defy all mortals with this open confession: I have robbed the golden vessels of the Egyptians to make a tabernacle for my God.

Thus ends a great episode in the story of astronomy because the period that followed, as represented by the work of Galileo, was a new beginning with the naked-eye observing of Tycho Brahe giving way to telescopic observing, and the guesswork and creative inquiries of Kepler giving way to the precise mathematical analysis of Newton.

CHAPTER 7

# Galileo, the Astronomical Telescope, and the Beginning of Modern Astronomy

*In questions of science the authority of a thousand is not worth the humble reasoning of a single individual.*

—GALILEO GALILEI

The works of Tycho Brahe and Johannes Kepler mark the end of the two great trends in astronomy that opened the doors to accurate observational astronomy, as exemplified by Brahe's work on the one hand, and theoretical astronomy as exemplified by Kepler's work on the other hand. Fortunately for observational astronomy, Galileo's astronomical telescope entered the stage just when Brahe had brought naked-eye observing to its limits of accuracy. Fortunately for theoretical astronomy, Isaac Newton's discovery of the laws of motion and the law of gravity entered the arena a few years after Kepler had completed his work on the motions of the planets. These two developments—the telescope and Newton's discovery of the laws of motion and gravity—raised astronomy to unaccustomed heights. No longer was it necessary to strain one's fallible sight to obtain and record results, and no longer, after Newton, did one have to cover hundreds of sheets of paper with the most careful and demanding computations. The orbits not only of the planets moving around the sun but also the orbits of stars around each other could be deduced in a very short time with the elegant Newtonian mathematics and his laws of motion and gravity.

Galileo Galilei was born in 1564 about 7 years before Kepler but he outlived Kepler by about 12 years, so that they were contemporaries during most of their professional lives and they both

Galileo Galilei (1564–1642) (Courtesy AIP Emilio Segrè Visual Archives)

greatly influenced each other. Like Kepler, Galileo began as a mathematician, starting his mathematical career at the age of 25 when he was appointed to the chair of mathematics at the University of Pisa. Though miserably paid, he was content to remain there for about 3 years because he had the leisure time to pursue his interests in mathematics and physics. During his few years at Pisa, Galileo turned to his true interest—the nature of motion and the motions of freely falling bodies. These investigations led him to the discovery or, more appropriately, to the founding of the modern science of dynamics. Indeed, his studies of the motions of bodies became the base on which Newton built his three laws of motion.

From Pisa Galileo went to Padua where he spent 18 years as professor of mathematics at the University of Padua. In addition to mathematics, he also taught astronomy there but devoted most of his spare time to constructing and studying the mechanics of such devices as pendula, metronomes, and the hydrostatic balance. His treatises on such instruments which he circulated and his lectures brought him to the attention of the outstanding scholars of the day. However, he was somewhat secretive about his important discoveries such as the law of falling bodies, the concept of inertia, and the laws of projectiles. He communicated these ideas, as well as his cosmological concepts, only to those persons like Kepler, who, he knew would accept them or, at the very least, consider them with an open mind.

Galileo first learned of Kepler through a common friend, Paulus Amberger, who delivered Kepler's book, *Cosmic Mystery*, to Galileo in 1597. Kepler, in a letter to his teacher, Maestlin, in 1597, states that he had sent a copy of his book "to a mathematician Galileus Galileus, as he signs himself." In acknowledging this gift in a letter, written within a few hours after Galileo had perused the book and perceived its full significance, he stated: "I indeed congratulate myself on having an associate in the study of truth, who is a friend of truth. For it is a misery that so few exist who pursue the study of truth and do not pervert philosophical reason." This book contributed to Galileo's belief that Copernicus was right; thus he became a convinced Copernican, although this letter to Kepler indicates that Galileo had already accepted the Copernican theory in his early twenties. In spite of Galileo's complete acceptance of the Copernican doctrine, he was loathe to publicize it, not because of his fear of persecution by the Church but because of his fear of ridicule. This fear is indicated in the final lines of the substantive part of this letter to Kepler where Galileo expresses his reluctance as follows:

> I have written [conscripsi] many arguments in support of him [Copernicus] and in refutation of the opposite view, with which, however, so far I have not dared into the public light, frightened by the fate of Copernicus himself, our teacher, who, though he acquired immortal fame with some, is yet to an infinitude of others (for such is the number of fools) an object of ridicule and derision. I would certainly dare to publish my reflections at once if more people like you existed; as they don't, I shall refrain from doing so.

In a reply to this letter in October 1597, Kepler exhorts Galileo to help spread the Copernican truth and not to fear the scorn of the ignorant, presenting himself as an example of one who prefers "the most acrimonious criticism of a single enlightened man to the unreasoned applause of the common crowd." Kepler went on to suggest very simple stellar observations which would prove the truth of the heliocentric theory. These were to be a series of observations of the apparent positions of the nearby stars. If those positions changed every 6 months this would be conclusive evidence that the earth moves back and forth relative to the stars as is required by the motion of the earth around the sun in a closed orbit. This was nothing less than a request that Galileo try to detect the semiannual parallactic shift of the stars.

In suggesting this exercise Kepler thought that all Galileo had to do was observe the apparent positions of the stars with an accuracy of a quarter-minute of arc. In this suggestion Kepler greatly underestimated the distances of the stars. Indeed, the stars are so far away that the semiannual shift of the nearest star is of the order of 0.75 of a second of arc, far too small for even Tycho Brahe to have detected with his naked eye. Not until 1840, some 200 years after Kepler's suggestion, was the first stellar parallax of a star measured by the German mathematician and astronomer, Bessel.

But Kepler was enthralled by the idea itself, pointing out in his letter to Galileo that "even if we could detect no displacement at all" the project was a "most noble one" which no one had undertaken before.

Not being an observer of the heavens like Tycho or a theoretician like Kepler, Galileo was content to accept the astronomical observations of Tycho Brahe and the Copernican doctrine as improved by Kepler's elliptical orbits and the three laws of planetary motion. Involved as he was in the study of motion and of mechanics and dynamics, in general, and knowing that he had little to contribute to astronomy at that time, he showed little interest in it, or, rather, did not pursue it with the intensity with which he pursued his interests in physics. During this period of his life he discovered one of the most remarkable laws in nature: all bodies starting from rest at the same point (the same height) above the surface of the earth fall with exactly the same speed in a vacuum regardless of their weights. Galileo did not state this as a natural law because he did not have the law of gravity (which Newton discovered later) to guide him. He simply stated it as an observation, after watching different bodies fall to the ground from the same height (from the top of the leaning tower of Pisa, it is said). Knowing that air (the atmosphere) resists the free motions of bodies, Galileo was amazed to discover that even with this resistance he could detect hardly any differences in the time that bodies of different weights took to fall to the ground. He properly concluded from these observations that if the atmosphere were absent, all the bodies would fall with exactly the same speed. This has been completely verified in many different observations and experiments since Galileo's time and is now accepted as a basic law. Indeed, Einstein based his general theory of relativity (his geometric formulation of the law of gravity) on this fact, so that today we accept gravity as a geometrical property of space-time rather than as a force.

Though Galileo's basic discoveries revolutionized the study of astronomy, he did not properly consider himself an astronomer but rather a "natural philosopher" (physicist) and mathematician. He contributed little to mathematics but completely altered the practice and theory of natural philosophy, rejecting the Aristotelian notions of opposites (e.g., "hot and cold," "light and heavy," "up and down,") and replacing such an abstract "guiding principle" with the investigation of nature by careful observation and measurement, with special emphasis on "measurement."

The need to make precise measurements of physical phenomena led him to the inventions of such instruments as the thermometer, the pendulum clock, and the hydrostatic balance. He also measured the acceleration of gravity by studying the time of descent of bodies on the inclined plane. By altering the slope of the plane by small amounts he deduced the acceleration of a freely falling body. His interest in projectiles led him to the discovery that all objects projected from the earth at various angles travel in parabolas whose geometric properties depend on the speeds and angles at which they are projected. This was the beginning of the modern science of dynamics.

Galileo's drive to discover the basic laws of nature was matched by his drive to educate the public about his discoveries. Knowing that his work was a direct threat to the dominance of Aristotelian principles, he was aware that he would arouse the strong opposition and enmity of the academicians of his day if he published his work. Nonetheless, that concern did not stop him and he became a pamphleteer, boldly challenging all those who could read to debate his discoveries and ideas. To win the authorities and influential people to his side, he was careful to dedicate his most controversial ideas to them with effusive praise. When he wrote for the general public, he wrote in the vernacular, writing in Latin only when he wanted to impress those he considered his intellectual equals.

He devoted his first articles and pamphlets to the descriptions of his experiments in physics. He was driven to write his first book (in defense of his giving priority to the invention of what he called "the geometric and military compass") which enjoyed a considerable commercial success. A student at Padua, Baldassar Capra, published in Latin, Galileo's description (written in the Tuscan dialect) of the compass and claimed it as his own. Galileo was so enraged at this attempted theft that in 1607 he wrote *A Defense Against the*

*Calumnies and Impostures of Baldassar Capra.* Here Galileo, accusing Capra of plagiarism and theft, exhibited his great skill as a polemist; he used this skill extensively in all his future literary works. Galileo brought a legal action against Capra, which he won, and so Capra's book was banned. Galileo's writings brought him fame throughout Europe and he was sought out by students everywhere.

During this early period he contributed little to astronomy in the way of original research or ideas, but he was a strong defender of the Copernican cosmology, which was later used against him in his trial before the Inquisition. Galileo had strongly supported Copernicus in his private communications to his correspondents. But despite his controversial pamphlets, he published no book specifically defending the Copernican doctrine for fear of ridicule, which can be a powerful deterrent to a sensitive mind.

Galileo's physical reasoning and his clever interpretation of observed phenomena are clearly manifest in his discovery of the law of inertia from his observations of sunspots. In his second letter to Mark Wilser on sunspots he states:

> For I seem to have observed that physical bodies have physical inclination to some motion (as heavy bodies downward) which motion is exercised by them through an intrinsic property without need of a particular external mover, whenever they are not impeded by some obstacle. And to some other motion they have a repugnance . . . and therefore they never move in that manner unless thrown violently by an external mover.

The "intrinsic property" Galileo refers to is what we now call the body's "inertia" or "mass." Further on in this letter Galileo states: "And it [a body] will maintain itself in that state [of motion] in which it has been placed; that is, if placed in a state of rest, it will conserve that; and if placed in movement toward the west (for example) it will maintain itself in that movement." This is essentially what we now call Newton's first law of motion.

The turning point in Galileo's pursuit of science and in the study of astronomy occurred in 1609 when he learned that a Dutch lensmaker (optician) Johann Lippershay had constructed a telescope. After reading the description of this instrument, Galileo built one himself, which was the first astronomical telescope. It did not take Galileo long to go from the description of the Lippershay telescope to the construction of his own model, which consisted of two lenses mounted in a long tube.

The basic optical principle of the telescope is quite simple. The lens mounted at the front end of the tube is called the "objective" of the telescope. It refracts (bends) the rays of light from a star so that they converge to a point near the rear end of the tube, thus forming an image of the star, or of any distant object. Since this image is very small, a second lens called an "eyepiece" is introduced at the rear of the tube to magnify the image. The objective (the front lens) is convex (both surfaces bulge out) but the "eyepiece" (the back lens) may be convex or concave (both surfaces curve in). In a convex lens the edge is thinner than the center; in a concave lens the edge is thicker than the center. From these descriptions we see that a convex lens must have one surface bulging out but the other surface may be convex, flat, or concave, as long as the round edge of the lens is thinner than its center.

Galileo constructed a telescope with a convex objective (an absolute requirement) and with a concave eyepiece; in modern telescopes the eyepieces are convex. Without the eyepiece a telescope is just a very large camera and one may, as is now done in all modern observatories, use it as such by replacing the eyepiece by a photographic plate. Three things, which Galileo understood, are important in constructing a telescope: its magnifying power (how much larger an object appears when viewed through the telescope as compared to its apparent size when viewed with the naked eye); its brightness (how much brighter the object looks when viewed through the telescope than when viewed with the naked eye); and its field of view (how much of an arc, defined as the field, the telescope covers, as compared with the naked eye.

Galileo knew these three properties of a telescope, which he called the "optic tube," and decided to construct one that would not make too great a demand on his limited mechanical and optical expertise and yet would reveal astronomical things and phenomena beyond anything that the naked eye could reveal. He thus first built a 3-power telescope which produced images that were three times larger than they appeared to the naked eye. Noting that the size of the image in the telescope increases if the length of the tube of the telescope is increased (which requires that the objective lens be thinner at its center and the surfaces of the lens flatter, that is, less convex), Galileo finally constructed a telescope with a 30-fold magnification (a 30-power telescope). He saw, however, that he could go no further because as the tube became longer, the field of view became smaller and the images were less clear since it was increasingly more difficult to hold the tube steady as it was made longer.

Galileo had to confront and solve another problem in constructing a telescope that could produce as bright an image as possible. This meant the objective (the front lens) had to be made larger, which taxed Galileo's optical technology. With all these restrictions it is a wonder that he had gone as far as he had and had built telescopes that revealed undreamed of celestial wonders. His main concern, then, was to inform the public of his great astronomical discoveries, and he began, on August 8, 1609, by inviting the Venetian Senate to look through his "spyglass" from the tower of St. Marco. By that time he had built a 9-power telescope and its effect on the senators was spectacular. The Senate immediately doubled his salary and made his professorship at Padua permanent. His next step was to publish the first of his books, *The Starry Messenger*, which revolutionized the study of astronomy.

Here, in Galileo's book, we see the revolution that made astronomy accessible to more than the few devotees such as Tycho Brahe who had developed the special aptitudes and patience that are required for the naked-eye study of celestial bodies and placed it in the hands of all who could construct or buy telescopes. In addition, the gap that existed between the observational and theoretical astronomers before Galileo was eliminated. As we have already noted, Copernicus was not an observer nor was Kepler. Thus Galileo was the first astronomer who combined observation and theory.

From the tone of his first book, Galileo prized his observations as much as, if not more than, the conclusions he drew from his observations. The contrast between Kepler and Galileo as proclaimers of their books—and therefore of their astronomical achievements—is indicated from the frontispieces of their respective first books. In 1609 Kepler published the *Astronomea Nova* the (New Astronomy) with the following frontispiece, as already noted, as translated by Stillman Drake: "A New Astronomy Based on Causation or A Physics of the Sky, derived from Investigations of the Motions of the Star Mars Founded on Observations of the Noble Tycho Brahe."

Here we see Kepler's self-effacement and his honesty in crediting Tycho Brahe's observations as the source of his theoretical discoveries as expressed in his three laws of planetary motion.

Coming now to Galileo's first book, *The Starry Messenger*, published in 1610, we read:

THE STARRY MESSENGER

> Revealing great, unusual and remarkable spectacles, opening these to the consideration of every man, and especially of philosophers and astronomers; as Observed by Galileo Galilei, Gentleman of Florence, Professor of Mathematics in the University of Padua With the Aid of a Spyglass lately invented by him, in the surface of the Moon, in innumerable Fixed Stars, in nebulae and Above all in FOUR PLANETS swiftly moving about Jupiter at differing distances and periods and known to no one before the author recently perceived them and decided that they should be named THE MEDICEAN STARS, VENICE 1610.

This frontispiece is notable for a number of things. First, it demands that Galileo receive full credit for his invention of the astronomical telescope, and all the observations he made with it. We feel that Galileo was justified in this claim even though he did not invent the telescope as an optical instrument. His application of the telescope to celestial observations was really "another invention." Certainly, his improvement of the telescope, bringing it from a 9-power instrument to a 90-power device, which no optician had considered before, constitutes an invention. Second, the list of his observations, stated with great self-assurance, left no doubt that Galileo considered his authority supreme and not to be questioned in the realm of observational astronomy. Finally, he dedicated *The Starry Messenger* to "The Most Serene Cosimó II De Medici" in the very flowery language that Galileo used in his pursuit of powerful public figures and patrons who might support his scientific ventures. Unlike Kepler, Galileo was never loathe to compliment most effusely those who could help him. Thus in his dedication to Cosimó de Medici he emphasized that he named the four satellites that he observed circling Jupiter "the Medicean Stars" in honor of the Medici family. The language of his dedication speaks volumes about Galileo's own personality:

> Indeed, the maker of the stars himself has seemed by clear indications to direct that I assign these new planets Your Highness' famous name in preference to all others. For just as these stars, like children worthy of their sire, never leave the side of Jupiter by any appreciable distance, so as, indeed, who does not know clemency, kindness of hearth, gentleness of manner, splendor of royal blood, nobility in public affairs, and excellency of authority and rule have all fixed their abode and habitation in Your Highness. And who, I ask once more, does not know that all these virtues emanate from the benign star of Jupiter next after God as the source of all things good.

Further on, in the introduction to the main body of the book Galileo repeats his list of discoveries: *Astronomical Message* which contains and explains recent observations made with the aid of a new spyglass concerning the surface of the moon, the Milky Way, nebulous stars, and innumerable fixed stars, as well as four planets never before seen, and now named The *Medicean Stars.*

This passage was written a year before the word "telescope" was introduced, which very quickly replaced the word "spyglass."

Seeking the approbation of his science peers and recognizing Kepler as "the Imperial Mathematician" of Europe, Galileo wanted, above all, Kepler's praise and so he requested the Tuscan ambassador to Prague, Julian de Medici, to inform Kepler verbally of his astronomical telescope and of his discoveries. Kepler's response was immediate; having no telescope of his own he accepted Galileo's claims on trust and wrote his support in the form of an open letter which appeared in the form of a scientific pamphlet which he titled "Conversations with the Star Messenger," which was printed in Prague and in Florence in Italian. In his most glowing language, Kepler exhorted the public to recognize Galileo's great discoveries. Referring to *The Starry Messenger*, he wrote that "it offered a very important and wonderful revelation to astronomers and philosophers, inviting all adherents of true philosophy and truth to contemplate matters of greatest import. Who can remain silent in the face of such a message? Who does not overflow with the love of the divine?"

To emphasize his support of Galileo, Kepler wrote to him that "in the battle against the ill-tempered reactionaries, who reject everything that is unknown as unbelievable, who reject everything that departs from the beaten track of Aristotle as a desecration . . . I accept your claims as true, without being able to add my own observations."

Galileo was quick to use this unqualified endorsement in his correspondence with various nobles, as in his letter to the secretary of state of Cosimó de Medici:

> Your Excellency, and their Highnesses through You, should know that I have received a letter—or rather an eight page treatise—from the Imperial Mathematician, written in support of every detail contained in my book without the slightest doubt or contradiction of anything. And you may believe that this is the way men of letters in Italy would have spoken initially if I had been in Germany or somewhere far away.

Returning now to *The Starry Messenger*, we see from his discussion of the four satellites of Jupiter that he considered that discovery to be of the greatest importance because their motions, as far as he was concerned, fully confirmed the Copernican doctrine. Stating explicitly that these satellites, "variously moving about most noble Jupiter as children of his own, complete their orbits with marvelous velocity—at the same time executing with one harmonious accord mighty revolutions every dozen years about the center of the universe; that is the sun." Galileo rejected the ecclesiastical doctrine that the earth is the center. This was the first published indication that Galileo accepted the Copernican system.

One can hardly treat *The Starry Messenger* as a book because it consisted of no more than 27 pages; it is thus a treatise or a pamphlet. Each of the discoveries announced in it could quite easily have been expanded into a thick tome, but Galileo was so astonished and overwhelmed by his discoveries that he could not resist the intense pressure to publish them immediately and receive his due credits and rewards before anyone else could do so. Galileo's book was as much a journal of his discoveries and was presumably written in the order in which the discoveries were made. First, he discussed the surface of the moon. Noting that the moon is "distant from us almost sixty earthly radii," but as seen through his telescope, "no farther away than two such measures so that its diameter appears almost thirty times as large" he stated that "the moon is not robed in a smooth polished surface but is, in fact, rough and uneven, covered everywhere, just like the earth's surface, with huge prominences, deep valleys and chasms." He then went on to describe the "bright" and "dark" spots on the moon's surface and correctly explained the variations in the appearance of the moon (its phases) as arising from the variations in the way the moon's surface reflects the sunlight that reveals it to the earth.

To enhance his description of the moon's surface at different phases, he complemented his word description with drawings which quite accurately show the moon as we actually see it through a telescope. The remarkable feature of Galileo's discussion of the lunar features as revealed by his telescope is its modernness. Reading *The Starry Messenger* we feel that we are reading an essay by a modern astronomer. Galileo would feel right at home and quite comfortable in the community of today's astronomers.

Galileo's description of the stars as revealed by the telescope followed this discussion of the moon. Interestingly, he began with his observation of the Milky Way, stating "it seems to me a matter of no small importance to have ended the dispute about the 'Milky Way,' by making its nature manifest to the very senses as to the intellect." He then went on to report that the Milky Way is not a nebulous gaseous structure like smoke but consists of innumerable "points of light" which he correctly interpreted as individual stars at such great distances that they appear faint. He considered each as a "sun" and so identified stars with the sun, appearing much fainter than the sun because they are at vast distances from us.

He noted in his discussion of the stars that even in his most powerful telescopes they appear as mere points of light, exhibiting no size; they do not appear as disks like the planets. He also recorded his amazement at the vast numbers of stars revealed by the telescope—far beyond anything the naked eye leads one to suspect. He stated this revelation as follows: "In order to give one or two proofs of their [the stars invisible to the naked eye] almost inconceivable number, I have adjoined pictures of two constellations. With these as examples you may judge of all the others."

Galileo followed this statement with a discussion and hand-drawn pictures of the belt and sword of Orion and the Pleiades. He then restated his analysis of the Milky Way, concluding that the galaxy is, "in fact, nothing but a [collection] of innumerable stars grouped together in clusters." To emphasize the analytical power of the telescope he states that "with the aid of the telescope this [galaxy] has been scrutinized so directly and with such ocular certainty that all disputes which have vexed philosophers through so many ages have been resolved and we are at last freed from wordy debates about it."

Galileo was also the first to discover gaseous nebulae in our galaxy, noting that they are distributed in regions outside the Milky Way and that a close inspection of these nebulae reveals the presence of stars within them. Galileo was particularly struck by the Orion nebula in which he found 21 stars and the Praesepe nebula in which he counted "40 starlets." Observing the sky every night, he soon discovered the difference between the fixed stars and those he called the Medicean stars (the moons of Jupiter) which do not remain fixed but, as he noted, revolve around Jupiter. This was the first clear statement in the story of astronomy that the fixed stars constitute a celestial system quite distinct from the system of planets and satellites.

Having devoted the 27 pages of *The Starry Messenger* to the surface of the moon, the Jovian satellites, and the fixed stars, Galileo announced his other astronomical discoveries, made later, in separate epistles, addressed to people (such as Kepler) he deemed worthy of his communications and confidences. He was particularly concerned that his correspondents understand his discoveries and credit him for them. As no science journals for publishing one's discoveries existed, and all kinds of claims and counterclaims about celestial phenomena were being made, Galileo wanted to be sure that the telescope would be recognized as the "ultimate authority" and the device that "would throw the superstitions into confusion." This communication was not received kindly by the Aristotelians and powerful Church leaders who were silently setting the stage for bringing Galileo to his knees and silencing him.

During this period Galileo continued using his telescope and announced three other important discoveries: the "strange" shape of Saturn; the phases of Venus; and the behavior of the spots on the surface of the sun. Of all of these discoveries, the apparent change in the shape of Saturn was the most mysterious. He saw that Saturn was visible as a disk but it appeared to have two handles, one on each side, which changed in size from day to day during the 3 months he viewed the planet, and then disappeared, only to reappear some time later. This baffling phenomenon puzzled Galileo who confessed that he was at a loss to offer an explanation. We know now that he was observing the rings of Saturn, which could not be distinguished from Saturn's bright surface with Galileo's telescope. Only the two edges of the rings, as seen against the black sky, were visible to Galileo, so he announced this discovery in the form of an anagram which, in translation reads: "I have observed the highest planet [Saturn] in triplet form [that is, the disk and the two edges of the ring]." A month later, he sent another anagram to Julian de Medici which was translated into the announcement that "the mother of love [Venus] emulates the shapes of Cynthia [the moon]." Galileo had discovered that Venus changes phases as it revolves around the sun just as the moon does in its revolution around the earth. The explanation is the same for both phenomena: we see both the moon and Venus owing to the sunlight that is reflected to us from their surfaces. The phases of Venus could not be the same as those of the moon if Venus were revolving around the earth. The changes in the phases of Venus prove conclusively that Venus revolves around the sun and not the earth. During Venus' motion around the sun, it appears as a crescent as it approaches and recedes from the earth (when it is between the sun and earth) and then when it is on the other side of the sun (the sun is between Venus and the earth) Venus appears full, with its entire disk visible. These discoveries were sent to the Jesuits at the Roman College as well as to Kepler and other astronomers and philosophers.

Galileo announced his work on sunspots in a series of letters to a Mark Welser who had written to Galileo to ask him about his opinion about "solar spots," as Welser called them. In this letter Welser states "you [Galileo] have led in scaling the walls [toward heaven] and have brought back the awarded crown [of knowledge]. Now others follow your lead with the greater courage, knowing that once you have broken the ice for them it would indeed be base not to press so happy and honorable an undertaking."

Welser's friend was a German astronomer, Father Christopher Scheiner, a Jesuit professor at the University of Ingolstadt. He had made a number of observations of sunspots with a telescope of his own and had considered his sunspot observations as new discoveries and asked Welser to publicize them; to conceal his identity for fear of clerical retribution he asked Welser to refer to him as "Apelles Latins post tabulum" (the author awaiting comment and criticism before revealing himself). Galileo was very happy to "comment and criticize," announcing his comments as the "*History and Demonstrations Concerning Sunspots and Their Phenomena*, contained in three letters, written to the Illustrious Mark Welser, Dumvir of Augsburg and Counselor to His Imperial Majesty by GALILEO GALILEI, Gentleman of Florence, Chief Philosopher and Mathematician of the Most Serene Cosimo II, Grand Duke of Tuscany, Rome 1613."

In his first letter to Welser, Galileo rejected certain conclusions drawn by Scheiner about the spots. He pointed out that the spots are not absolutely dark but dark only by comparison with the brilliance of the surrounding solar surface, noting that they are "real objects and not mere appearances or illusions of the eye or telescope." He concluded further that the spots are part of the sun and not outside it, for, as he pointed out, the apparent motions of the spots across the sun from "west to east" conform to the solar rotation. He noted also that the shapes and sizes of the spots change from day to day, ultimately disappearing and then reappearing in definite cycles. In this letter Galileo clearly expressed his philosophy and feelings about scientific research: "As your Excellency well knows, certain recent discoveries [of mine] that depart from common and popular opinions have been noisily denied and impugned, obliging me to hide in silence every new idea of mine until I have more than proved it."

In his second letter to Welser, Galileo discussed and described in greater detail the geometry of the spots, their apparent sizes, their clustering, their thickness, and their relationship to each other as they appear and disappear. He also described how the spots can best be observed by projecting the image of the sun produced by the telescope onto a "flat white sheet of paper about a foot from the concave lens."

Galileo's third letter to Welser discussed not only the sun spots but also Venus, the moon, the Jovian satellites, and the strange appearance of Saturn.

These three letters in a sense summarize Galileo's great discoveries which changed the study of astronomy forever; never in the history of astronomy has any single astronomer made as many discoveries, each of a revolutionary nature, as did Galileo. He set a record for discoveries which will never be equaled or even challenged. Galileo performed another great service for astronomers and scientists, in general, by insisting that the Scriptures must play no role in the pursuit of science. His point of view and philosophy about the separation of science and religion was expounded in a letter he sent to the Grand Duchess of Tuscany. By this time, Galileo, though supported by many, was under attack by the Aristotelians and clerics who saw his doctrines as a direct threat to their power and positions. They therefore turned to the Bible to find arguments against his support of the Copernican cosmology and, as Galileo stated in his letter, "have endeavored to spread the opinion that such propositions in general are contrary to the Bible and are consequently damnable and heretical." Galileo was clever enough to see that he could defeat his attackers only by showing first that the Bible cannot and, indeed, should not be used as a measure of the truth of scientific enquiries and, second, that none of his own discoveries conflicted with the Bible.

Like all good debaters he was very skillful in demolishing the arguments of his opponents, but he could not defend himself from arrest and trial by the Roman Inquisition. This was preceded by the publication of Galileo's last great work, *The Dialogue on the Two Chief Systems of the World.* Written in his mother tongue (Italian) for everyone to read, it had received the imprimatur of the "Master of the [papal] Palace" and was the first book of popular science ever written. This was a layman's encyclopedia of all of Galileo's discoveries and his explanation of the natural phenomena they represented. Here he boldly defended the Copernican system and the doctrine of the motion of the earth around the sun.

Only with Galileo's death in 1643 did the struggle between science and religion end. The persecution of Galileo by the Church marked the last unsuccessful gasp of the clerics' efforts to prevent the flowering of the new science brought forth by the works of Copernicus, Kepler, and Galileo. Ultimately these new ideas swept aside the increasingly obsolete objections voiced by their critics. Although the Church was not happy to lose its primacy in matters relating to the description of creation, there was increasingly little to be done about it.

CHAPTER 8

# The Newtonian Era

*I do not know what I may appear to the world; but to myself I seem to have been only like a boy playing on the seashore, and diverting myself in now and then finding a smoother pebble or a prettier shell than ordinary, whilst the great ocean of truth lay all undiscovered before me.*

—ISAAC NEWTON

By the time Galileo died in 1643, the year Isaac Newton was born, astronomical research had been irreversibly altered. On the one hand, Kepler's discovery of the three laws of planetary motion deduced by appropriately applying algebra and trigonometry to Tycho Brahe's observations of the motions of the planets (particularly the motion of Mars), indicated that the motions of the planets are governed by some universal principles. The search for these principles then became the dominant focus of the theoretical astronomers of the day. On the other hand, Galileo's brilliant application of the telescope to celestial observations forever changed observational astronomy: all the pre-Galilean observatories, best exemplified by Tycho's Brahe's observatory, Urania, on the Island of Hven, with their quadrants, cross-staffs, triquetrums, and armillaries, were cast aside. Post-Galilean observatories were dominated by the telescope, which ruled observational astronomy during that period.

Anyone who wanted to become an observational astronomer could do so by either building or buying a telescope and mounting it in some convenient spot. Thus the spirit and essence of Galileo's work were continued. However, one cannot say as much for Kepler's revolutionary work because no one understood, before the formulation by Newton of his laws of motion and the law of gravity, how Kepler's laws of planetary motion are related to more basic universal laws. Both Kepler and Galileo had contributed dramatically to the revolutionary change in astronomy but it remained for Newton to bury scholasticism, Aristotelianism, and clericalism forever as true paths to the understanding of the universe. However, some 20-odd years elapsed before Newton began his momentous work.

Astronomy was not dormant during this period because various monarchs, in the tradition of King Frederick II of Denmark who funded Brahe's observatory Urania, and Emperor Rudolf II of Bohemia who funded Brahe's observatory at Benathky, began to encourage and fund astronomical research and the construction of observatories and telescopes in their countries. Most notable among these was Louis XIV (the "Sun King") of France, who established the French Academy of Sciences which, in time, became the international bureau of standards for the various physical quantities (e.g., centimeters, grams, degrees, etc.) that are used in science today. The French Academy of Sciences attracted scientists from all over the world, among them astronomers such as Johannes Hevelius of Danzig, Giovanni Cassini of Bologna, Olaus Roemer of Denmark, and Christian Huygens of Holland.

With the great publicity that followed Galileo's exploits with his telescope, it was quite natural that members of the French Academy and other astronomers should concentrate on constructing larger and larger telescopes, with the hope of increasing their magnifying power and thereby seeing the images produced by these telescopes in greater detail. Thus gigantic telescopes up to 150 feet long were built. Because constructing metal tubes of this size to house the lenses was beyond the capabilities of metalworks in those days, the builders used a framework of metal instead of a tube to support the front lens (the objective). With these large, rather clumsy devices, it became very difficult to direct the telescope toward any given celestial object.

Aside from this practical difficulty associated with the construction of such large telescopes, other problems arose. The telescopes could not be held steady enough to allow the viewer to see any image clearly. Moreover, the lengths of these telescopes severely restricted the field of view; only a tiny fraction of the sky could be seen in a single viewing. Another difficulty was presented by the lens of the telescope itself, which, by its very nature, produced flawed images and limited the amount of light that could be brought to a focus (the image). The flaws inherent in a lens are called aberrations, and two kinds in particular destroy the quality of the image produced by the objective of the telescope. The first of these aberrations is called chromatic aberration and the second is called spherical aberration.

We can understand why a lens produces chromatic aberration (the edge of the image is surrounded by the colors of the rainbow) if we understand that each half of the lens is essentially a thin prism that breaks up white light into its constituent colors. Spherical aberration arises because the rays of light that pass through the lens near its perimeter are bent more than those that pass through the lens near its center so that these different rays (edge and center rays) are not brought to a single point focus. As telescope technology evolved, these aberrations were corrected in various ways, which we discuss later.

Johannes Hevelius of Danzig was particularly interested in the moon and constructed one of the first private observatories to study the lunar surface; he produced more than a hundred prints of its surface which he had engraved on copper plates. In studying the lunar surface he discovered what we now call "the librations of the moon." As the moon revolves around the earth, it also rotates around its own axis. Because it rotates around this axis at the same rate as it revolves around the earth, it completes a single rotation in the time that it makes one revolution; thus the same face of the moon is always turned toward the earth so that the observer on the earth can see only one face of the moon. But the observer can see a bit more than one side because the moon oscillates back and forth a little around its axis. These lunar oscillations are called librations of the moon. Hevelius also studied sun spots, cataloged many stars, discovered four comets, recorded the phases of Saturn, and was one of the first to observe the transit of Mercury across the face of the sun.

The French Academy met in a baroque palace, which would have remained just that if its future director Giovanni Cassini had not insisted that it also serve as an observatory to accommodate

Cassini's telescope. But he could not convince Louis XIV to go to the additional expense of altering the palace, so its desired functions had to be pursued outside the palace, where Cassini set up his instruments and made his observations. He made important discoveries about Mars such as its polar "ice" caps and its rotation period (about 24 hours—very close to the earth's period). His greatest discovery, however, was that the "handles" that Galileo observed around Saturn do not form a continuous structure but consist of a set of rings separated from each other by gaps. Cassini discovered the largest of these divisions, called the "Cassini division." In 1850 the observer Bond discovered another division and today we know, from space probes, that the rings contain many such divisions. Cassini also measured Jupiter's period of rotation (about 10 hours), discovered the dark band across Jupiter's equator, and found four other Jovian moons which are much fainter than the four moons Galileo had discovered.

During this pre-Newtonian period advances in astronomy remained in the realm of observation and measurement, with very little time or effort devoted to theory, that is, to the understanding or explanation of celestial phenomena in terms of basic laws or principles. But the observations and measurements became more sophisticated and accurate. Thus the French Academy set as one of its goals the accurate determination of the circumference of the Earth, which, as we have already mentioned, Eratosthenes had done in about 250 BC. The method used by the French academicians is essentially the same as that used by Eratosthenes: measuring the length of a degree on the earth's surface. This was done by measuring the distance one must move along a meridian for the direction of a plumb line (the vertical) to change by one degree. This distance multiplied by 360 gives the earth's circumference which these academicians found to be longer by about 240 miles than Eratosthenes had figured—24,647 miles. With this accurate measurement of the earth's circumference they went on to measure the mean distance of the sun, which is about 93 million miles, and the distance of Mars from the sun.

The known distance of the sun from the earth and the eclipses of Jupiter's moons enabled Olaus Roemer to measure, for the first time, the speed of light. The determination of the nature of light and, particularly, the measurement of its speed, had eluded scientists for thousands of years. To many people it appeared that light traveled instantaneously from point to point in space. To others, a minority, it seemed just as reasonable that light took time to move from point to point. This belief was fully confirmed by Roemer, who used a very ingenious analysis involving the eclipses of the moons of Jupiter, to measure, for the first time, the speed of light. He noted that when the earth in its orbit is receding from Jupiter, the measured interval between successive eclipses by Jupiter of any of its moons increases. In other words, the time, as measured by an observer on the earth, moving away from Jupiter, for the moon to reappear after it passes behind Jupiter is longer than it would be if the earth were standing still with respect to Jupiter. He reasoned that this delay is caused by the earth's motion, as the earth moves away from Jupiter. The distance the light from the reappearing moon has to travel to reach the earth increases so that the time between successive eclipses, as observed on the moving earth, increases. Roemer very carefully measured the total time delay, taking into account the total number of eclipses from the first one he observed when the earth is closest to Jupiter (the earth and Jupiter are on the same side of the sun with the earth between the sun and Jupiter) and the last eclipse when the earth is at its greatest distance from Jupiter (the sun is between the earth and Jupiter). He found that this total delay is 1000 seconds. Reasoning correctly that this 1000-second delay arises because the total distance the light from the reappearing moon has to travel to reach the earth after this last eclipse has increased by the diameter of the earth's orbit (186,000,000 miles) he divided this distance by the total delay time—1000 seconds—to obtain 186,000 miles per second for the speed of light.

Roemer also noted that the interval between successive eclipses decreases (and by the same amount as the previous increase) when the earth begins to approach Jupiter. This phenomenon (the increase or decrease in the period of a periodic event of which the variation in the period of the eclipses of a Jovian moon is a special case) is called the Doppler effect, named after Christian Doppler, the nineteenth century Austrian physicist. He discovered the change in the frequency (or wavelength) of a wave (e.g., light or sound) as measured by an observer moving with respect to the source of the wave.

Roemer did not limit himself to the study of light or to observing the Jovian moons; he greatly increased the usefulness of the telescope as an instrument for measuring and cataloging the positions of celestial bodies by anchoring the telescope to the floor of an observatory in such a way that it could be rotated only along the meridian (the north–south circle). This circle thus became known as the "meridian circle." With the telescope mounted in this way, he could determine what we now call the "longitude" and "altitude" of a celestial body. The altitude (the angular distance of the body above the observer's horizon) was given by the tilt of the telescope with respect to the horizon (the horizontal). The longitude of the object was given by the time it took the earth to rotate the telescope eastward so that the object appeared exactly at the center of the eyepiece of the telescope. This measurement depended on having an accurate clock. In time, such clocks were designed and built.

Because Roemer's telescopic mount was quite restrictive in that one could not rotate the telescope to point to any desired region of the sky, the Roemer mount was later improved by introducing another degree of freedom in the rotation of the telescope to allow it to be rotated in a plane parallel to the earth's equator. This mount, now called the equatorial mount, is standard in all observatories today.

With its powerful Academy of Sciences attracting scientists from all over Europe, France became the center of scientific research. But this dominance gradually changed as the foreign members of the French Academy such as Christian Huygens felt the need to encourage and promote science in their own countries. Huygens himself was the youngest member of the Academy but he contributed greatly to all branches of science, particularly to the wave theory of light. He spent only a few years at the French Academy, and then returned to Holland. As Huygens was Newton's contemporary, we discuss his contributions to astronomy together with those of the other scientists of that period after we discuss the work of Newton himself.

Considering how astronomy had developed before Newton came on the scene, one would hardly have expected the center of astronomical activity and research to shift to England. After all, Kepler and Galileo, the great precursors of the seventeenth century astronomers, were German and Italian respectively. Why then was an Englishman, Isaac Newton, the dominant figure of eighteenth and early nineteenth century physical astronomy? We do not imply here that England had no tradition of important science before Newton. To understand the English tradition we can consider such figures as the thirteenth century scientist Roger Bacon, the great sixteenth century essayist Francis Bacon, the sixteenth century physicist William Gilbert, and the sixteenth century physician and anatomist William Harvey to see the breadth and variety of England's pre-Newtonian scientific tradition. Nevertheless, very little significant work in astronomy or physics was done in Britain before Newton.

Roger Bacon, an English monk, was educated at Oxford and Paris, and taught at the University of Paris. After returning to England he entered the Franciscan order and began his experimental research and his astronomical observations. In his *Opus Magnus* he outlined his philosophy, emphasizing the need to study nature with an unbiased mind and to apply mathematics to the analysis of one's observations.

Francis Bacon, though not a scientist, influenced scientists in England by his brilliant essays in philosophy and logic. Two of his works *The Advancement of Learning* (1605) and *Novum Organum* (Indications Respecting the Interpretations of Nature)(1620) were particularly important in that they emphasized the acceptance of accurate observations as the only path to an understanding of nature. Thus accurate observations and measurements must, according to his philosophy, replace all prejudices (religious or otherwise) and preconceived notions. He categorized these prejudices as "idols of the tribe," "idols of the cave," "idols of the marketplace," or "idols of the theater."

William Gilbert, whose work was praised by Galileo, was a physician and physicist who carried out the first important experiments in electricity and magnetism, noting the way different substances, after being rubbed in certain ways, attracted other lighter substances. He called this phenomenon "electric" and was the first to introduce the concept of an "electric force." His greatest contribution was his experimental demonstration of the existence of a terrestrial magnetic field. To describe this field he introduced the concept of north and south magnetic poles. He also contributed to astronomy by championing the Copernican system and postulating that the fixed stars are not all at the same distance from the earth. His book *Magnets, Magnetic Bodies, and the Great Magnet of the Earth, New Natural Science*, was the first great scientific work published in England.

William Harvey, a contemporary of William Gilbert, was neither a physicist nor an astronomer, but he contributed to the growth of science in England by insisting that all aspects of nature must be the concern of scientists. Devoting himself to the study of all kinds of life forms, he discussed the circulation of blood, demonstrating that the flow of blood is sustained by the pumping action of the heart. That careful research could predict the existence of a pump within the human body was so amazing to the seventeenth century savants that the public was ready to accept the pursuit of science as almost as sacred as the pursuit of the Holy Grail. Though Harvey contributed nothing to astronomy, his research and philosophy showed that seventeenth century England was as welcoming and accepting of science as any of the continental countries.

Separated physically from the continent as well as intellectually, it offered great encouragement to science on all levels. It is no wonder, then, that England under Newton, became the center of research not only in experimental science (physics) but also, owing to Newton's great mathematical powers, in theoretical physics. This primacy, of course, influenced the development of astronomy enormously.

Because Newton made basic discoveries in mathematics, physics, and astronomy, he was an astronomer, mathematician, and physicist rolled into one. Here, however, we consider primarily his contributions to astronomy, pointing out how his mathematical genius and his important discoveries in physics led him to his astronomical work, which changed astronomy from a data gathering activity to a precise science. Newton was not an observer of the heavens and probably seldom looked through a telescope, but his contributions to the optics of the telescope led to great improvements in telescope construction and in the quality of the images produced by telescopes. Before we discuss these topics, however, we describe Newton's great mathematical and physical discoveries, starting with his physics and proceeding to his mathematics, which stemmed from the physics.

Newton's most remarkable mental aptitude was his ability to concentrate on a problem that interested him without giving up on it until he found a solution—if it was at all solvable. Even when he was not working on the problem directly, it was always on his mind. In his description of his mental process, he compared it to digging away at a cement wall. All he needed to pierce the wall was an initial crack, which he could then enlarge by chipping away at it. Though not a spectacular classroom student during his boyhood, Newton was always performing all sorts of simple experiments on things all around him to satisfy his curiosity about "how things worked." His curiosity was boundless and he thought constantly of measurements that might give him a deeper insight into nature than one can obtain from the superficial appearance of things. Another of his remarkable characteristics was his deep desire for privacy and his reluctance to discuss his ideas with anyone; he insisted on "full leisure and quiet."

His boyhood fancies gave way to great and profound ideas after he had enrolled at Cambridge in 1661 at the age of 18. Newton completed his bachelor's degree in 1665 when he was ready to astound the world of science and mathematics with the announcement of his many simultaneous discoveries. During the 2 years 1665 and 1666, Newton made a number of theoretical and experimental discoveries that many consider to be the greatest intellectual accomplishment of all time. Among his greatest discoveries were his summing of infinite series, representing binomials as series (e.g. the binomial theorem), trigonometry, the differential calculus (the fluxions), the "theory of colors," the integral calculus, the law of gravity, and the laws of motion. Though we do not know

Sir Isaac Newton (1642–1727) (Courtesy AIP Neils Bohr Library; Burndy Library)

in what order he made his great discoveries, it is clear from his own evidence that he considered the laws of motion as the basis of all scientific (that is, physics) discoveries. This is indicated by his Preface to the first edition of his *Principia* in which he stated that "the whole burden of philosophy [natural laws] seems to consist in this—from the phenomena of motions to investigating the forces of nature and from these forces to demonstrate the other phenomena." This immediately presents to us the revolutionary nature of Newton's departure from the Aristotelian approach to the study and the nature of motion. Whereas Aristotle described motion as a property of the kind of matter one contemplates (e.g., heavy bodies by their very nature always move toward the center of the universe (earth) whereas light bodies like fire, whose motion is contrary to that of the heavy, move to the extremity of the region which surrounds the center), Newton did not differentiate between "heavy" and "light" bodies, stating that the way all bodies move depends on the forces acting on these bodies.

Nor did Newton limit himself to the study of motions of bodies toward or away from the center of the earth. Instead of relating the motion of a body to some inherent characteristic of the body, he related its state of motion to the force acting on the body, concluding that one can deduce the nature of the force acting on the body from the change in the state of motion of the body. Not limiting himself to motion toward or away from the center of the earth, he pointed out that a given force applied to any body (light or heavy) can make the body move in any direction, depending on the direction of the force.

This concern with the relationship between the motion of a body and the force acting on it led Newton to a deep analysis of the nature of motion itself and of how to represent this motion mathematically. It was clear to him that one must differentiate between "straight line motion at a constant speed" and nonrectilinear motion at varying speeds produced by forces. He noted that the motions of the moon and the planets deviate from rectilinear motion and concluded that some kind of force acts on these bodies which emanates from the earth (in the case of the moon) and from the sun (in the case of the planets). With these ideas slowly maturing in his mind, Newton was about to formulate his three laws of motion, but before he could do so, he had to draw a clear distinction between the speed of a body and what we now call the velocity of the body. The speed is merely the time rate of change of a moving body (how fast it is moving) measured along the path of its motion at any moment; whereas the velocity gives another important bit of information—the direction of the body's motion at that moment. To Newton, this distinction meant that the thorough specification of motion requires the simultaneous knowledge of a body's speed and direction at each movement. This concept of the instantaneous specification or knowledge of the motion of a body led Newton to the calculus, which is really the algebra of infinitesimal quantities.

We can best understand Newton's discovery of the calculus (the theory of "fluxions" as he called it) by considering first the rectilinear motion of a body at constant speed in a given direction. To this end we consider the body as moving along a line, which we may call the $x$-line and represent the body's position along the $x$-line by giving the distance $x$ from some fixed point 0 on the $x$-line. Because $x$ changes (increases from moment to moment), Newton called this rate of increase (or decrease) of $x$ its "fluxion" and represented it by putting a dot over $x$, viz., $\dot{x}$. This was the beginning of Newton's discovery and development of the differential calculus. In introducing the concept of the "fluxion" which, later, the philosopher and mathematician Gottfried Leibnitz called the derivative of $x$ with respect to time, Newton related the calculus to "rates of change," that is, speeds or velocities. He had struck on the "fluxions" concept as a way to understand the motions of bodies, particularly the motion of the moon, because he saw that a proper analysis of motion requires some kind of mathematical technique that permits one to determine or represent the instantaneous speed of a moving body.

How then does one express the speed of a body at any particular moment? Newton saw that just dividing the distance $x$ a body moves in a time $t$ by the time, that is, using the formula $x/t$, gives the average speed over the distance $x$ during the time interval $t$, but does not give the speed at any moment during that time or at any point within the distance $x$. To obtain the instantaneous speed Newton introduced the idea of looking at the body for a very short time interval $\Delta t$ (the symbol $\Delta$ in front of $t$ means "very small") during which the body moves a very short distance $\Delta x$, and then dividing $\Delta x$ by $\Delta t$ to obtain $\Delta x/\Delta t$. Does this give the instantaneous speed? Clearly no, because the speed can change (even if only by a small amount) during the time $\Delta t$. To meet this objection, Newton proposed that $\Delta t$ be made infinitesimally small and called it the "fluxion of $x$" written as $\dot{x} = \lim_{\Delta t \to 0} \Delta x/\Delta t$, in a shorthand mathematical form which was later offered in a more useful form by Leibnitz as $dx/dt$ called the derivative of $x$ with respect to $t$. Underpinning Newton's mathematics was his decision to examine the consequences of allowing $\Delta t$ to go to 0, that is, to become infinitesimal. This very simple concept is the essence of differential calculus; it rapidly evolved from Newton's initial fluxion idea of a time rate of change to the broader concept of the rate of change of one quantity with respect to another quantity on which the first quantity may depend in any way. Mathematicians call this a functional dependence, stating that the first quantity is a "function of the second quantity." The weight of an infant, for example, depends on its age; his weight is therefore a function of his age.

Newton's concern with the motions of celestial bodies (the moon, planets, etc.) and the motion of the earth around the sun, forced him to consider the changes of the velocities of these bodies from moment to moment. He therefore had to extend the "fluxion" concept to the velocity itself and consider the fluxion $\dot{v}$ or $\ddot{x}$. In terms of fluxions, this concept is then the "fluxion of a fluxion," but in terms of Leibnitz's concept of a derivative it is the derivative of a derivative written as $dv/dt = d^2x/dt^2$. Newton's "fluxion of a fluxion" of a moving body is what we now call the "acceleration" of the body. Newton's introduction of the acceleration of a body as the basic element in its motion was a revolutionary advance in all phases of science—particularly in the study of dynamics—because it led Newton to his three laws of motion which stand at the very summit of scientific discovery and which launched astronomy from guesswork and speculation to being a precise science.

To see how Newton arrived at his laws of motion we start from his acceptance of Galileo's concept of inertia (mass) as that property of a body which causes the body to persist in its state of motion (rest or uniform motion in a straight line) unless prevented from doing so by some external agency. Newton accepted this hypothesis as a law of nature and called the external agency "force." This, indeed, became Newton's first law of motion, with force introduced as a specific physical entity. Newton was wise enough to see that he could advance the science of dynamics enormously without specifying in any detail the nature of the force acting on the body. Indeed, he did not even attempt to define the concept of force in any way, choosing to present it simply as the cause of the change in the state of motion of a body.

His introduction of the first law of motion and his recognition of the role that force plays in changing the state of motion of a body led him to a profound insight into the motion of the moon around the earth and the motions of the planets around the sun. These motions were changing continuously and this implied, according to his first law, the action of a force. But before he could specify the nature of this force, he had to find the general relationship between the force acting on the body and the change in the body's state of motion. This led him to his second law of motion, which is one of the greatest discoveries in the history of science.

Newton proceeded first by carefully defining the concept of acceleration which Aristotle and his disciples did not consider at all in their study of motion. They believed that the speed of a body was the only important thing. Galileo was aware of the acceleration of a body and, in fact, was the first to measure accurately the acceleration of a freely falling body (now called the "acceleration of gravity"). Newton was the first, however, to define acceleration as the "time rate of change of velocity." This was a very important step for Newton, who was deeply involved in studying the motion of the moon, a body whose velocity changes continuously so that only by considering this continuous change of the moon's velocity (acceleration) could Newton develop a sensible theory of the lunar motion around the earth.

Newton began his momentous work on motion by defining the motion of a body, as we have said, as the "time rate of change of its velocity" expressed in terms of the infinitesimal changes of its velocity during infinitesimal time intervals. In this way, Newton extended his fluxion concept to velocity and acceleration. Thus the acceleration, in Newton's terminology, is the fluxion of a fluxion, which Newton expressed by placing two dots over the changing quantity. Thus if $x$ is the changing position of a body moving along a line $x$, then its velocity $v$ is $\dot{x}$ and its acceleration $a$ is $\dot{v}$ or $\ddot{x}$. In modern usage, restating what we have already said, we write $v$ as $dx/dt$ and the acceleration $a$ as $dv/dt$ or $d^2x/dt^2$ and say that $v$ is the first derivative of $x$ with respect to time and $a$ is its second derivative. Newton clearly understood that acceleration must take into account the rate of change of the direction of motion of a body as well as the rate of change of its speed. This distinction was particularly important to Newton in his study of the lunar motion for he saw that even if the moon were moving at constant speed in a circle around the earth, it would still be accelerated because its direction of motion would still be changing continuously.

Newton's next step in his formulation of the laws of motion was his recognition that the acceleration of a body can be produced only by the "action of a force" on the body. This was one of the greatest (if not the greatest) advances in the history of science because it altered science from a purely speculative branch of thought to a precise intellectual discipline. In introducing force as the acceleration-producing agent, Newton took a giant step from Galileo.

Newton continued by expressing the relationship between force and acceleration in a precise mathematical form which led him to the profound perception that, in addition to force and acceleration, one other physical entity is involved in the relationship—the mass (or the inertia) of the body. We can see this from our own experiences in throwing or pushing bodies. The faster we want to make a body that we are pushing, the harder we must push—which means that the greater the acceleration we seek to impart, the harder we must push. We express this idea symbolically by writing $F$ (force) is proportional to $a$ (acceleration). If we wish to double the acceleration we must double the force, and so on. We also know from our experience that it is easier (requires less force) to throw a baseball than to push a massive rock. Indeed, as we have already stated we find we must exert twice as much force to throw a 10-pound weight as to throw a 5-pound weight. This means that the force that must be exerted to accelerate a body by a given amount is proportional to the mass ($m$) of the body. Newton combined this proportionality of force and mass with the previously stated proportionality of force and acceleration into the equation $F = ma$; force $F$ equals the product of the mass $m$ and acceleration $a$ of the body. This equation marked the beginning of modern science because it is the essence of dynamics. We note, however, that it (the equation) can be applied to the study of dynamics only if the nature of the force $F$ is known (if it can be expressed mathematically).

Here, too, Newton started it all by introducing the force of gravity as a mathematical formula that involves space and matter. His observation that since a freely falling body (e.g., an apple falling from a tree) falls with increasing speed, that is, with accelerated motion, its motion is governed by his second law of motion. This means, therefore, that a force is acting on it to produce its accelerated motion. To Newton, this reasoning meant only one thing: that the force which he called gravity emanates from the earth so that one may say that the earth is pulling downward on the object. He then extended this idea to the motion of the moon, arguing that the same force that causes a body near the earth to fall to the earth is the same as the force that causes the moon to move in its orbit around the earth. The only difference between the pull of the earth on a gram of the freely falling apple and its pull on a gram of the moon, as Newton saw it, is not in any difference in the nature of the force but in its magnitude. He reasoned that the force on the moon is weaker, gram for gram, than the force on the apple.

In this reasoning Newton proposed two important ideas: the force of gravity decreases with distance but increases with mass, so that the total force of the earth on the moon is very large, but the force on a single gram of the moon 240,000 miles from the earth is much weaker than the force on a gram of an apple a few feet above the surface of the earth. Newton understood that a formula for the earth's force of gravity on any object must contain the mass of the object and its distance from the earth. To obtain such a formula which describes not only the pull of the earth on a body but the pull of the sun on the planets and the pull of one star on another, he proposed the revolutionary idea that every object in the universe exerts a force of gravity on every other object. His next step was then to consider two objects separated by a given distance, pulling on each other.

To eliminate all extraneous elements such as the sizes and shapes of the two bodies from their gravitational interaction, he introduced the unphysical concept of a mass point (a given mass concentrated in a point) and pictured two such mass points, one of mass $m_1$ and the other of mass $m_2$ separated by the distance $r$. He then made two additional assumptions: that the force of gravity between the bodies acts only along the straight line connecting the two bodies and the gravitational force is symmetric between the masses of the two bodies. Thus the pull of the mass $m_1$ on $m_2$ is exactly equal to and opposite to the pull of $m_2$ on $m_1$. This is also the essence of Newton's third law of motion which is called the law of action and reaction and states that forces come in pairs. If two bodies pull or push each other, each one experiences an identical pull or push. This law of action and reaction applies to all forces, gravitational or not. The response of each of these two bodies to the equal pull or push depends, of course, on its mass: the more massive the object, the smaller its response. With these ideas to guide him, Newton wrote down a remarkably simple algebraic formula for the force of gravity ($F$) between the two mass points $m_1$ and $m_2$: $F_{gravity} = Gm_1m_2/r^2$. The formula immediately tells us that the force is symmetrical between the two mass points and that it does not matter whether we speak of the force of one object on the other or vice versa. The presence of the square $r^2$, the square of the separation of the two mass points, tells us that the gravitational force falls off as the distance increases. If the distance $r$ is doubled, the force is one-fourth as large; if the distance is tripled, the force is one-ninth as large, and so on.

The factor $G$ in this formula, known as "Newton's universal constant of gravitation" is an extremely small number (of the order of one hundred millionths) so that, under ordinary circumstances, the gravitational force between two ordinary bodies (even huge boulders) is extremely weak. But under certain conditions (e.g., white dwarfs, pulsars), the force of gravity can overwhelm all other forces in nature, so that once a compact sphere (e.g., the core of a supernova) begins to collapse, there is very little to stop it.

Having written down his second law of motion and his law of gravity, Newton reasoned that he could apply these two laws to calculate and even predict the orbit of the moon around the earth. His point of view, correct in every respect, was that the moon revolves around the earth as though it were constantly falling freely toward the earth combined with its lateral motion (e.g., at right angles to the line from the earth to the moon) thus producing its observed motion.

To check this prediction by applying his laws to the motion of the moon, Newton first had to show that his law of gravity in the form he had written it applies to large spheres like the earth and moon even though they are clearly not mass points. Using his calculus he proved that his law of gravity can be applied exactly as written to such spheres because, gravitationally, they behave as though their masses were concentrated at their centers. But he also had to know the correct distance between the center of the earth and the center of the moon. Using the best known value as given by the French Academy, Newton calculated the moon's acceleration (produced by the earth's pull) and, from that, the moon's speed. This calculation did not quite agree with the moon's observed speed, and, so, reluctantly, Newton set aside his calculations. However, some 16 years later, when Newton was informed of a more accurate value of the distance between the earth and the moon, he obtained excellent agreement between the moon's observed speed and his calculated speed. This was the birth of modern theoretical astronomy for it showed that the combination of Newton's laws of motion and his law of gravity, applied to celestial bodies, can predict the motions of these bodies exactly.

Newton next became interested in the motions of the planets, including the earth, around the sun, reasoning that the sun's gravitational force on them governs their motions just as the earth's gravitational pull on the moon governs the moon's motion. The pull of the sun on any planet decreases as the distance of that planet from the sun increases, and does so as the square of that distance increases. Therefore, the distant planets are moving more slowly than the nearby planets, as predicted by Kepler's third law of planetary motion. Moreover, the strict application of Newton's laws of motion of the planets shows that Kepler's third law is not quite correct but a very good approximation to the actual motions of the planets. This shows the great power of a law of nature (law of gravity) combined with logic (mathematics) to reveal unknown truths about nature. Because Newton's laws of motion and law of gravity are universal, they can be applied as, indeed, they were, to stars (binary stars) and to the universe, as a whole. This was the beginning of the science of cosmology because one could now try to explain the distribution of the stars in the sky as a direct consequence of Newton's laws. It is no wonder, then, that we now consider Newton's work and his contributions to dynamics as a scientific revolution of the most profound kind. He changed not only the way people look at the universe but also the way they solve the problems presented by the universe.

Newton applied his analytical technology to the earth itself showing, for example, that the periodic rise and fall of the ocean tides can be explained by the differential gravitational pull of the moon on the surface waters (lunar tides) as the earth rotates on its axis. He also showed that the combination of the "centrifugal force," stemming from the earth's rotation, and its self-gravity produced a nonspherical earth, one that is flattened at its poles. We leave this discussion of Newton's work on gravity with a few remarks about his emphasis on the mysterious nature of gravity. Emphasizing his rejection of any preconceived notions and his strict adherence to scientific proof, he stated in his famous *Principia:* "Hypotheses non fingo (I frame no hypotheses)." And in answering a letter sent to him by the first Boyle Lecturer, the famous seventeenth century classical scholar Richard Bentley, Newton wrote: "You sometimes speak of gravity as essential and inherent to matter. Pray do not ascribe that notion to me, for the cause of gravity is what I do not pretend to know."

Newton's contributions to astronomy were not limited to his theoretical work, as manifested by his laws, for his work in optics led to the discovery of the spectroscope and to the introduction of the reflecting telescope. The spectroscope evolved from one of the most famous discoveries about the nature of light. We all know "color" is one of the properties of objects that make the world all around us exciting and enjoyable to look at and one of its obvious manifestations is in painting and film, but until Newton's work on optics little was understood about color. In passing white light through a prism and thus separating this light into a continuum of constituent colors, from red to violet, Newton showed that light consists of an array of colors, which he called the optical spectrum, and that these colors are primary in the sense that each of these colors remains unaltered when passing through another prism. Newton emphasized that these colors are not a property of the prism but a basic characteristic of the light itself. In time, physicists and astronomers recognized that an instrument, now called the spectroscope, can be used to analyze the light coming from a star (such as the sun). In this way, scientists can learn a great deal about the star itself (e.g. its chemistry, its temperature, etc.). Considering the innumerable applications of the spectroscope to the study of stars and to celestial phenomena in general (as we show in later chapters) the advance of astronomy owes more, for such very little cost, to the spectroscope than to any other astronomical instrument. Never have so many discoveries been made with such little effort and such small cost.

Having discovered that a piece of glass, in the shape of a prism, spreads white light passing through it out into a spectrum of colors, Newton explained the colors surrounding the image produced by a convex lens as a prismatic effect. He noted that a convex lens cut through its center may be pictured as already mentioned as two prisms cemented together at their bases, so that they spread white light, passing through their edges into a color spectrum. This undesirable color halo surrounding the image in a Galilean telescope is called chromatic aberration. Newton knew that a concave lens placed behind the convex lens of the Galilean telescope, would eliminate the color spectrum, but he did not pursue this idea further because he thought that the concave lens would also eliminate the image. In this idea he was mistaken as was shown by the British optician John Dolland who in 1757 combined a convex and concave lens into a single lens, called an achromatic doublet. Dolland discovered that by using a special kind of glass called flint glass and the proper concave shapes for the surfaces of the concave lens, one can produce a doublet lens that not only eliminates the chromatic aberration, but also another optical defect called spherical aberration. These optical discoveries opened the doors to the construction of large telescopes, called refractors, with doublet objectives. The sizes of such telescopes are strictly limited, however, by the weights of the objectives; for that reason refractors gave way, in time, to reflectors, the first of which was constructed by Newton in 1668.

Newton turned his attention to designing and constructing reflecting telescopes when he discarded the idea of a lens doublet consisting of a convex and concave lens. Knowing that a concave mirror reflects rays of light to make them converge to form an image with no color fringes, he reasoned that the undesirable chromatic aberration produced by the refraction of the light passing through the first lens of a refracting telescope is not present in a reflecting telescope. Such telescopes, therefore, were to be preferred to refractors. Moreover, since large mirrors can be made much thinner and, hence, much lighter than large lenses, very large reflecting telescopes can be constructed.

Though reflectors are free of chromatic aberration, a spherical mirror is not free of spherical aberration because the rays of light reflected from the rim of the mirror do not come to the same focus as the rays reflected from the center of the mirror. Moreover, it is more difficult to look at the image produced by a reflector telescope than a refractor telescope. The first difficulty, the spherical aberration, was eliminated by changing the spherical reflecting surface into a parabolic surface. Reflecting telescopes are therefore called parabolic reflectors, which constitute the modern telescopes now found in all large observatories.

In this chapter we have discussed in some detail Newton's direct contributions to physics and astronomy, but have said little about his great influence on the growth of science and on the general community of scientists. We need only say in connection with these points that Newton changed forever the way scientists pursue their profession. No longer was the collection of observational data the main purpose of science. However important the culling and organizing of data were and still are to the development of science, Newton's work showed that the more important objective was to understand these data in terms of the basic laws of nature and to use these data to confirm or disprove the assumptions that they are laws. Newtonian science elevated scientists to a new status: they became the predictors and purveyors of new knowledge.

CHAPTER 9

# The Rise of Modern Astronomy

*Every great scientific truth goes through three stages. First, people say it conflicts with the Bible. Next they say it had been discovered before. Lastly, they say they always believed it.*

—LOUIS AGASSIZ

To evaluate the full impact of Newton's work on astronomy is difficult, if not impossible, but we are not off the mark if we say that Newton ushered in the rise of modern astronomy. His discoveries in mathematics, physics, and optics were incredible stimuli to the rapid development of theoretical astronomy, observational astronomy, and to the construction of observatories, either as adjuncts to universities or as independent national institutions. In particular, Newton's discovery of the differential calculus and its enormous usefulness in expressing in manageable mathematical forms (differential equations) the motions of bodies acted on by forces, loosed a veritable torrent of activity in the mathematical analysis of the dynamics of the solar system and stellar systems. These studies were no longer the exclusive domain of astronomers for they were open to anyone who knew some calculus and knew how to apply it to dynamical analysis. Clearly the trained mathematicians were most eager to contribute to these activities, and, as we shall see, they became so productive in this area that celestial mechanics became the dominant field of study for theoretical astronomers who depended on mathematicians for their guidance during those early years.

Observational astronomy, observatories, and telescopes grew and evolved together. Naturally, considering Newton's great influence, we can understand why England led in these developments. The first great national observatory at Greenwich, founded by King Charles II in 1675, has been in the forefront of astronomical research for more than 300 years. At the same time the office of the Astronomer Royal and the Royal Astronomical Society were established. John Flamsteed was appointed the first Astronomer Royal and he was followed by Edmund Halley, famous for his discovery of the comet named for him. He was in turn succeeded by James Bradley. The continental countries quickly followed England's lead, constructing telescopes and observatories of their own. But England had an additional motivation that contributed to the rapid construction of observatories: the Royal Navy's need for accurate navigational instruments and techniques. As the leading maritime European power at the time of Newton, with commercial interests all over the world, England's devotion to astronomy and to the Greenwich observatory was prompted, in part, by its search for an accurate navigational technology based on accurate observations of the positions of groups of well-known stars. This search ultimately led to the development of a highly accurate type of navigation called celestial navigation. The navigator on a ship can use this technique only if he has available a table of the positions of the stars in observable constellations and an accurate clock. Indeed, the Greenwich Observatory was constructed primarily to supply navigators with accurate stellar tables.

To prepare an accurate table of the positions of stars the astronomers of the Greenwich Observatory described a system of circles on the sky similar to the system of circles (latitude and longitude) on the earth. One can then describe the position of any star in the sky by giving its position relative to these celestial circles. For a navigator to determine his position on the earth, he compared the positions of groups of stars as he located them relative to his position on the sea with the positions of these same stars as given by the stellar table. Using an accurate clock he could then determine his own latitude and longitude on the earth.

The construction of accurate clocks (chronometers) went hand in hand with developing the technology needed to construct observatories. Astronomers found it necessary to separate ordinary clocks used to give the time of day (the solar time) as measured by the apparent position of the sun as seen by an observer on the earth from astronomical or "sidereal time" as measured by the rising and setting of the "fixed stars." The observatories today therefore have two kinds of clocks: a solar clock which gives the solar or ordinary time of day and a sidereal clock which gives sidereal or astronomical time.

The difference between these kinds of time arises because of the motion of the earth around the sun. Owing to this motion, which is an eastward motion (the motion is in the same sense as the rotation of the earth from west to east) the sun appears to move eastward, whereas the stars—at vast distances—appear fixed. The true period of rotation of the earth (the time for the rotation with respect to the fixed stars) is not 24 solar hours but 23 hours, 56 minutes, and a few seconds. The difference of 4 minutes between this period of rotation (the sidereal period) and the 24 hour solar day arises as previously described because the solar day is defined as the period (time interval) between two successive passages of the sun across our meridian. But owing to the apparent eastward motion of the sun, it appears to pass across our meridian 4 minutes later than the true period of the earth's rotation. Put differently, we say that the diurnal (daily) rising of the sun is delayed by 4 minutes compared to the diurnal rising of any star. If a given star on a certain day is on the observer's meridian together with the sun (the star, of course, will not be visible), the star will be back on the meridian 23 hours, 56 minutes (solar time) later, whereas the sun will be back on the same meridian 24 hours later.

The final use of the observatory data and the navigator's data to determine the position of the ship requires the solution of what is known as the "navigational" or "astronomical triangle," the three vertices of which represent, respectively, the position of the ship, the geographical position of the stars (or sun) and the earth's north or south pole. Putting all of these things together required accurate measurements so that positional astronomy (preparing accurate tables of stellar positions) became a major task of the early observatories. Such tables were and still are known as nautical almanacs, which are prepared and issued for general use every year.

Although observational astronomy—the main concern of the observatory astronomers—was the dominant astronomical activity during the Newtonian years, the Newtonian principles (the laws of motion and the law of gravity) and the calculus beckoned physicists and mathematicians, as well as theoretical astronomers, to the pursuit of theoretical astronomy. The emphasis in those early days was on the study of planetary motions and on attempts to deduce them mathematically from Newton's laws. One can see that if an object (e.g., a planet) was launched some billions of years ago at a given distance from the sun in some direction tilted at some angle to the line from the sun, the object must have moved in an orbit which was some mathematical combination of its initial launch velocity and the change in this velocity produced by the sun's gravitational pull.

If no sun had been present at the moment of launching, the object, owing to its inertia, would have moved off in a straight line instead of moving in a closed orbit around the sun. With the direction and the speed of launch just right the planet under the sun's action moved in one of the Keplerian orbits. This description, though reasonable, is a far cry from deducing the planet's orbit mathematically. This deduction thus became the goal of some of the British physicists and astronomers who formed an admiring coterie around Newton, all of whom were members of the Royal Society. Both the Royal Society and the French Academy were fathered, in an intellectual sense, by Francis Bacon, who in his fable, *The New Atlantis* (published shortly after his death in 1626), proposed that a "House of Solomon," whose adherents were to pursue "the new experimental philosophy," be established. Even before Bacon's proposal appeared in print, a group of philosophers in Rome established in 1600 the Academia dei Lincei, of which Galileo was a member. This was followed in 1657 in Florence by the Academia del Cimento which was founded by the two Medicis, Grand Duke Ferdinand II and Leopold, who had been tutored by Galileo. These groups were forerunners and prototypes of the Royal Society and the French Academy.

As the first Astronomer Royal, Flamsteed laid the foundation for the vast scientific structure that we now call observational or positional astronomy, which deals primarily with measuring accurately the positions of stars. In addition to meeting the needs of navigational astronomy, this was extremely important to the development of dynamical astronomy because astronomers following Flamsteed showed, by comparing the positions of the same stars over time, that the stars are not fixed but move about the evening sky. This discovery stemmed from later observations that the positions of the stars had changed. Flamsteed himself had measured the positions of about 3000 stars. Considering the relatively inaccurate instruments Flamsteed used, we consider his accomplishments in observational astronomy to be nothing short of astounding.

Edmund Halley, Flamsteed's successor, improved on Flamsteed's observations of stellar positions, becoming the first astronomer to demonstrate that "fixed stars" are not fixed but move with respect to each other and with respect to the solar system. Before becoming the Astronomer Royal in 1678, he went to the Island of St. Helena to catalog stars visible only in the southern latitudes. In his ocean journeys he discovered that his magnetic compass did not always point due north and from that observation deduced that the north magnetic pole does not coincide with the geographic North Pole.

From a comparison of his measured positions of stars with those of Flamsteed, Halley proved by direct observations a conclusion he had already drawn from a comparison of ancient astronomical manuscripts listing stellar positions with those that his contemporaries were preparing using the best instruments then available—that the stars are actually moving objects. In particular, he found that Sirius, Aldebaran, and Arcturus had shifted from their positions listed by Aristarchus.

All of these deductions were a natural consequence of his acceptance of Newton's laws, and his reasoning that the stars, owing to their mutual gravitational interactions, must be moving about, and that they are held together as a group by the universal force of gravity. Propelled by these ideas he decided to apply Newton's laws to the motions of comets and proved that the comets he studied were, like the planets, moving in elliptical orbits around the sun. He applied his analysis of comets, in particular, to the famous comet that appeared in 1687 and which now bears his name, correctly calculating its period as 76 years, from which he deduced that the same comet had appeared in 1066, 1531, and 1607. He went on to predict that it would reappear in 1758, as it did.

By comparing the times of occurrence of ancient eclipses with the time of occurrence of similar eclipses in the seventeenth century, Halley deduced that the length of the day had increased since biblical times and that the moon's distance from the earth had increased, while at the same time, its speed in its orbit around the earth had decreased.

Halley was a devoted disciple of Newton and advanced his ideas and theories whenever and wherever he could. Concerned that the Newtonian truths might be lost, he persuaded Newton to write his famous *Principia*, which Halley published at his own expense. Halley also oversaw the physical production of the book and spent many hours reviewing its proofs. Indeed, it may be said that Newton's *Principia*, perhaps the single most influential scientific work of all time, would never have seen the light of day had it not been for Halley's insistence that Newton expound his revolutionary ideas about motion and gravity.

The third Astronomer Royal, James Bradley, who took over management of the Greenwich Observatory from Halley, was also a follower of Newton. His great contribution to observational astronomy was his insistence on the need for accuracy in such observations; he pointed out that such stellar observations are affected by certain properties of the earth. If the earth were not rotating or revolving around the sun but were a perfect sphere with no atmosphere, then the observed position of a star would be its accurate position relative to the earth. But the earth's atmosphere and its various motions introduce errors in the observations. These errors arise from the refraction of the light from the star by the earth's atmosphere, the aberration of starlight stemming from the earth's motion around the sun, and the shift of the earth's north celestial pole owing to what we now call the precession of the equinoxes.

Because the earth's atmosphere decreases the speed of light passing through it (called the refraction of light), the path of a beam of light from a star is bent when it passes into the earth's atmosphere. The direction of the beam is thus altered so that to an observer on the earth receiving this beam the position of the star appears to be altered; the star thus appears to be in a position different from the true position. Bradley drew up tables of this phenomenon for different seasons of the year and different stars.

The aberration of light is produced by the motion of the earth and hence of the observer, transverse to the direction of the star; this motion makes the light from the star appear to come from a direction displaced by a small amount in the direction of motion of the observer (i.e., of the earth) from its true direction. This means that the telescope pointing toward the star along the star's true direction will not produce an image at the center of the telescope; the telescope must be tilted in the direction of the earth's motion to produce an image at the center of the eyepiece. Bradley discovered this by noting that the same stars seemed to shift back and forth parallel to the earth's motion every 6 months. Bradley was not looking for this effect, which he called the aberration of light, but rather for the parallactic shift of the stars which stems from the displacement of the earth relative to the stars by 186 million miles owing to the earth's motion around the sun. Picturing the sun as fixed we see that each star appears to shift its position by a very small amount (a very small angle). This parallactic shift is so tiny even for the closest star that Bradley could not have possibly measured or even observed it with the crude astronomical instruments available to him. Indeed, the parallactic shift of a star (61 Cygni) was first measured in 1838 by the German astronomer Bessel.

The difference between the apparent shift of a star's position stemming from the aberration of light and that stemming from the star's parallax, is that the aberration is produced by the velocity of the earth transverse to the direction to a star. The aberration depends only on the velocity of the earth and not on the motion of the star or the position of the earth. Bradley discovered that the angular displacement measured in radians of the star produced by the earth's motion equals (to a very good approximation) the velocity $v$ of the earth divided by the speed of light $c$, which gives the aberration angle $v/c$. Finding this angle to be about 20 seconds of arc (1/10000 of a radian) Bradley calculated the speed of the earth in its orbit to be 1/10000 times the speed of light or about

18.6 miles/second. He then calculated the circumference of the earth's orbit and, thus, the radius of the earth's orbit.

The precession of the equinoxes as already stated had been discovered by Hipparchus in 125 BC, who noticed that the year of the season (the tropical year—the interval between two successive appearances of the sun at the vernal equinox—that is, the beginning of spring) is shorter by about 20 minutes than the length of the year as measured with respect to the fixed stars (the length of the sidereal year). Bradley emphasized the great importance of this difference in observational astronomy since the observed positions of the stars are measurably affected by this precession. At the same time Bradley pointed out another dynamic property associated with the earth's axis, which must be taken into account in making accurate measurements of stellar positions. The precession of the equinoxes arises because the celestial tip of the earth's axis, which is tilted by 23.5 degrees with respect to the normal to the ecliptic (the line at an angle of 90 degrees with respect to the plane of the earth's orbit) describes a circle, moving westward around the earth's North Celestial Pole. But the earth's axis also wobbles in a north–south direction at the same time. This is called nutation.

Taking all of these discoveries into account we may say that Bradley was the father of precision in observational astronomy. The great importance of this precision in the continual advance of astronomy cannot be overemphasized. Only if an astronomer's observations are very accurate can he discover phenomena that require changes in our formulations of the basic laws of nature. We shall see in a later chapter how the proof of the validity of the general theory of relativity, perhaps the greatest single creation of the human mind, depended on the incredible accuracy of astronomical observations.

Not all of Newton's contemporaries worshipped him as devotedly as Halley and Bradley. Robert Hooke was one of the those who even went so far as to claim (unjustly) prior discovery of some of Newton's discoveries; he even claimed that he had prior knowledge of the law of gravity, but he could point to nothing that he had written or said that can be taken as a formulation of the law of gravity (the inverse square law). Hooke was a very ingenious experimenter with a brilliant mind who dabbled in many things including philosophy. He made some important discoveries in the structure and properties of matter, including Hooke's law of the elasticity of solids: the amount by which a solid can be stretched in a given direction is proportional to the force acting on the body in that direction. This law is generally expressed by the statement that "strain is proportional to stress." But Hooke's work did not advance astronomy to any great extent, even though he was a fervent supporter of it.

Newton turned the interests of many of his intellectual contemporaries toward mathematics, physics, and astronomy. Sir Christopher Wren was perhaps the most notable among them. He showed his remarkable aptitude in science and mathematics at the age of 14 when he began inventing various scientific devices and developing simple proofs of theorems in geometry. Indeed, he was so precocious that he was appointed professor of astronomy at Gresham College, London in 1657 at the age of 25. Three years later he accepted the Savilian professorship of astronomy at Oxford. The importance of this appointment for the story of astronomy is that it dignified the study of astronomy as a full-fledged domain of scientific pedagogy with its own faculty and retinue of students. Wren is more famous as an architect than as an astronomer, but even as an architect he contributed to astronomy in his design of the Greenwich Observatory in 1675.

Turning away from Newton's immediate contemporaries we come to Thomas Wright, who may be called the father of galactic astronomy. As a sailor he had ample time to study the heavens, particularly the Milky Way or galaxy. Many observers before Wright had discovered that as one turned toward the Milky Way the number of stars concentrated in a given area of the sky increased. But Wright was the first, in 1740, to suggest that the stars in the Milky Way and those in our own celestial neighborhood (the visible stars that stand out as distinct individuals owing to their proximity) as first discovered by Galileo, form a single system shaped like a lens, thick at the center and thin at the edges. Wright concluded, very boldly, that the stars do not form a spherical system but rather a somewhat flattened system. His only error was his assumption that our solar system is at the center of this system. As astronomers discovered in the modern astronomical era we are far from the center—indeed, our solar system is at the edge of the galaxy, some 30,000 light years from its center. This same idea was promulgated some years later by the German philosopher Immanuel Kant in his book *General Natural History and Theory of the Heavens* (1755). Kant went so far as to suggest that many galaxies like ours exist in the universe, each consisting of numerous stars. We may consider this prediction as marking the beginning of modern cosmology.

Because all astronomical knowledge comes from the light emitted by the stars, it was inevitable that, in time, Newton would attempt to construct a theory about the nature of light (i.e., the constitution of light). Knowing that light "travels in straight lines," at least as far as he could discern from the apparently sharp edges of the shadows cast by objects that are illuminated by the parallel rays of light, he reasoned, from his laws of motion, that light consists of particles or "corpuscles" as Newton called them. This became known as the "Newtonian corpuscular theory of light."

In promulgating this theory Newton had to account for the refraction (the bending of the path) of light when it passes from one medium into another (e.g., from air or from the vacuum into water or glass). He did this by stating that a corpuscle of light, on entering a denser medium, is pulled into the medium by the constituent particles (atoms) of the medium. Thus, the corpuscles speed up owing to being pulled as they enter the medium, change their direction of motion—producing refraction. Newton thus concluded that light travels faster in a denser medium (e.g., glass) than in air. Since the path of red light is bent least, he concluded that red corpuscles of light travel faster than blue ones.

This theory of light was challenged by Newton's great Dutch contemporary, Christian Huygens, who proposed the wave theory of light, which marked him as one of the outstanding physicists of all time. Not only was Huygens an excellent mathematician and physicist, but he was also a very good astronomer. He developed a new method of grinding and polishing lenses which produced more accurate and better lenses. He also designed a new type of eyepiece for telescopes so that he achieved sharper optical definition, leading him to the discovery of a satellite of Saturn and to a more accurate description of Saturn's rings than had been possible previously.

Huygens was led to his wave theory of light by combining two of his observations: (1) that light travels extremely rapidly with a speed a "hundred thousand times greater than that of sound," (2) that light is some kind of motion but not the motion of particles of matter. From his knowledge of the great speed of light he correctly reasoned that light is the "propagation of motion itself" not of matter. He then drew an analogy between the propagation of sound and that of light, concluding that light, like sound, is a wave and that the different colors that Newton discovered can be explained by the different wavelengths of light. Huygens developed in detail his theory of the propagation of waves of light which is accepted today. According to Huygens' theory, light rays do bend slightly around corners so that very close and careful observations of the edges of shadows should reveal that these edges are not perfectly sharp. This phenomenon, called the "diffraction of light," was later fully confirmed. Moreover, according to Huygens' theory, the waves of light advance more slowly through a dense medium than through a vacuum or a rarefied medium. This was also confirmed experimentally later so that the Newtonian corpuscular theory of light was abandoned.

The wave theory of light (called physical optics) enables us to understand better the formation of images in telescopes, which would be rather mysterious without the wave theory. It was known for a long time that the image of a point source of light (e.g., a star) is not a point but a disk, which is not the image of the true disk of the star itself but is formed at the center of the focal point of the front lens (objective) of the telescope by waves of light passing through the outer part of the objective interfering with waves passing through the center part of the objective. This explanation was first presented and properly calculated by G.B. Airy in 1834, the Astronomer Royal at the time. The disk is therefore called the Airy "spurious disk." The formula he deduced for the diameter of this disk shows that the larger the diameter of the objective and the shorter the wavelength of the light, the smaller the "spurious disk" and the sharper the image. Therefore to obtain sharper images of stars, astronomers began to work with increasingly large telescopes.

Newton's influence on his contemporaries extended beyond his immediate circle and beyond pure mathematics and astronomy. For the first time scientists began to set up experiments to lead them to a deeper understanding of the nature and properties of matter. As we have already noted, Robert Hooke studied the elastic properties of matter which, in time, led to what we call "solid state physics." Although Hooke's discoveries had no direct bearing on astronomy, they convinced others that careful experimentation and measurements reveal laws of nature. Robert Boyle was a contemporary of Newton, for example, who stands high in the ranks of such experimenters. His greatest discovery, now known as Boyle's law of gases, describes how the pressure of a gas, confined to a given volume and kept at a constant temperature, is related to the volume of the gas. The law states that if $P$ is the pressure of the gas and $V$ is its volume, then the product $PV$ remains constant as the volume changes. Simply stated, this means that the gas pressure increases if the volume is decreased (the gas is compressed). It may appear, at first, that this has little to do with astronomy, but, in time, as astronomers began to study the nature and structure of stars, the laws of gases, of which Boyle's law is a special case, became extremely important in pursuing such studies since stars are very hot globes of gases. Boyle thus took the first step, without knowing it, in the development of astrophysics, which is now one of the dominant branches of astronomy.

Boyle was interested primarily in the behavior of gases kept at constant temperatures so that Boyle's law, though a step, is not applicable to the study of stellar structures because the temperatures of stars increase from the surfaces to their centers. The French physicist Edme Mariotte (1620–1684) was the first to point out the limited application of Boyle's law to the behavior of gases under real circumstances, arguing that a correct law of gases must take into account the change in temperature of gases; such a law must be an algebraic relationship tying together the temperature, volume, and pressure of a gas. Some years later the French physicist and chemist Joseph-Louis Gay-Lussac and Alexander Cesar Charles, independently discovered the complete law of gases, which is a cornerstone of modern astrophysics. This law states essentially that if the volume, pressure, and temperature of a given amount of gas change together in any way, the pressure multiplied by the volume, divided by the temperature, remains constant (does not change). If the letter $P$ stands for pressure and $V$ and $T$ stand for volume and temperature respectively, then the gas law of Charles and Gay-Lussac is written, algebraically as $PV/T$ = constant. This remarkably simple law is of extraordinary importance to astrophysics.

Newton's contribution to astronomy went far beyond his discoveries of the laws of motion and gravity because he changed the whole spirit and tenor of the study of celestial bodies and even of space and time. This was a giant step away from the Galilean era. After Newton, no one dared to challenge scientific doctrine or discoveries, no matter how far they departed from religious doctrine or scholasticism. But the study of physics itself was altered; no longer could the physicist merely state a hypothesis; every formulation of a new principle was, after the appearance of Newton's *Principia*, to be subjected to a critical analysis in the spirit of Newtonian principles. At the same time the Newtonian philosophy stimulated scientists to investigate all kinds of celestial phenomena and to explain the behavior of all the bodies in the universe and, indeed, the universe itself. Nothing now was beyond understanding. Euphoria had taken over, imbuing all who read the *Principia* with a sense of tremendous intellectual power—mathematical analysis became the "name of the game." One no longer had to hypothesize because Newton himself had stated, "I frame no hypotheses," in his dismissal of preconceived notions and questions about the nature and the cause of gravity.

Newtonian philosophy and its insistence on precise mathematical demonstrations in general were so overwhelming and convincing that few post-Newtonians deviated from the Newtonian method of solving astronomical problems; that is, expressing the problem mathematically and solving the mathematical equations entailed. This approach was particularly attractive to mathematicians and physicists in general, for it now became possible as they saw it to do astronomy with pencil and paper. This approach, of course, applied only to positional astronomy, to stellar and planetary motions and stellar distributions. Constructing theoretical models of the interiors of stars was beyond the Newtonian laws of motion and gravity. These laws would, however, play enormous roles in stellar modeling, but something else beyond gravity was required for the construction of models of stars, which we consider in a later chapter.

Accepting this somewhat restricted role for their discipline in astronomy, mathematicians were only too happy to contribute whatever they could to the advance of astronomy. Many of them saw this as a fairly direct road to fame which might escape them if they devoted themselves to pure mathematics, which few people understood. But demonstrating that mathematics is the bridge from fantasy to reality appealed enormously to the French mathematicians who were beginning to dominate the field of mathematics, as we shall find.

Newton himself had indeed deduced mathematically Kepler's three laws of planetary motion, but only by considering the motion of a single planet revolving around the sun. This is known as the two-body gravitational problem, which does not take into account the gravitational interactions among the planets themselves. Solving the general gravitational problem of many interacting bodies, known as the $n$-body problem, became the great challenge to those astronomers, mathematicians, and physicists who followed Newton. Solving such a problem would have applications not only to planetary motions but also to aggregates of stars such as stellar clusters and galaxies. Although the post-Newtonian French mathematicians were the dominant moving spirits in these activities, other continental mathematicians began to enter the fray. We may therefore say that astronomy was the universal tie that bound together mathematicians throughout the world. This was also true of physicists, who saw astronomy as an inexhaustible source of problems and the cosmos as a vast laboratory which dwarfs any manmade laboratory. We shall see, as our story goes on, that the truth of this perception has become increasingly more evident, particularly today, when physicists are studying phenomena which involve energies far beyond anything available in laboratories on earth.

CHAPTER 10

# Post-Newtonian Astronomy

*Science is the topology of ignorance.*

—OLIVER WENDELL HOLMES

Newton's discoveries and his formulations of the laws of motion and the law of gravity, together with the calculus, produced a revolution in the practice of astronomy, particularly in the development of theoretical astronomy. Even observational astronomy, which can be pursued without any theory about motion and gravity, was greatly stimulated, for the observational astronomer was now challenged either to prove or disprove the Newtonian laws. Any observed deviations from these laws would redound greatly to the fame of the discoverer and challenge the theoreticians further to explain any observed deviations. But to this end the Newtonian laws had to be reexpressed or reformulated in the most useful mathematical forms possible. These mathematical forms are what we now call the "differential equations of motions" which are, in fact, algebraic equations of infinitesimals. Interestingly enough, these theoretical developments, in their most useful forms, stemmed not from British mathematicians but from the French school of mathematics led by such great mathematicians as Joseph Louis Lagrange, Pierre Simon Laplace, Alexis Claude Clairaut, Claude Jean D'Alembert, Pierre Louis Morean de Maupertuis, and Simeon Denis Poisson. Among these we also include the great Swiss mathematician Leonhard Euler. In studying the contributions of this remarkable coterie to astronomy, we note that this was not a one-way street, for astronomy stimulated its members to the discoveries of new mathematical techniques. Such new techniques were required to solve the problems presented by astronomy because these problems are inherent in the very equations that must be used to describe astronomical phenomena.

These are not ordinary algebraic equations that deal with finite quantities but, as stated, differential equations that deal with infinitesimals. Mathematicians were thus forced to develop a new branch of mathematics called "the theory of differential equations." Today, differential equations are the basic mathematical tools for probing all branches of astronomy from astrophysics to cosmology.

Among all those mathematicians who contributed to this work, we must rank Leonhard Euler at the very top. Probably the most productive mathematician of the eighteenth century, if not of all time, with a phenomenal memory and with the ability to see instinctively the solutions to the most complex mathematical problems almost at a glance, he is credited with some 500 books and 890 mathematical papers. Among these is his famous treatise on Newtonian dynamics, *Analytical Mechanics,* which presented the first systematic treatment of motions of mass points acted on by mutually interacting forces such as gravity. Much of Euler's mathematical work was devoted to lunar theory, a large section of which deals with what is known as the three-body problem. Newton and his contemporaries had already shown that the two-body gravitational problem can be solved exactly and that the solution leads to Kepler's three laws of planetary motion. The two body problem can be stated simply as follows: what are the orbits of two mass points interacting exactly in accordance with Newton's three laws of motion and his law of gravity?

Newton obtained Kepler's laws (elliptical orbits) with very little effort, but he could not solve the general three-body problem, which deals with three mass points interacting gravitationally. Euler confronted this problem, which has still not been solved, in his treatment of the lunar problem, the three bodies in this case being the earth, the moon, and the sun. In his work on the lunar theory Euler laid the mathematical foundation for almost all future work that dealt with orbital theory. In particular, he initiated the theory of "successive approximation" and the theory of "perturbations," which encompass more than astronomical problems. Among his astronomical treatises we note his work on celestial mechanics, which stemmed from his analysis of the motions of the planets and comets. He is justly famous in mathematical, physical, and astronomical circles for his discovery of what is now known as the "Euler equation" which is a generalization of Newton's second law of motion. It is applicable not only to the motion of a mass particle but the motion of collections of particles. The Euler equation is really a set of partial differential equations, as they are called, which Euler deduced by the application of a new branch of calculus called "the calculus of variations" which is useful and, indeed, indispensable in every branch of science. Euler also developed a set of equations, again an extension of Newton's laws of motion, to describe the motions of rigid bodies. These equations are indispensable in the study of the rotation of bodies such as the earth, the sun, and the planets.

As previously stated, mathematicians were greatly influenced by Newton's work, primarily by Newtonian mechanics and dynamics. Applying this new mathematics to the most important astronomical questions, mathematicians, for the first time, demonstrated the great power and usefulness of mathematics as the key to the doors behind which new truths in astronomy and physics can be discovered. Euler, following Newton, was the first to show that many such doors can be opened by the proper application of the mathematical key. Mathematics in itself cannot reveal new truths, but combined with or applied to basic physical laws such as the Newtonian laws of motion and gravity, it can perform wonders, as demonstrated by Euler and his successors such as Joseph Louis Lagrange.

Lagrange is most famous for his *"Mécanique Analytique"* in which he extended Euler's dynamical equations in a form which is still used extensively in physics and astronomy. For the first time the concept of energy was introduced (specifically in the study of the motions of the planets). Indeed, Lagrange pointed out that Newton's laws need not be introduced specifically; one can replace them by introducing a quantity that is now known as the Lagrangian of a system of bodies. The "Lagrangian" consists of two kinds of energy of a body: the kinetic energy and the potential energy. The kinetic energy of a body (its energy of motion) is just the quantity 0.5 of $mv^2$, where $m$ is the mass of the body and $v$ is its speed. The kinetic energy is always positive. The potential energy of the body, which is always taken as negative, is the energy associated with the force of the gravitational field on the body in which the body moves. The total energy of the body is then the sum of these two energies, which may be negative, zero, or positive.

With the aid of Lagrange's equation we can then show, as Lagrange did, that only if the total energy of a body, revolving around another one gravitationally, is negative, can the two bodies stay bound to each other. The orbit is then an ellipse. If the total energy is zero, the orbit is a parabola and the two bodies cannot stay together. If the total energy is positive, the orbit is a hyperbola and, again, the bodies fly apart. Since the orbit of any one of the planets in the solar system is closed, the planet remains attached to the sun; its total energy is thus negative.

We can understand this controlling role of the total energy if we note that the kinetic energy of a body is a measure of its ability to escape from a bond (such as gravity) because it is proportional to the square of its speed. The potential energy, on the other hand, is a measure of its gravitational bond, which must, if it is to remain in orbit, outweigh its tendency to run away. The binding energy (negative) of the body must exceed its dispersive energy (positive) if the body is part of a permanent structure (e.g., the solar system). The introduction of the energy concept in dynamics led to the principle of conservation of energy, the first of a series of conservation principles which are extremely important in astronomy. This principle tells us that the total energy of a planet moving around the sun must remain constant.

Lagrange applied his dynamic theories to the analysis of many bodies interacting gravitationally with each other. As we have described, this is known as the *n*-body gravitational problem, which, as Lagrange immediately saw, cannot be solved by standard gravitational theory. He began constructing approximate mathematical methods for obtaining solutions to any desired degree of accuracy. These methods, taken together with Euler's work, constitute what we have already called "perturbation theory" or the theory of "successive approximations." With modern electronic computers performing the numerical calculations involved in perturbation theory and successive approximations, finding approximate solutions can be simple and quick.

Giving up on finding a general, complete solution to the $n$-body gravitational problem, Lagrange turned his attention to finding a complete solution to the three-body gravitational problem. This problem also proved to be intractable; even now no general solution of the three-body problem is known. A simple example shows why this problem is so difficult. As the three bodies in our example we take the sun, the planet Jupiter, and a pebble moving under the gravitational attraction of the sun and Jupiter. Because the mass of the pebble is negligible, it has no gravitational effect on the sun or on Jupiter so that Jupiter moves around the sun in its Keplerian orbit as usual. But the pebble can move in an infinitude of orbits, depending on the initial conditions of its motion; these orbits can range from a closed orbit around Jupiter to a complex orbit that weaves in and out between Jupiter and the sun. To find a single algebraic expression from which all such possible orbits can be deduced is well-nigh impossible.

From these daunting problems Lagrange turned his attention to what is now known as the restricted three-body problem. In this problem the three bodies, such as in the sun–Jupiter–pebble example above, are restricted to moving in the same plane. Lagrange solved this problem quite easily, obtaining two general solutions: (1) the three bodies lie on a line which is spinning around an axis perpendicular to the line; and (2) the three bodies are at the three apices of an equilateral triangle. This second solution was completely verified with the discovery of two sets of Jovian or Trojan asteroids which are at the same distance from Jupiter as from the sun and equal to Jupiter's distance from the sun. One set of these asteroids is in advance of Jupiter in its orbit around the sun and the other set follows Jupiter. This is probably the first example in astronomy of a theoretical prediction that was confirmed by observations much later.

Lagrange's final contribution to astronomical theory dealt with the stability of the planetary system. Although the gravitational interactions of the planets among themselves are negligible compared to the overall solar gravitational action, one might expect these interplanetary interactions to be cumulative and, ultimately, after millions of years, to destroy the planetary order. But, as Lagrange proved, this is not so; the perturbations produce small periodic changes in the planetary orbits, but the planets always remain near mean positions which define their orbits. The solar system is, accordingly, highly stable.

Next after Euler, Lagrange was the most famous of the eighteenth century mathematicians whose contributions to theoretical astronomy were prodigious. Following him were d'Alembert, Clairaut, and Laplace. D'Alembert dabbled in many mathematical activities, but, unable to resist the great attraction of astronomy, he devoted himself to studying perturbation theory, the three-body problem, and to Newtonian dynamics. He is famous for the "d'Alembert principle" which enables one to solve the complex dynamical problems by reducing the many forces acting on a collection of bodies to a single force. Using his own principle d'Alembert produced the first analytical solution of the problem of the precession of the equinoxes based on Euler's theory of the motions of rigid bodies. These contributions are contained in his book *Treatise on Dynamics*. Elected "perpetual secretary" of the French Academy in 1754 he became the most influential man of science in France. He wrote on many scientific topics, both practical and theoretical, devoting most of his best efforts to astronomy, particularly the three-body problem.

Alexis Claude Clairaut was next in order of brilliance among this incredible quintet of theoreticians. Having published his first mathematical paper ("The Differential Geometry of Space Curves") at the age of 18, he was elected to the French Academy by the time he entered the university. Interested as he was in space geometry, he turned his attention to the shape of the earth, which he analyzed in his book *The Theory of the Shape of the Earth* in 1743. In this book, which was primarily devoted to the equilibrium of fluids and the gravitational attraction of ellipsoids of revolution, Clairaut proved that, owing to its rotation, the earth is not a sphere but an ellipsoid, flattened at its poles. He showed that the earth's shape is determined by its surface gravity, which depends both on its ellipticity and on its internal constitution. As the earth's rotation produces a centrifugal force on its surface, the earth's shape at any point on its surface depends on how large this "centrifugal force" is compared to the force of gravity at that point. Clairaut derived the formula for the oblateness of the earth (the flattening at its poles) in terms of its surface gravity. This must be credited as one of the great successes of gravitational theory.

Clairaut achieved another success when he confirmed, by detailed calculations, Halley's prediction that Jupiter's gravitational pull on Halley's comet delays its periodic return. Clairaut had predicted that the comet would return in the early part of 1759, somewhat later than one might have expected. It returned on March 13, 1759, a month earlier than Clairaut's calculations had predicted. This was a remarkable achievement when one considers the almost primitive level of mathematical calculations in those days. Clairaut also contributed extensively to perturbation theory, to lunar theory, and to the three-body problem.

Clairaut's scientific enterprises are associated with the activities of another famous mathematician, Pierre de Maupertuis, who, under the auspices of the French Academy, conducted an expedition to Lapland in 1735 to measure the length of a degree. If the earth is flattened at the poles, the length of a degree must be larger at the poles than the length of a degree at the equator. The length of a degree means the distance a vertical line must be displaced along a meridian for its direction to change by one degree. If the earth were a perfect sphere, this would be the same at every point on the earth's surface. If, however, the earth is an oblate spheroid, the length of a degree increases as one moves from the equator to the pole. Maupertuis, accompanied by Clairaut, set out to prove this and was successful in this undertaking.

Though Maupertuis did not contribute much to astronomy, he discovered a remarkable principle which made him famous and which is a powerful adjunct of and useful addendum to Newton's laws of motion. This principle, known as the "principle of least action," is the first of a series of such minimal principles. It states that a certain entity called "action" associated with the motion of a particle or a group of particles must be a minimum along the actual path of the particle. In a sense, then, this principle determines the path along which a particle must move when acted upon by a force. According to this point of view, a planet moves along an ellipse because its action along the ellipse is smaller than along any other path. To define the action of a body Maupertuis introduced the concept of the momentum of a particle (mass point) which he defined as the mass $m$ of the body times its velocity $v$ ($mv$). He then defined the action of the body moving a tiny distance $d$ as $mvd$, and stated his principle of least action as follows: The path along which a particle (e.g., a planet) moves among all possible paths is that one along which the total action is smaller than along any other path.

This remarkable minimal principle has expanded enormously since it was first proposed by de Maupertuis in its simple form. This principle probably stemmed from a different minimal principle proposed by the great French mathematician Pierre de Fermat (still famous for "Fermat's last theorem"). Fermat was looking for some general principle that would unify the laws of the universe and he believed he had found it in what he called the "principle of least time" as applied to the path along which a ray of light moves in going from its source to any other point. He argued that if paths are drawn from some point $A$ from which the light originates to some distant point $B$, the light, in moving from $A$ to $B$, moves along that path which it traverses in the shortest time (hence, the principle of least time).

This principle has predictive power even if points $A$ and $B$ are not in the same medium. If $A$ and $B$ are both in air or in a vacuum, or in any other medium (e.g., glass or water) the path of light from $A$ to $B$ is a straight line because the speed of the light does not change along its path. But if $A$ is in air and $B$ is in water below the surface at some distance away the light path cannot be a straight line because the speed of the light beam is smaller in water than in air. It is then easy to show that the path of the light beam must be a broken path if it is to satisfy Fermat's principle of least time. This is completely in accordance with the observed refraction of light when it passes from one medium (air) into a denser medium (glass or water) as we have already mentioned in our previous discussion of Bradley's contributions to astronomy.

Euler restated the principle of least action in a form from which Newton's laws of motion can be deduced so that this principle seems to be more fundamental than Newton's laws. In the early nineteenth century the great Irish mathematician and physicist William Hamilton extended the principle and reformulated it, enlarging it to include energy and time in such a way that it is now the basis of the most recent developments in theoretical physics and astronomy.

Though Maupertuis, on the basis of the success of his expedition to Tornea in Lapland became famous, his devotion to his principle of least action led him to religion and metaphysics and he contributed little more to science. He became president of the Berlin Academy and a favorite of Frederick the Great, spending many years at Frederick's court. The attention of the French circle of astronomers and mathematicians now shifted to Laplace, whom we may consider as the father of cosmogony and, in a sense, of cosmology. Laplace was the last of the late eighteenth and early nineteenth century group of mathematicians who contributed extensively to astronomy. Though primarily a mathematician with special expertise in probability theory (his book *Theorie Analytique des Probabilités*, which was published in 1812, is still a standard text), he turned to astronomy as the greatest mathematical challenge of his day. In his five-volume work *Traite de Mécanique Veleste*, written between 1779 and 1825, he systemized all the mathematical work that had already been done on Newtonian gravitational theory. He also wrote a history of astronomy, *Exposition du System du Monde*. First attracted to the dynamics of the solar system, he progressed from an analysis of the perturbations of the planets, proving that the solar system is highly stable, to his magnum opus, the origin of the solar system, based on his "nebular hypothesis" of the origin of stars.

It was only natural for mathematicians, philosophers, physicists, and theologians to turn to cosmogony (which deals with the origin of the cosmos) and cosmology (which deals with the evolution of the cosmos) with the acceptance of Newtonian gravitation as the basis of the dynamics which governs the entire universe. Newton's laws of motion and his law of gravity are inconsistent with an unchanging universe; if we accept Newton's laws, we must reject the pre-Newtonian concept of a fixed, unchanging universe, and accept instead a universe that is in a state of flux, which we should be able to describe completely in accordance with Newtonian theory. In promulgating his nebular hypothesis of the origin of the solar system, Laplace went far beyond anything that had been done with Newtonian dynamics up to that time. This required boldness of thought and great mathematical skill, both of which Laplace had in abundance. But the nebular hypothesis was not entirely original with Laplace; the German philosopher Immanuel Kant had proposed a similar idea in 1755, as had the Swedish scientist Emanuel Swedenborg in 1734. The difference between Laplace's proposal and those of Kant and Swedenborg is that Laplace's work, still extant, was developed in a rigorous mathematical way whereas Kant and Swedenborg did nothing more than propose their speculative ideas.

Laplace was led to his nebular hypothesis by carefully analyzing four observed properties of the solar system: (1) all the planets and the moon revolve around the sun in the same direction; (2) all the planets, as observed then, rotate around their own axes in the same direction as they revolve (from west to east with the exception of Venus as observed from a point above the earth's North Pole); (3) the planets all move around the sun in very nearly the same plane (the plane of the ecliptic); and (4) the orbits of all planets are almost circles (very round ellipses). Laplace, greatly puzzled by these "coincidences," refused to accept them as such, for his mathematical skill in probability theory, applied to these collective phenomena, showed him that they could not have been fortuitous. His calculations show that the chance that the four properties of the planets listed above resulted from pure coincidence is negligible. He therefore dismissed pure chance (random coincidence) as the explanation of these dynamic phenomena and looked for a rational explanation of them.

In going over each of these phenomena carefully, he saw that he could make them all fit together and explain them as stemming from a single cause if he pictured the sun, planets, and moon originating from a single, slowly rotating cloud which he called a nebula—hence, the "nebular hypothesis." We can see how bold and revolutionary this hypothesis was at the time for it completely destroyed the belief held then that solar material and planetary material are different. Stars, planets, moons, rocks, plants, and animals are made of the same stuff and they all came from the same primordial material. Laplace was far ahead of his time, for the conclusions that stemmed directly from his hypothesis was that all stars were formed in the same way, and that there is a system of planets associated with each star.

The nebular hypothesis was not immediately accepted by Laplace's contemporaries because the mathematical skills required to show in detail just how the sun and planets in the solar system took their present shapes and sizes, and why the planets move in orbits that are spaced in a fairly regular pattern around the sun, were not available. Why, for example, are no planets found in orbits between Mars and Jupiter or in an orbit closer to the sun than Mercury's orbit. Another difficulty associated with the nebular hypothesis stems from the distribution of the mass in the solar system. We know that more than 99 percent of the total visible mass in the solar system is contained in the sun, but most of the rotational motion (angular momentum) is contained in the planets. This is difficult to reconcile with a rotating nebula, whose outer regions, containing little mass, were revolving much more slowly than the inner regions.

Though accepted quite readily in its early years, the nebular hypothesis lost favor in time, but it regained favor in the twentieth century and today, though still presenting very challenging problems, it is universally accepted as the only viable theory of the birth of stars and planets.

Laplace's boldness in proposing new theories and hypotheses knew no bounds for he promulgated what we must accept as the first statement of a unified theory of science. Expressed in his own words it must have shocked the eighteenth century intelligentsia, in spite of their acceptance of their age as the "age of reason," for it was a direct acceptance of what we would now call "mechanical materialism," which Laplace stated as follows:

> An intelligence which, for a given instant, knew all the forces by which nature is animated, and the respective positions of the beings which compose it, and which, besides, was large enough to submit these data to analysis, would embrace in the same formula the motions of the largest bodies in the universe, and those of the lightest atoms: nothing would be uncertain to it, and the future as well as the past would be present to its eyes. Human mind offers a feeble sketch of this intelligence in the perfection which it has been able to give to Astronomy.

For many years this doctrine was at the center of many controversies among mathematicians, physicists, and astronomers owing to its uncompromising determinism, but became purely academic after 1927 with Werner Heisenberg's discovery of the uncertainty principle which tells us that we cannot simultaneously have knowledge of a particle's position and motion (momentum).

Before retiring from his great creative work in mathematics and astronomy, Laplace performed one other great service for astronomy and astronomers by simplifying the mathematical technology for solving complex gravitational problems. He did this by enlarging on the concept of the "gravitational potential," which Euler had already discussed briefly in his treatment of hydrodynamics. To understand the importance of Laplace's work in this area we recall that Newtonian gravitational theory brought with it a very mysterious feature—action at a distance—that no one, not even Newton, understood. In fact, Newton himself was greatly puzzled by gravity's action at a distance, which he expressed as follows in a letter to Richard Bentley: "It is inconceivable that inanimate brute matter should, without the mediation by something else, which is not material, operate upon and affect other matter without mutual contact."

This troubling feature (action at a distance) of Newtonian gravity was somewhat ameliorated later by the introduction of the "gravitational field" generated by a mass; the gravitational field was (and still is) pictured as extending to infinity, but with its intensity falling off in all directions, becoming zero at an infinite distance. The field was then described by giving its intensity and direction at each point. Another mass (particle) placed at any point in the field would then respond to the field at that point by moving in the direction of the field with an acceleration determined by the intensity (strength or magnitude) of the field. The field concept in itself did not simplify the gravitational many body problem; it just shifted it from one point of view to another. To see this we consider a given body in the gravitational field produced by a number (more than one) of other bodies at various distances and in different directions from the given body. To find the action of these combined fields on the given body at its position, we have to add all of these separate fields and that is, in general, extremely difficult because the fields all point in different directions. Combining these directions to obtain the single direction of the sum of all the fields at the point is prohibitively difficult.

Euler and Laplace overcame this difficulty by introducing the "gravitational potential" at any point and in any combination of gravitational fields. This greatly simplified gravitational theory because the potential is just a number without any direction and it is fairly easy to calculate it at any point for a particle and for many particles. If a given particle is at a point in a gravitational field of $n$ different particles, we calculate the $n$ different potentials of the gravitational fields of the $n$ particles at the given point and add them arithmetically to obtain the potential at that point of the combined fields. Both Euler and Laplace discovered the general equation for the potential of the general gravitational field. This equation eliminates the troubling concept of "action at a distance." Laplace's equation for the potential shows that a particle in a gravitational field always moves from a point where the potential is high to a point where it is lower. Knowing the value of the gravitational potential at each point of a gravitational field one can easily deduce, mathematically, the magnitude and direction of the gravitational field at that point.

What Newton had begun was now complete, or so it seemed to those who followed Laplace. With the elegant mathematical techniques developed by Euler, Lagrange, Clairaut, d'Alembert, Maupertuis, and Laplace to handle all kinds of gravitational problems, there seemed little more to be done than to scan the heavens, study the positions and motions of the stars, and collect ever more observational data to lend increasing support to Newtonian theory. Recognizing this apparent truth the observational astronomers began to take over to show that the mathematicians were wide off the mark in their assessment of what the future of astronomy was to be. New observational techniques—in particular, photography and the use of spectroscopy, were used to introduce new branches of astronomy such as astrophysics, galaxy and stellar cluster astronomy, cosmology, and cosmogony. With the deduction of Kepler's three laws of planetary motion from Newton's laws of motion and gravity, observational astronomers lost interest in observing the planets and shifted their greatest attention to the stars. Of course, solar system observations went on, but not with the excitement that followed Galileo's construction of the first astronomical telescope. The solar system still presented astronomers with questions that the French mathematicians could not answer.

We mention a few of these questions to show how much remained to be explained in our solar system: the nature and structure of the sun; the spacings of the orbits of the planets; the differences between the four small inner planets and the large massive outer planets; the nature and origin of the comets; the nature of the rings of Saturn, etc. In principle, if we completely understood the solar system, we could understand the universe itself. This idea was beautifully stated by the late eighteenth century British poet William Blake:

> To see a World in a Grain of Sand
> And a Heaven in a Wild Flower
> Hold infinity in the palm of your hand
> And Eternity in an hour

To the late eighteenth and early nineteenth century observational astronomers, making meaningful observations beyond the solar system meant constructing large telescopes and large observatories to house them. In this resurgence of observational astronomy, the lead, under the direction of the British astronomer, Sir William Herschel, was taken by England. As we shall see in the next chapter, the first exciting new astronomical discovery made by Herschel was not about the stars but within the solar system itself: the discovery of the seventh planet—Uranus—beyond Saturn. But the study of the stars, the galaxies, and space itself offered a great deal more excitement than the study of the solar system.

CHAPTER 11

# The Beginning of the New Age of Astronomy: Beyond the Solar System

*The true worth of an experimenter consists in his pursuing not only what he seeks in his experiment, but also what he did not seek.*

—CLAUDE BERNARD

In 1757 a 19-year-old musician from Hannover, Germany, settled in England as a music teacher and organist. Coming from a highly cultured and gifted German-Jewish family of musicians—his father was a kappellmeister—and having won a prestigious prize for his organ playing, William Herschel, at the age of 36, became the music master at Bath and a popular conductor. At Bath he conducted Handel's oratorios with a large chorus and a hundred-man orchestra. He continued his devotion to music by playing the organ in a church, playing first violin in a theater, giving private recitals, and teaching many students. But even in the midst of these time-demanding musical activities (some 14 hours a day) it became clear that his deepest interests were undergoing a profound change. He was transferring his most creative efforts from music to astronomy. Although he had become a respected composer, he was devoting more and more of his time to astronomy.

He had begun to read mathematical treatises—particularly those dealing with calculus—and devouring astronomical literature. In the spirit of his devotion to music he threw himself into astronomy, not as an inactive bystander, amateur, and admirer of those who had gone before, but as a very active participant. He began by setting up an observational program. Buying small telescopes he transformed his house into part observatory and part music studio.

Because only very small telescopes were available commercially, he decided to construct a large telescope for his own use. This entailed setting up a forge and an optical shop for polishing and testing metal mirrors. He became so proficient at polishing mirrors for telescopes that he began to construct telescopes for sale.

Unable to perform all the activities required by his extensive program, he brought his sister, Caroline, and brother, Alexander, over from Germany to live with him and help him in all his projects. Caroline became his assistant in astronomy and Alexander his commercial assistant. Within a few years the Herschel brothers produced 200 4-inch telescopes, 150 6-inch telescopes, and 80 9-inch telescopes. These numbers are the diameters in inches of the mirrors ground by William and Alexander. Large diameter mirrors are desirable for two reasons: the amount of light entering the telescope increases as the square of the mirror diameter and the larger the diameter, the sharper the image produced by the mirror. Thus a 6-inch diameter mirror allows four times as much light to enter the telescope as a 3-inch mirror does, and a 9-inch mirror allows nine times as much light to enter as a 3-inch diameter mirror.

Caroline became William's observing assistant, spending many hours peering through the best and largest of his telescopes. Devoting herself to searching for comets, at her brother's suggestion, she became the first female astronomer. In all, she discovered eight comets, adding to the fame of the Herschel family. With Alexander and Caroline well integrated in his numerous activities, William finally settled down in Slough. Having married into a wealthy family, his wife's income made it possible for Herschel to fulfill his greatest desire: to construct the largest telescope ever built. This telescope became the sensation of the day and attracted visitors from all parts of England. These visitors enjoyed not only the view of the telescope but also the charm of Herschel himself.

A great teacher and lecturer, Herschel was described by those who heard him as a "delightful, extremely modest man, for all his vast knowledge; candid as a child, delicately tactful and considerate; he makes everything extremely clear . . . and puts over his own ideas with indescribable charm. He knows the history of all the heavenly bodies to the furthermost boundaries of the Galaxy."

William Herschel (1738–1822) (Courtesy AIP Emilio Segrè Visual Archives; E. Scott Barr Collection)

Although Herschel became world famous owing to his discovery of the seventh planet Uranus, astronomers do not consider that to be a discovery of basic astronomical importance. Anyone who had surveyed the sky with a large telescope as patiently and as assiduously as Herschel had, would have, in time, picked up Uranus, which Herschel did in 1771. Of course Herschel used a special observational technique which he called sweeping the sky. Just as the surface of the earth can be divided in strips bounded by arcs of longitude so, too, can the sky. Herschel surveyed strips of the sky bounded on the east and west by the two meridian circles separated by 2 degrees. He then systematically searched each such strip for unusual objects which had never before been observed, and he found a very interesting object in one of these strips.

An inexperienced observer might have dismissed such a fuzzy object as being a comet or another unimportant celestial body, but Herschel, from his deep knowledge of the apparent motions of comets and planets, knew at once that he had discovered a new planet. From its apparent motion, its brightness, and its apparent size, he deduced that the new planet was more distant than Saturn but roughly about one-third its size. Herschel first suggested the name Georgium Sidus, in honor of George III, the British king at the time, but soon settled instead on the name Uranus.

Herschel's well-deserved reputation as an innovative astronomer rests on his stellar work, the new observational techniques he introduced, and the types of celestial objects on which he concentrated: nebulae, stellar clusters, and double stars. He studied double stars because he was convinced, and wanted to prove, that many double stars (two stars that appear to be close together) are, indeed, gravitationally bound to each other and thus form what astronomers call a "binary system." He observed in many of these double stars that the two components shift their positions with respect to each other, twirling around a common center of mass like two dancers on a ballroom floor. Herschel had begun studying double stars with the hope that they might lead him to a procedure for measuring their parallaxes and thus their distances, which was still beyond the astronomical technology at that time.

By the parallax of a star we mean the apparent change in a star's position against the more distant background stars when the earth changes its position in its orbit around the sun. This stellar position is defined by the direction of a straight line from the earth to the star. As the earth moves in its orbit, the direction of this line changes so that the star appears to change its position. The farther the earth moves along a straight or nearly a straight line (in a fixed direction), the greater the star's parallactic shift is, if, indeed, it can be observed at all. The star's parallax is defined as the angle between the two straight lines drawn first from the earth's initial position to the star and then from the earth's final position to the star. The star's parallax is thus the angular change (apparent change in direction) in the star's apparent position. Since the stellar distances are vast and accurate angular measurements are difficult, astronomers base their parallax measurements on the maximum displacement (separation) between two points that the earth experiences in its orbit around the sun. This is just the diameter of the earth's orbit around the sun (now called one astronomical unit) which is 186 million miles.

Herschel, of course, knew that all attempts to measure this angle for any star, beginning with those observed by Bradley, had failed because this angle (the parallax) even for the nearest star is exceedingly small. He then thought he might get around this problem by finding the parallaxes of double stars by using the orbits of the two individual stars around their common center of mass. If Herschel could have determined the true distance of each star in the binary system from their center of mass he could then, from their angular separation, have found their true distance (actually the distance of their center of mass) from the earth. But he was not successful in this endeavor because he did not know the total mass of the two stars in the binary system and this was information that he could not obtain with his astronomical technology.

But Herschel's failure to measure the parallaxes of binary stars did not deter him from studying these very interesting objects in detail. He first showed by a general analysis of the gravitational interaction between two stars that they could form a bound system moving in Keplerian orbits around a common center of mass, just the way a planet moves around the sun. He measured the periods of many different binaries and showed that they range from one to a few hundred years and that their separations may range from tens of millions of miles to hundreds of millions of miles. To calculate these distances from the observed angular separation he assumed that each star in a binary system has the mass of the sun. By the time he had finished studying double stars he had discovered some 800 binary star systems.

From the study of binary stars Herschel went on to the study of groups of stars called stellar clusters, consisting of stars that move together through space, relative to the sun, like a gravitationally bound system. He first discovered triplet systems of stars, then quartets, and finally groups (clusters) consisting of dozens of stars. The Pleiades are a very striking example of such a cluster. Seven stars in this cluster are visible to the naked eye, but even with a small pair of binoculars one can pick up some 50 stars in this cluster. Herschel discovered clusters containing hundreds and even thousands of stars and showed that certain clusters look like a globe of stars, so tightly packed with stars that the ones in their centers cannot be seen as individuals. He estimated that the total numbers of stars in such clusters (now called "globular clusters") may be as large as 100,000, all held together by gravity.

Like all great scientists, Herschel was very imaginative, bold, and speculative, all within the constraints imposed on his thinking by Newtonian mechanics and dynamics. In his thinking the laws of nature came first and were not to be denied or contradicted. But these laws allowed him to roam intellectually to his heart's content, and roam he did. He began to study distant galaxies and, quite correctly, proposed that they are all "island universes" like the Milky Way. This was the beginning of rational cosmology, the attempt by astronomers to understand and describe the structure and manifestations of the entire universe in terms of Newton's laws. Observing the galaxies distributed throughout the space that was observable to him at the time, he reasoned correctly that they could not be fixed or suspended in space like lamps or candelabra attached by rope to the ceiling of a cathedral. The mutual force of gravity would cause them to move together, ultimately collapsing into a single sphere of matter. This would be the end of the universe. Because he had no way of measuring the motions of the distant galaxies, this hypothesis was reasonable but very speculative.

Herschel paid considerable attention to the stars in the Milky Way and the structure of the Milky Way. He spent hours cataloging the stars in the solar neighborhood, noting how the stellar population increased dramatically as he turned his telescope toward the Milky Way itself. Using statistical methods he drew the first diagram of the Milky Way, showing it as an elongated ellipsoidal structure, thick at the center and thinning out at the edges (although he incorrectly placed the solar system near the center). This diagram was a most remarkable achievement and synthesis.

In his study of the stars near the solar system in the Milky Way, he drew a very important conclusion from the observed apparent motions of these stars. The Greenwich Observatory astronomers, by studying the observed changes in the positions of these some 40 stars over a period of years had concluded correctly that they are moving. These apparent motions of stars are called stellar proper motions. Herschel readily accepted this concept but, in studying these observations carefully, he discovered that they revealed a peculiar feature: the stars seemed to be moving in all directions away from the sun, as though they were avoiding the sun. Thinking about it Herschel rejected the idea that all these nearby stars were actually moving in this way and introduced the very revolutionary idea that part of the apparent motions of the nearby stars is due to the motion within the galaxy of the entire solar system. This fit very nicely into his Newtonian scheme of things: the solar system is moving among the stars owing to the gravitational pull that these stars exert on the sun and planets in the solar system. His detailed analysis of the stellar and solar motions led him to the conclusion that the solar system is moving toward the constellation Hercules.

To verify his hypothesis that the motions of the stars and the solar system within the galaxy are produced by stellar gravity, Herschel decided to count the stars in the entire galaxy but to do so by replacing an actual count of each star by a statistical count, arguing that he could achieve his goal by counting only the nearby stars and then simply assuming that these sample counts would do for all parts of the galaxy. He called these sample counts "standard star fields" which were assumed to be the same everywhere in the galaxy. This is not quite correct but at that time it was a very heuristic idea which stimulated great interest in galactic research. He lent great support to these often speculative ideas by the thoroughness of his stellar observations which marked him as probably the greatest star observer of all time. With his largest telescope he was counting as many as 100,000 individual stars per hour, picking up stars that were so faint that they were barely visible. Of course, he did not count each star but he sampled small areas of the sky and from these samples estimated the total number in the field he was studying.

Herschel was quick to emphasize that his purpose was not to become the "champion star counter" but to draw important conclusions about the structure and dynamics of the galaxy. He expressed his philosophy clearly: his purpose was not "to accumulate observations" for their own sake but to use these observations as the foundation for constructing and developing theories. He emphasized that the role of the scientist—particularly the astronomer—is to generate such theories even if the probability for them to be correct might be low. Even if the imagination of the astronomer in constructing a model of the universe might lead him to "overstep the boundaries set up by nature," he must still pursue his search with the hope of arriving at the truth. In this philosophy Herschel was expressing the doctrine that guides scientists today, and he went far beyond what most of his contemporaries were willing to accept as a guiding principle. In spite of his glorification of speculation and imagination, he insisted on strict adherence to the policy that all such intellectual wanderings must be checked by the facts. The stars themselves were to be the final arbiters. This may appear to go counter to Newton's statement: "I frame no hypotheses" but it is really in the spirit of Newton in that Herschel's hypotheses never denied basic theorems (e.g., gravity).

Without going into all the details of Herschel's other achievements, we list them briefly:

1. By comparing the brightness of distant stars with nearby ones he set up a distance scale for stars.
2. He was the first astronomer to use the light year (the distance light travels in one year—about 6 trillion miles) consistently as an astronomical yardstick.
3. He was the first to present the idea that sun spots, by affecting the solar luminosity, can be correlated to harvests on the earth.
4. He was the first to study the variations in the intensities of the different colors of the light emitted by the sun, noting that the intensity varied from color to color. He did this by using different colored filters. This led him to the discovery that some of the solar radiation has no color at all—such radiation is called infrared.
5. He studied the lunar craters and the heights of lunar mountains, finding their correct heights by measuring the lengths of their shadows during certain phases of the moon.
6. Finally, his great success in extending his observations to great depths in space altered observational astronomy for all time and demonstrated the importance of working with large telescopes.

What place in the story of astronomy do we reserve for this remarkable genius? If Newton was the monarch, then Herschel was the crown prince who fulfilled the great promise held out by Newton. His life marked a watershed; astronomy was altered beyond recognition after he completed his work; it was changed into a rigorous science that set the standard for all other sciences. Moreover, it encouraged amateur astronomers in all walks of life to try to emulate Herschel's great success.

The Herschel name did not disappear from the roll of active astronomers after William died, for his son, Sir John Herschel, continued his father's work by applying his father's methods to the stars in the southern sky. Sir John led an expedition in 1848 to the Cape of Good Hope. But he is best known for introducing an accurate scale for expressing the brightnesses of stars (the magnitude scale) which was applied and used very loosely until a more rigorous magnitude scale was introduced in 1854.

William Herschel's influence on astronomy, particularly on stimulating observational astronomy, was immeasurable, for he attracted the wealthy amateurs. But this restricted observational research to the few that could afford to construct telescopes and the observatories to house them. The most famous such "wealthy amateur" was William Parsons, the third earl of Rosse in Ireland, who, in 1845, built the largest telescope of that time for his own use. The telescope boasted a mirror 6 feet in diameter and was mounted at the front end of a 60-foot tube which was controlled by a system of cranes, pulleys, and cables. Rosse set out to examine in greater detail than William or John Herschel could, the many nebulae (galaxies) that William Herschel had discovered and cataloged. To Herschel these very distant objects appeared nebulous, without structure, and he was unable to resolve them into their constituent elements (stars). He therefore, quite properly, called them nebulae.

As Rosse had no photographic equipment to record them (the nebulae), he sketched them, showing their spiral structures; from that time on these objects have been called "spiral nebulae." When more advanced observational technology, particularly photography, resolved the "spiral nebulae" into individual stars, astronomers concluded that the "spiral nebulae" are distant galaxies like our own and so the name "galaxy" is applied to each of these nebulae. Rosse is most famous for having detected and sketched the detailed spiral structure of what he called the "Whirl Pool Nebula" (now known as the M51 galaxy). His sketch shows a concentrated "core" or "nucleus" with spiral arms emanating from the core, and also a smaller "satellite galaxy" connected to the "whirlpool" by an extension of the outermost spiral arm.

While ever larger telescopes were being constructed by wealthy amateurs, a new technology—astronomical photography—burst upon the world in 1870 and was to have as great an effect on nineteenth century astronomy as Galileo's telescope had on seventeenth century astronomy. The advent of photography freed observational astronomy from its dependence on the uncertainty of the observer's naked eye. In photography we need not rely on the interpretation of the image formed on the retina of the observer's eye. It supplies the certainty of a photographic record that all observers can accept as an "objective truth." If one photograph of a celestial object does not suffice, many photographs can be taken and compared with each other.

Photography improved almost every aspect of observational astronomy. In measuring the parallax of a star, the astronomer must measure the change in the apparent positions of a given star observed 6 months apart. Without photography he must note its position at the start of his observations of the star and compare that with its apparent position at the end (6 months later) of his observations. But when he employs photography, this arduous task is reduced to comparing two photographs of the same star taken 6 months apart. These photographs are taken with the telescope itself which becomes a very long focal length camera. The comparison of the two apparent positions of the star is then obtained by simply superimposing the two photographic plates and measuring the change in the positions of the image of the star on the two plates.

This same procedure can also be used to determine the motions of stars relative to our solar system. Photographs of a given star taken over a period of years and compared with each other give us a good measure of the motion of the star relative to the solar system.

Photography very quickly permitted astronomers to measure stellar brightnesses and, from these measurements, to determine stellar luminosities very accurately. We recall that the Greek astronomer Hipparchus introduced the concept of the magnitude of a star to classify stars according to their apparent brightnesses. This idea was carried on by Ptolemy who formulated the magnitude concept more precisely by arranging the visible stars into six magnitude classes; but since this was all done visually, the magnitude scale was not precisely defined so that assigning a magnitude to a star was pretty much of a hit or miss procedure. This changed with the introduction of photography because one could now measure the brightness of a star precisely by measuring the impression the light from a star made on the photographic plate.

To pursue this subject further and to clarify the basic ideas involved, we must move forward to the first decade of the twentieth century during which time the concept of the photon (a quantum of light) was developed owing to the pioneering work of Max Planck and Albert Einstein. We consider now the formation of an image on a photographic plate as the action of photons of light on the thin gelatin on the surface of the film. This gelatin consists of minute particles of silver halides, which, on absorbing the light, are precipitated as dark grains in the emulsion, thus forming an image on the film. Each photon of light interacts with a single silver halide molecule so that the image of the star becomes increasingly darker and larger as more light strikes the film.

The photographic film is superior to the eye in measuring the brightness of a star, because the retina of the eye on which an image of the star is formed keeps no record of individual photons striking it, whereas a photographic film does. The response of the brain to the stimulation produced on the retina is to an integrated effect which is recorded by the brain as very bright, very faint, or something in between, so that when we look at two different stars we note that one is brighter than the other, but by how much, we can give only a fuzzy estimate. But if we expose to two different stars the same kind of photographic film, for the same length of time, using the same telescope, we can measure the brightness difference between the two stars by actually counting the number of grains of silver halide in the image of each star. This can be shown to be equivalent to measuring the size of each image on each photographic plate. Thus photographic astronomy, for the first time, permitted astronomers to introduce a precise magnitude scale of brightness. This was done by the British astronomer Norman Pogson in 1856.

Before we discuss and define the Pogson magnitude scale, we must define the concept of the apparent brightness of a star. We emphasize the qualifying adjective "apparent" to differentiate clearly between the light that we receive on the earth from the star and the light the star pours out in all directions, which we later relate to the luminosity of the star. We now define the star's apparent brightness as the amount of light from the star in 1 second that strikes 1 square centimeter of surface area held perpendicular to the line from the star to the earth. We label this with the small case $b$ to distinguish it from what astronomers call the absolute brightness $B$, which is related to the star's luminosity. These concepts are related to photometry, the study of which was begun by physicists at about the time when photography was introduced in astronomy, so that the two disciplines developed together. Returning now to Pogson, we note that he applied photometric measurements to determine by how much the brightness of stars of a given magnitude as proposed or defined by Hipparchus and Ptolemy differ from the brightness of stars assigned to a different magnitude. He then discovered that, on the average, stars of the first Ptolemic magnitude are brighter by a factor of 100 than stars of the sixth Ptolemic magnitude. He therefore introduced a precise magnitude scale by defining a first magnitude star as one that is brighter by a factor of 100 than a star of the sixth magnitude. In other words, the magnitude scale of brightness is just an arbitrary scale of numbers which is so defined that if the two numbers assigned to two different stars differ by one unit, the brightnesses of the two stars differ by a factor of about 2.512; this means that each magnitude step upward (e.g., from 2 to 3) means a drop in brightness by a factor of very nearly 2.512. Thus if we take five steps from a first to a sixth magnitude star, the brightness decreases by 2.512 raised to the fifth power, which is nearly equal to 100 ($2.15 \times 2.15 \times 2.15 \times 2.15 \times 2.15$).

The reader should note that the word "magnitude" in its astronomical usage has nothing to do with the size of the star; it was introduced by Hipparchus to express his belief that bright stars are more important than faint stars. He expressed this anthropomorphic feeling about stars by calling the 20 brightest stars he could find "stars of the first magnitude" (importance). This is why the magnitude scale is inverted numerically with small numbers assigned to bright stars and large numbers assigned to faint ones. With very faint stars being discovered with larger telescopes after Galileo, the magnitude scale had to be enlarged to take account of stars much fainter than those just visible to the naked eye, so that magnitude classes beyond Hipparchus's 6, 7, 8, 9, etc., had to be introduced. This was particularly so with Herschel's discovery of the faint nebulae which are given large magnitude numbers. At the present time magnitudes as large as 25, 26, 27, etc., have to be assigned to the most distant objects visible through the very large modern telescopes.

Just as the magnitude scale had to be extended beyond 6 to include very faint objects, so it also had to be extended to smaller and negative numbers (e.g., 0, –1, –2, –3, etc.), to include objects brighter than the typical (average) first magnitude star Hipparchus and Ptolemy had used in their classification. Finally, we note that today magnitudes of stars can be measured with an accuracy of one part in a thousand so that one finds magnitudes such as 4.536 listed in the literature. Since the Pogson scale, like any other scale, is purely arbitrary, it can only be used if it has a zero point or a point which relates the magnitude to a measurable brightness. This point on the Pogson scale is associated with the star Polaris (the North Star) which was arbitrarily assigned the apparent magnitude 2. This number gives a physical meaning to the Pogson scale for it permits us to assign an apparent magnitude to any star just by comparing its brightness (the amount of light it sends us per second) with that of Polaris. On this scale, the sun's apparent brightness is –26.87, a very "large" negative number. On the same scale, the apparent magnitude of the full moon is –12.6 and that of Venus and Jupiter at their brightest is –4. Among the visible stars Sirius is the brightest with an apparent magnitude of –1.5. The apparent magnitude of any celestial object is represented by the lower case $m$.

As photography advanced and the emulsions and films improved, astronomers discovered that various emulsions respond differently to different colored light and differently from the reaction of the human retina to the same radiation. It was, therefore, necessary to distinguish between the visual apparent magnitude of a star, written as $m_0$, and the photographic magnitude written as $m_{ph}$. Knowing that the emulsion on film responds more readily to blue and violet light than to red light, whereas the normal eye responds more readily to red than to blue light, astronomers devised a very simple measure of the color of a star by comparing the visual magnitude of a star with its photographic magnitude. The difference between these two measurable numbers was introduced as the "color index" of the star. These developments did not occur overnight but evolved steadily as astronomical photography grew.

The next step after the introduction of the Pogson apparent magnitude scale was the use of a simple procedure to obtain the absolute magnitude written as $M$ to differ from the apparent magnitude $m$. This was necessary because the knowledge of $M$ leads us to the luminosity of the star. If the star's $m$ and its distance are known, its $M$ and therefore its luminosity can be calculated. The star's $M$ is defined as the apparent magnitude it would have if it were at a distance of 32.6 light years or 10 parsecs, where the parsec is defined 3.26 light years.

Herschel's striking success in his observational program greatly stimulated observational astronomy on the continent, particularly in Germany, which had developed a flourishing optical industry. Joseph von Fraunhofer, born in 1787, was foremost among the opticians and lens designers who began their activities which, in a relatively short time, projected Germany into world leadership of the optical industry. Fraunhofer began his career as a glass grinder apprentice at the age of 14, and he quickly began to dominate the industry. Owing to his intuitive grasp of optical theory, his understanding of mechanics, and his skill as a designer he became director of the famous optical institute in Benedictbeuern in 1812, at the age of 25, and in 1823 became a member of the Academy of Science in Munich and its "conservator" of physics. Aware of Herschel's great accomplishments, Fraunhofer set himself the goal of constructing a telescope superior to Herschel's telescope. Instead of building a reflector with a mirror as its objective, he decided to build a refractor which required the grinding and polishing of a lens doublet, an achromat, to eliminate the chromatic aberration; he housed his achromat in a tube weighing many tons. To permit this tube with its lens in front to be manipulated (rotated) easily, he mounted it on ball bearings, a design which set a completely new trend in telescope design. In constructing this telescope with its doublet objective, Fraunhofer set the style for the construction of all future refractors. Two famous German telescopes, based on Fraunhofer's design, were built shortly before his death—the 11-inch Königsberg telescope and the 10-inch Dorpot telescope.

If Fraunhofer had done no more than revolutionize the art and science of building telescopes, he would still be famous, but most of his fame among astronomers rests on his discovery of the famous dark lines (the Fraunhofer lines) in the solar spectrum. We are all acquainted with the sun's rainbow and with the way the light from the sun that strikes a prism at a certain angle is spread out into a band of colors. This band, called the solar spectrum, was known to the early Greeks, to Galileo, to Newton, and to Herschel. And yet none of these early scientists nor Newton, who had studied the solar spectrum carefully, had detected anything unusual about it. Even Herschel, who might have been expected to discover anything unusual and exciting in the sun's spectrum, failed to do so. The reason for these failures is that the scientists who predated Fraunhofer did not think of examining the spectrum in detail, whereas Fraunhofer did.

To examine the solar spectrum under magnification Fraunhofer designed a special instrument, now called a spectroscope, which consists of three parts: (1) a small tube with a narrow slit in front and a lens in back; (2) a prism on a small mount; and (3) a small telescope mounted on a turntable behind the prism. This device (spectroscope) is mounted in the tube of the telescope behind the objective, close to the eyepiece. Light from the sun enters the front tube of the spectroscope but only a narrow beam (defined by the slit) gets through and enters the prism. This beam, now bent by the prism, is inspected by the small telescope behind the prism. By rotating this telescope one can then examine each color of the spectrum, which is really an image of the slit in that color. Using this spectroscope Fraunhofer discovered one of the most important phenomena in astronomy: the light from the sun does not consist of an unbroken band of colors: The band, the "continuous spectrum," is present, but superimposed on this band like slits in a wooden fence are literally thousands of dark lines called absorption lines. Each line is identified by its wavelength, or frequency. The wavelength is measured by the position of the line in the spectrum; those in or near the red part of the spectrum have long wavelengths and those near the violet end have short wavelengths. Here "long" and "short" are relative terms since the wavelengths at both ends of the spectrum are very tiny compared to the ordinary distances we deal with in our daily lives. Owing to the minute wavelengths of the different colored lights our eyes respond to, physicists and astronomers introduced the Ångstrom (Å) as a unit of wavelength named after the Swedish astronomer and physicist Anders Jonas Ångstrom (1814–1874) of the University of Uppsala. He was director of the Royal Society of Sciences in Uppsala and a pioneer in the study of spectra.

The Ångstrom, a unit of length, is 1/100,000,000th of a centimeter. The wavelengths of the rays are in the range of about 7000 Å for red light and about 3500 Å for violet rays. Thus the visibility of the human eye spans about one octave in the electromagnetic spectrum. The wavelengths of infrared rays are longer than those of the red rays and the wavelengths of ultraviolet rays are shorter than those of the violet rays. If we picture a light wave as a train of oscillations moving through space, with the wave crests moving along a line like the crests of a water wave, the wavelength of this train is the distance between any two successive crests; this wavelength is the same all along the wave.

Instead of describing the wave in terms of its wavelength, we can describe it in terms of its frequency: the number of wave crests that pass a given point per second. The speed of the wave is then given by the product of its wavelength and its frequency. This is just the speed of light—very nearly 30 billion centimeters per second—which is the same in a vacuum for all colors. From this speed and the known wavelength of any color we can calculate the frequency of that color, which is about 500 trillion vibrations per second for red rays and about 1000 trillion vibrations for violet.

Fraunhofer understood the nature of the dark lines in the solar spectrum and reasoned correctly that they are produced by the solar atmosphere as the light coming from what he pictured as the "surface" of the sun (now called the photosphere) passes through the atmosphere. When the solar radiation leaves the solar photosphere, it consists of a continuous band of colors, but each color interacts with the solar atmospheric atoms (or molecules) in such a way that it either passes through unaffected or suffers some kind of absorption. The lines thus produced by the interactions are therefore called "absorption lines." Fraunhofer saw that these lines are not completely black (completely devoid of light) but appear so only by comparison with the continuous background spectrum; if this background spectrum were removed, the absorption lines would be visible as bright lines.

Fraunhofer observed thousands of lines but had time to map and catalog only 57 percent of the most intense lines. Since then tables of thousands of these lines have been prepared but many thousands more remain to be studied. Fraunhofer also spent time studying bright "emission" spectral lines such as those produced when salt is thrown into a flame and emits a very brilliant yellow light. Fraunhofer examined the yellow light with his spectroscope and discovered it consists of two lines called the sodium D lines because they are produced by the sodium atoms in the salt molecule. Because the wavelengths of the sodium D lines are very nearly equal, they are called a "doublet."

Fraunhofer was quick to see that his discovery of the solar absorption lines would have a profound influence on astronomy for it was the first step in the chemical analysis of stars, each of which exhibits a spectrum similar but not identical to the solar spectrum. The variations in the spectrum of different stars must then reveal differences in their chemistry. This was the beginning of astrophysics.

Fraunhofer's colleagues recognized the great significance of his work during his lifetime. The esteem in which Fraunhofer's contemporaries held him is best revealed by the inscription on his gravestone: "He brought the stars closer." We append here our own commentary on Fraunhofer's great contribution to science and to the overall advancement of knowledge. The spectroscope is probably the greatest scientific invention of all time. It has enabled us to penetrate atoms, molecules, galaxies, human cells, and the universe itself. Never has so much been achieved by so small an investment in technology.

While Fraunhofer was opening the door to astrophysics, other astronomers were enlarging the exploration of the solar system with particular emphasis on the study of comets, meteors, meteorites, asteroids, sunspots, and the lunar surface. The most dramatic event in these studies occurred in 1801 when the Italian monk, Giuseppe Piazzi (an amateur stellar observer), discovered the first of an ensemble of a new kind of celestial body which he called "asteroids" or "minor planets."

A study of the distances of the major planets Mercury, Venus, earth, etc., from the sun, reveals a remarkable and simple numerical relationship among these distances. In 1766 the German astronomer Daniel Titius of Wittenberg noted that if to each of the integers 0, 3, 6, 12, 24, 48, . . . (note the doubling) the number 4 is added to obtain the integers 4, 7, 10, 16, 28, 52, . . . and each of these is divided by 10 to obtain the sequence 0.4, 0.7, 1, 1.6, 2.8, 5.6, . . . these are very nearly equal to the distances of the planets from the sun expressed in astronomical units (the mean distance of the earth from the sun, which is thus taken as 1). In 1772, J.E. Bode, a professor of astronomy at Berlin, published this numerical sequence, which is now known as the "Bode–Titius law" of planetary distances. It was first considered a "law" when the mean distance (20 A.U.) of Uranus, shortly after its discovery, was found to fit this sequence extremely well. However, as no planet, at that time with a mean distance of 2.8 A.U. was known to exist, doubt was cast on the "Bode-Titius law." But then came Piazza's discovery of the first asteroid Ceres with a mean distance of 2.8 A.U. from the sun.

The story of Piazzi's discovery is interesting because he announced it in a letter to Bode. But the letter was delayed for almost 2 months, after which time Ceres was no longer visible. Fortunately, the great German mathematician, physicist, and astronomer Karl F. Gauss had developed a mathematical technique for determining the complete orbit of a body in the solar system if any three points of its orbit were known. Piazzi had, luckily, made three distinct observations of Ceres before it disappeared. Using these observations, Gauss calculated the orbit very accurately and Ceres was found very near the point predicted by Gauss, crossing in front of the constellation of Virgo, on December 7, 1801. This was all very exciting for its mean distance from the sun is 2.8 A.U. in accord with the Bode–Titius law. However, the mass of Ceres is so small, less than one ten thousandth of the earth's mass, that one could hardly call it a planet. But shortly after Piazzi's discovery of Ceres, Heinrich Olbers discovered the asteroid Pallas in 1802; two other asteroids, Juno in 1804 and Vesta in 1807, were then discovered. By 1890, more than 300 asteroids were known and their orbits plotted. The mean distance of this group from the sun is almost exactly 2.8 A.U., so that the Bode–Titius law was supported, based on the argument that the asteroids (which are now estimated to exceed 100,000 in number) are the remnants that never coagulated gravitationally into a planet. These asteroids have played an important role in astronomy for their orbits are an excellent test of Newtonian gravitational theory.

While some of the professional and amateur astronomers were drawn to the search for asteroids, others were devoted to the search for comets, which led to some interesting discoveries. In particular, it was found that the orbits of comets can be drastically altered by the gravitational attraction of the massive planets Jupiter, Saturn, and Uranus, and that a comet is ultimately destroyed by the tidal action of the sun. An example of the first phenomenon is displayed by the orbit of what we now call Encke's comet, even though it was actually discovered by Jean Louis Pons in 1818. Johann Franz Encke, an observer at the University of Göttingen observatory, calculated the orbit of the Pons comet and found that the entire Pons ellipse is not much larger than the orbit of Mars and that its period is 3.3 years. This comet's orbit departs so drastically from the usually larger elliptical orbits of most observed comets that Encke and astronomers following him concluded that its original orbit was greatly altered by the gravitational actions of Mars, Jupiter, Saturn, and Uranus.

A good example of the second phenomenon (the tidal destruction) was the behavior of Biela's comet in 1846. An Austrian army officer, Baron Wilhelm von Biela, had discovered a short period comet (6 years) and Olbers, calculating its orbit, predicted accurately that its future closest approach to the sun might result in its tidal dissection. This was verified some 14 years later in 1846 when it returned as two distinct comets, moving in tandem around the sun. Since this comet was never seen again, it was argued that it had suffered many solar disruptions and its material had been dispersed into tiny remnants distributed over the entire previous orbit. If this were so and if this orbit intersected the earth's orbit, many of these remnants would penetrate into the earth's atmosphere to produce a meteoric shower and such a shower or swarm would reappear periodically. The first such shower was observed in 1833 and we now encounter one such shower every month of the year. The most dramatic such shower occurs each year about August 11. This shower is known as the "Perseid shower" because all the meteors seem to diverge from a point in the constellation of Perseus. This simply means that all the particles in the swarm are moving parallel to each other, as expected, and hence seem to converge to or diverge from an infinitely far off point, that is, from the vast distances of a constellation.

Olbers became very interested in comet tails and made the first attempt to explain their structure and behavior in terms of the solar radiation. He suggested that the solar radiation exerts a pressure against the individual particles that constitute the tail of a comet, causing the tail to trail the comet as the comet advances toward the sun but to be in advance of the comet as it recedes from the sun. This is what is observed so that Olbers' explanation was accepted for many years but discarded in recent years because calculations show that the radiation pressure by itself is too small to account for the behavior of a comet's tail. Today we explain the tail by taking into account the solar wind of ionized atoms; unknown to Olbers.

During this time astronomers, particularly amateurs, became feverishly interested in Mars owing to the announcement by the Italian astronomer Giovanni Schiaparelli in 1877 that he had observed very long straight furrows or ravines which he called "canali" on the Martian surface. This word was unfortunately translated as "canals"—artifacts of intelligent beings. Since Schiaparelli was a very competent, highly respected astronomer, who had correctly explained meteor showers or swarms as remnants of comets that had been destroyed by the tidal action of the sun, his statement about "canali" on Mars was misinterpreted and blown up by the media out of all proportion to its true significance. His "canali" became true canals (i.e., artificial) in newspapers all over the world. This produced a Martian craze with observatories devoting themselves almost entirely to the study of Mars. Schiaparelli had also observed the white polar caps of Mars which change with the Martian seasons, which one would naturally expect if the Martian polar caps were frozen water. This added fuel to the popular excitement about "intelligent life" on Mars, for it was argued that the canals had been constructed to bring water from the polar caps to the arid equatorial regions. And so the debate about life on Mars continued until well into the twentieth century.

During this period of very important astronomical activities, certain advances in physics and optical technology were emerging which were to have a profound effect on the future of astronomy. To begin with, the controversy between the adherents of the Newtonian corpuscular theory of light and Huygens' wave theory was settled in favor of the wave theory with the discovery of the diffraction and interference of light. Newton had proposed the corpuscular theory because he had no evidence that light can bend around corners the way sound does. But this bending (called diffraction now) was observed by a young French engineer, Augustine Fresnel, who noted in 1816 that after passing through a narrow slit a beam of light spreads out very slightly. He observed further that if two such slits are parallel and placed next to each other, the two beams formed after passing through the slits recombine to form a spectrum. This can only be understood if the oscillations of the waves of light leaving the two slits reinforce each other at some point and cancel each other at other points to produce the spectrum.

This conclusion was strengthened further by the discovery of the English physicist Thomas Young, who showed that if by the proper arrangement of mirrors a beam of light is broken into two beams and then recombined, the two beams interfere to produce a pattern of bright and dark bands. Finally, the French physicist A.H. Fizeau measured the speed of light in water and showed that light slows down when it passes from a less dense to a denser medium which the wave theory supports but the Newtonian corpuscular theory does not.

On learning of Fresnel's discovery of the diffraction of light, Fraunhofer immediately saw that he could replace the prism in his spectroscope by what we now call a diffraction grating to obtain a spectrum. He reasoned that he could simulate slits by scratching a set of very close parallel lines on a flat piece of clear glass. The light could pass only through the clear spacings between the scratches. These clear spaces would then behave like slits and produce a spectrum. This is exactly the way things went so that the diffraction grating (either a transmission or a reflecting grating) replaced the prism with results far superior to those produced by a prism. Today very good diffraction gratings are produced with spacings between neighboring lines smaller than one-thousandth of a millimeter. Fraunhofer introduced another advance in astronomical spectroscopy with his "objective prism." He did this by placing a very large prism in front of the front lens (the objective) of the telescope so that the point images of the many stars that the telescope produced without the "objective prism" were now spread out as spectra, so that the spectral lines of many stars can be viewed simultaneously with the objective prism.

We come now to the astronomer, Friedrich Wilhelm Bessel, who in 1834 brought astronomy to its most precise state. Bessel, a reformed entrepreneur, an amateur astronomer, a disciple of Olbers, and a mathematician, who was taught by Gauss, gave himself over entirely to astronomy. So devoted was he to astronomy that the King of Prussia appointed him to the newly created state observatory in Konigsberg. His rare combination of theory and practice (precise observation) marked him as so outstanding that in a short time he was launched to the very top of his profession; he became the greatest German astronomer and set the pattern of astronomical observations for all other astronomers. In fact, other astronomers were put on notice that inaccuracies in observations would no longer be tolerated by Bessel. In specifying the positions of stars, all measurable terrestrial motions would have to be taken into account, not only the earth's rotation and revolution around the sun. He insisted that slight variations in the ambient temperature of the telescope, very slight vibrations of the observatory and variations in the observational idiosyncrasies of the individual observer, must be taken into account when evaluating the correctness of any observational data.

Bessel stated his philosophy about astronomical observations in terms of his view of the nature of the telescope: that it is a "twice built" instrument in that it is a structure designed and constructed in the optical and machine shops and, second, that it is a working tool in the hands of the professional astronomer. He was deeply concerned about extreme accuracy in reporting the positions of stars because he had decided to become the first astronomer to measure the parallax (its semiannual shift in position owing to the earth's motion) of a nearby star. But to do this he had to choose a star, which, he had reason to believe, is close enough for him to succeed. How then, without knowing the distances of the stars beforehand did he choose one to work on? He used the apparent motions of the stars (their proper motions) to guide him.

A few words about the observed stellar motions will guide us in understanding Bessel's approach. Our experience teaches us that we can distinguish between nearby and distant objects when we are moving in a train or an automobile by noting that fixed nearby objects seem to pass us much more rapidly than the distant hills. Bessel reasoned that this is true of stars also; owing to the motion of our solar system among the stars within our galaxy, the nearby stars should appear to move past us faster than the more distant stars. Herschel had already argued that the motion of our solar system is reflected in the apparent motions of the nearby stars. Bessel accepted this idea and found that the star 61 Cygni in the "Swan" appears to travel very rapidly—about 5 seconds of arc per year—which is the apparent displacement of a point which shifts its position by 1.4 inches if viewed from 1 mile. We say that a displacement of 1.4 inches subtends an angle of 5 seconds of arc at a distance of 1 mile. This large apparent motion of 61 Cygni encouraged Bessel to believe that 61 Cygni is close enough to have a measurable parallax. The periodic, semiannual parallactic apparent motion stemming from the earth's revolution around the sun is superimposed upon the star's proper motion. Bessel found that this parallactic displacement is about 0.3 seconds of arc. From this finding and his knowledge of the mean distance of the earth from the sun, Bessel calculated the distance of 61 Cygni to be about 11 light years.

Reasoning that the very brightest stars are the closest, the astronomer F.G.W. Struve following Bessel measured the parallax of Vega, the brightest summer sky star, in Lyre, and found its distance to be no less than 27 light years. These efforts marked the beginning of a vast new astronomical enterprise—the preparation of tables of stellar parallaxes and distances. Astronomy had entered a new domain, that of cosmology. Since Sirius is the brightest winter sky star, its parallax and distance were easily measured; its distance was found to be 9 light years. Some years later the parallax of Alpha Centauri, the closest star, was measured—0.75 seconds of arc—so that its distance of about 4.5 light years or 26 trillion miles could be calculated. Noting that looking at any star transports us back in time, we can identify any star with any great historical event such as, for example, the American Revolution, by associating the distance in light years with the historical event that occurred when the light reaching our eyes left the star.

We leave this chapter with a brief discussion of the measurements of stellar motions which began during this fruitful astronomical era. Since absolute motion has no absolute meaning we may take any system (e.g., the solar system) as our frame of reference. This means that we do not separate the motion of the solar system from the motions of the stars. Later if we discover how our solar system is moving with respect to the center of our galaxy, we can subtract this motion from the observed stellar motions relative to the center of the galaxy. All that Bessel and his contemporaries could do at the time with their very accurate telescopic observations was to note the change in the position of a star over a number of years. Doing this in annual intervals for any given star eliminates the semiannual parallactic apparent motion. This gave these mid-nineteenth century astronomers a true picture of the apparent transverse motion of the star, that is, the motion of the star across the line of sight. But it told these astronomers nothing about the star's radial motion, its motion away from or toward the solar system. But a remarkable optical discovery made about that time was the key to the solution of the problem of stellar radial motions. In 1842, as already noted, the Austrian physicist Christian Doppler, examining the spectra from various sources of light, discovered that the color (wavelength) of the light from the source changed slightly if the source was moving toward him or receding from him as compared to the wavelength if the source was not moving. Moreover, he found that this was so whether he was moving or the source was moving. He explained this effect—known as the Doppler effect—by reasoning that if a source of light and observer are approaching each other, the waves of light are crowded together so that the observer finds the wavelengths shortened and the light is therefore bluer. If, on the other hand, the observer and source are receding from each other, the waves are stretched out (wavelengths are longer) and the light is redder.

For astronomy this was one of the most important discoveries of the nineteenth century. The Doppler effect is expressed by a simple algebraic formula: if $\lambda$ is the wavelength of a light wave, $v$ is the relative velocity of the source of the wave with respect to the observer, and $c$ is the speed of the wave, then the wavelength $\lambda$ is increased or decreased by the fraction $v/c$ for a moving source. As an early example of the usefulness of this simple formula in astronomy, we note that the British astronomer William Higgins discovered that the spectral lines of Sirius are shifted toward the violet (wavelengths are shortened) indicating that Sirius is approaching us. The technique for observing these spectral changes was later (in 1890) greatly improved by Herman Vogel at Potsdam who perfected the photographic technology of spectroscopy. As we shall see, the thorough analysis of the motions of the stars in our galaxy was essential to an understanding of the structure and rotation of our galaxy, which we discuss later.

CHAPTER 12

# Astronomy as a Branch of Physics

*The universe is not hostile, nor yet is it friendly. It is simply indifferent.*

—JOHN HAYNES HOLMES

No one today doubts that physics and astronomy are interrelated and contribute to each other in very significant ways. But only in the last half of the nineteenth century was the indispensability of physics to astronomy clearly recognized. This, of course, is not surprising because Newtonian physics had its birth in astronomy with Newton's theoretical deduction of Kepler's three laws of planetary motion from pure physics (Newton's laws of motion and his law of gravity). This work showed the predictive power of the proper combination of basic theory and mathematics and this greatly encouraged the continuing close collaboration between these two disciplines.

For more than a century after Newton's great work, physics was pretty much limited to gravitational dynamics, and its greatest application was to the motions of celestial bodies, particularly those in the solar system. But this restricted role of physics changed dramatically in the last half of the nineteenth century with the discovery of the electromagnetic theory of light. From that time on, astronomy became increasingly dependent on physics. Indeed, the branch of astronomy that arose from the intimacy between physics and astronomy, now called "astrophysics," could not have emerged from Newtonian physics alone. The physics of optics and electromagnetism, purely a nineteenth century physics, was required.

Having discussed some of the laws of optics previously, we now turn to electromagnetism, limiting ourselves to the basic ideas which finally led to James Clerk Maxwell's magnificent synthesis: his electromagnetic theory of light. Electricity in its most elementary form was known to the ancient Greeks. Indeed, the word "electric" comes from the Greek word "electrum" for amber. The Greeks knew that if amber is rubbed by fur, it acquires the ability to attract small bits of matter. In fact, Thales, one of the seven wise men of Greece, first observed this phenomenon in 600 BC. This discovery was handed down from generation to generation, but nothing new was added to it until about AD 1100. when the English court physician William Gilbert found that many substances have the same property as amber and he coined the word "electric" to describe this phenomenon. About 50 years later the word "electricity" came into general use to describe all such phenomena. But different substances behave electrically only if they are rubbed with certain substances. Thus whereas amber must be rubbed by fur, glass must be rubbed by silk.

The rubbing operation is not the essential feature in this process; the actual feature is the surface contact between two dissimilar appropriate surfaces. The essence of this difference in producing electricity was discovered when it was found that the electricity produced on the amber when rubbed with fur is different from the electricity produced on glass when rubbed with silk. Careful observation showed that two electrified rods (rubbed with fur) repel each other, as do two electrified glass rods. But when an electrified amber rod is placed next to an electrified glass rod, they attract each other. This was correctly explained by the hypothesis (now a known fact) that two kinds of electricity (electric charge) exist in nature and that similar electric charges repel each other whereas two dissimilar electric charges attract each other.

The concept of electric charge, however, was not introduced until Benjamin Franklin correctly suggested that lightning is produced by the flow of electric charges through the atmosphere. In fact, he conclusively demonstrated the flow of charge by his famous (and nearly fatal) kite experiment. He called one kind of charge positive and the other negative. The electric charge produced on amber is now called negative and that on glass positive (opposite to Franklin's designation). Franklin's kite/lightning experiment almost led him to the discovery of electric current. Further experiments with the production of electric charges by rubbing two different substances together showed that each of the two substances acquires a charge but the charges are of opposite sign (positive or negative). Thus amber acquires a negative charge when rubbed with fur while the fur acquires a positive charge. When glass is rubbed with silk, the glass acquires a positive charge and the silk acquires a negative charge.

All of these observations and qualitative experiments with electric charges finally led to the branch of physics called electrostatics, the main goal of which, in the late eighteenth and early nineteenth century, was to introduce precision and quantitative methods in the study of electric charge. The first step in this heroic and revolutionary work was taken in 1785 by the French physicist and engineer Charles Augustin de Coulomb who, in that year, discovered what we now call Coulomb's law of force between electric charges. Coulomb, well acquainted with Newton's laws, particularly his second law of motion and his law of gravity, reasoned that if he could concentrate an electric charge on each of two small spheres he could measure the force (either attractive or repulsive) between them and from that determine how the force depends on the magnitude of each charge and the distance between the two "point charges" as he called them. He suggested that this force depends on the inverse square of this distance (like Newton's law of gravity between two mass points) and on the product of the quantities of the two charges involved. He saw that this experiment would also enable him to define a unit quantity of charge which would lead to a method of measuring quantity of charge in general.

Before discussing Coulomb's law further, we consider the whole question of units in astronomy. Today almost everything in our lives has been quantified so that we have no trouble understanding such quantities as the prices of various products, the rate of inflation, the unemployment "rate," the intelligence quotient (IQ), etc., even though these are rather crude measurements. As science evolved from its amorphous structure during its early Greek phase to its precise Newtonian phase, the introduction of precise units for all the measurable quantities used became imperative. But this does not mean that standard units (i.e., the same units in all countries) were used. The French Academy in the late eighteenth century, as we have already stated, led the way to the use of a standard universal system of measurements by introducing the metric system of weights and measures. In this system, now universally accepted, the unit of length, the meter (100 centimeters), was defined as the "one ten millionth of the distance from the equator to the pole measured along a meridian (a great circle)." The unit of time was introduced as the second, given as a definite fraction of a year.

With the spread of navigation over the globe, precise time and space units were used extensively, but with the introduction of Newtonian mechanics and dynamics, precise units for force and mass were required. Here we must be careful to distinguish between mass and weight, which common usage had tended to confuse, and, even today, most people confuse these two concepts by giving the weight of a body when asked to give the mass of the body. This confusion has been compounded by the introduction of the same units (grams) for mass and weight. People commonly speak of a mass of 1 gram and a weight of 1 gram, meaning that a mass of 1 gram weighs 1 gram.

We easily avoid this confusion by starting from Newton's famous second law of motion $F = ma$, where the three quantities $F$, $m$, $a$, are precisely defined as force, mass, and acceleration, respectively. Because acceleration is defined as the time rate of change of velocity, and velocity is the time rate of change of distance (centimeters per second), then acceleration is centimeters per second squared. This gives us the units for acceleration. We now come to the units for force, $F$, which was also set by the French Academy of Science. This unit is called the "dyne" which was defined as the millionth part of the atmospheric pressure defined as dynes per square centimeter.

With this unit of force, the gram (as a unit of mass), is defined as the amount of mass that acquires an acceleration of 1 centimeter per second squared (second per second) when acted on by a force of 1 dyne. When we apply Newton's law $F = ma$ to a freely falling body of 1 gram, we have the formula for the weight (a force) of a body. If $g$ is the acceleration of gravity (the acceleration of a body falling freely [in a vacuum] on the surface of the earth), then its weight equals its mass times $g$ or $W = mg$. Since $g$ is 980 cm/sec$^2$, then the weight of a body of mass $m$ is 980 m dynes, so that one gram weighs 980 dynes.

With this discussion of units used in physics and astronomy, we now return to Coulomb's law of force between two electric charges separated by the distance $r$. He wrote the law of force as $F = q_1q_2/r^2$, where $q_1$ is the magnitude of one of the charges and $q_2$ is the magnitude of the other. Stated in words, this equation says that the force between two charges $q_1$ and $q_2$ equals the product of the charges divided by the square of the distance $r$ between them. From this he defined the unit charge $q$ as that amount of charge which exerts a unit force (1 dyne) on an equal charge $q$ at a distance of 1 centimeter. This was the beginning of electrostatics.

The next great development in electricity came with the discovery of electric current by Galvani in 1791 and Volta in 1800. Luigi Galvani was an Italian physiologist and professor of anatomy at the University of Bologna who discovered accidentally that a frog's leg in a metal dish containing vinegar twitched when a metal scalpel touched it. He incorrectly concluded that the muscles and nerves in the leg had generated the current; his friend Count Alessandra Volta, a professor of physics at the University of Paris corrected him, pointing out that the electric current was produced by the combination of vinegar, the metal dish, and the metal scalpel. Analyzing this phenomenon very carefully, Volta concluded that he could produce an electric current by immersing two different metals (e.g., zinc and copper) in an acid medium and connecting them by a wire. This device, called a voltaic pile, did produce an electric current and so a new phase of electricity was born. The voltaic pile was later refined, becoming the voltaic cell and, finally, the electric battery (or cell) which we now know and use extensively.

The next important development in this story of electromagnetism was the discovery by the Danish physicist Hans Christian Oersted in 1819 that a magnetic field is produced by an electric current. Magnetism was known to the Chinese, who, as early as 1100 BC, had discovered that a piece of magnetite suspended to rotate freely always turned in such a way as to point north–south along a meridian. The ancient Greeks and Romans, many years later, learned that the iron ore ferric oxide—called the lodestone—attracts small fragments of iron and aligns itself along a meridian (north–south). The lodestone thus became the basis of the magnetic compass. The Greeks had also discovered that the two ends of the lodestone behave differently; one of the ends always points to the earth's North Pole and the other end always points to the earth's South Pole. Owing to this behavioral difference between the two ends, one was called the north pole of the lodestone and the other the south pole. These names were later carried over to bar magnets. With the construction of such magnets, physicists began a detailed study of the force produced by a magnetic field. Such forces are called magnetic forces; physicists soon discovered that the force between magnets has the same mathematical form as the force between electric charges. In other words, the force between magnets obeys Coulomb's law of force, with the electric charge replaced by magnetic strengths. Physicists also quickly discovered that two north poles repel each other as do two south poles, but also that north and south poles attract each other.

With all of these discoveries about electric charges and magnetic poles, no one had the faintest notion that electricity and magnetism are related. This relationship was discovered accidentally by Oersted while he was lecturing on electricity to a class at the University of Copenhagen. This was the beginning of the science of electromagnetism and electromagnetic technology, both of which were to play very important roles in astronomy.

Oersted's discovery was the first half of a remarkable symmetry in nature, that of electricity and magnetism; the second half was discovered some 20 years later by the experimental physicist Michael Faraday. Oersted's discovery is essentially that electric charge plus motion generates magnetism. This is obvious if we keep in mind that an electric current is a flow of electric charges; hence the magnetic field accompanying an electric current is produced by the motions of the individual charges in the current. Michael Faraday was thoroughly devoted to the concept of symmetry in nature, and he was convinced that the electricity–magnetism relationship discovered by Oersted is a two-way street. Being a tenacious experimentalist, it did not take Faraday long to discover that if a magnet is moved in and out of a loop of wire (an electrical conductor) an electric current flows in the loop. This is called electromagnetic induction. Thus Faraday completed the electric–magnetic symmetry circle that Oersted had begun. We may summarize these discoveries as follows: If an electric charge is placed next to a magnetic pole, the charge and pole do not acknowledge or recognize each other, but if either one moves, almost instantaneous recognition or acknowledgment occurs. Thus, if the electric charge moves it creates a magnetic field to which the magnetic pole responds and vice versa. This may appear fanciful, reducing the Oersted and Faraday discoveries to motions of charges and poles, but this became fact when the American physicist Henry Augustus Rowland, in 1880, working with individual electric charges proved this to be true. The Dutch physicist H.A. Lorentz later deduced, from basic electromagnetic theory, his famous formula for the force exerted by a magnetic field on a moving electric charge.

The final synthesis of these discoveries into one of the most beautiful creations of the mind was produced by the British theoretical physicist James Clerk Maxwell when he formulated the equations of the electromagnetic field. These equations lead to a wave equation for the electromagnetic field which is the basis of the electromagnetic theory of light. According to these equations, an oscillating electric charge generates a wave consisting of an oscillating electric and oscillating magnetic field (oscillating at right angles to each other) which are propagated together through space at the speed of light.

To clarify this further, we describe the simplest example of the generation of electromagnetic waves using an elementary electric oscillator. We consider two parallel metal plates with a small space between them; in electricity two such plates are called a condenser or a capacitor because they are used in all kinds of electrical devices (e.g., radio and television sets) to store electric charges. We now picture the plates charged with equal but opposite electric charges. If an electric conductor (a metal wire) connects the two plates, current flows from one plate to the other. But instead of dying out immediately, the current in the wire oscillates back and forth, sending out electromagnetic waves. These were investigated experimentally in great detail and with great precision by a young German physicist, Heinrich Hertz, who was a professor of physics at the University of Bonn from 1889 to his death in 1894. There he performed his revolutionary experiments which proved definitely that Maxwell's equations and his electromagnetic theory are correct.

Hertz showed that if an electric spark is produced between two small spheres, the spark behaves like an oscillator, sending out electromagnetic waves, which, as Hertz decisively proved, behave in every way like waves of light. He also demonstrated that such waves can be received by loops of wire constructed like the loop that produced the waves. This was the beginning of "wireless transmissions" (telegraphy) which was developed technically into radio in 1899 by the Italian electrical engineer Guglielmo Marconi. The great importance of all this for astronomy was that for the first time astronomers began to think of modeling stars, the essence of astrophysics. After all, if the radiation from the stars is electromagnetic, it appeared that all one had to do to model a star was to write down the set of equations, based on the Newtonian laws of motion and gravity and Maxwell's laws of electromagnetism that govern the internal structure of a star. But this was easier said than done because Newton's and Maxwell's laws alone cannot lead to a complete description of the internal structure of a star. Thus it is easy to show that the vibrations alone of electric charges in stellar interiors cannot possibly account for the vast amounts of radiant energy stars emit. Nor can these theories alone account for the presence of the Fraunhofer lines in stellar spectra. As we shall see, some 50 years were to elapse before the revolutionary discoveries in physics that were to come opened the door to an understanding of stellar physics.

But another important development in theoretical physics became the basis for astrophysics. This was thermodynamics, which grew out of the work of a group of brilliant physicists: Julius Mayer and Rudolf Julius Clausius in Germany, Nicholas Sadi Carnot in France, and William Thomson, first Baron Kelvin in England. Strictly speaking, we should not include Mayer in this group because he was a medical doctor and not a physicist, but Mayer's work led to the development of the science of thermodynamics, the mastery of which is absolutely essential to the understanding of stellar structure. We are also stretching the definition of "physicist" by including Sadi Carnot, who was primarily an engineer interested in the operation of heat engines. But this interest led him to the discovery of an important property of heat.

Heat was used through the ages for all kinds of useful purposes from cooking to providing warmth, but the study of heat as a branch of physics was not pursued in earnest until the physical nature of heat was understood. This understanding came primarily from Julius Mayer's insistence that heat is a form of energy that must be taken into account in the formulation of the principle of the conservation of energy. Until Mayer's definitive analysis of heat, most physicists believed that heat is a kind of "caloric" substance which can flow from one body to another. This contradicts the well-known phenomenon that the friction between two surfaces moving in contact with each other produces heat; the faster the bodies move and the stronger the friction, the greater is the amount of heat produced. This is exactly the definition of the energy produced by a force displacing a body or setting it in motion, so that heat is, indeed, a form of energy. We show the importance of Mayer's identification of heat with energy, now called the first law of thermodynamics, by considering a gas confined to a cylinder with a freely moving piston on top. Those of us who drive cars are quite familiar with pistons and cylinders, without which our gasoline motors could not work. Physicists describe the physical conditions of the gas in the cylinder by assigning three measurable physical parameters to it: pressure $P$, temperature $T$, and volume $V$. The quantities $P$, $V$, $T$ completely determine the conditions in the gas. Indeed, the law of gases which we discussed previously tells us that no matter what we do to the gas—heat it, cool it, compress it, or expand it—without altering the quantity of gas in the cylinder, the algebraic combination $PV/T$ (pressure times volume divided by temperature) cannot change. A change in any one of these three quantities produces compensating changes in the other two. The importance of this for astrophysics becomes apparent because stars are spheres of gases. As a result, we must apply the laws of gases and thermodynamics to the study of stellar interiors. We do not try to study the entire gas sphere that we call a star but instead small elements of it which can be represented by our cylinder of gas. To do this properly we must limit the size (the volume) of the stellar element (cylinder) of gas so that the conditions within it (pressure and temperature) are the same throughout it.

We now apply Mayer's first law of thermodynamics to our cylinder of gas when we heat the bottom of the cylinder with a flame, which we can control so that we can measure the amount of heat that enters the cylinder. At the same time we keep a thermometer in contact with the gas and keep a fixed set of weights on the piston so that the pressure in the gas remains constant no matter how much heat enters the gas. We then discover that two things change: the piston rises and the temperature of the gas increases. How are we to interpret these changes? Only one interpretation is possible: the heat entering the gas does work (it raises the weights on the piston) and it increases what we call the internal (molecules) energy of the gas. These phenomena had been observed for many years but no one had interpreted them correctly. Only after Mayer's remarkable deductions was it clear that heat, as a form of energy, can be made to do work. It soon became clear that heat engines could be built to do the backbreaking physical work of men, women, and even animals. Indeed, this was the beginning of the Industrial Revolution and, economically speaking, the end of slavery, for human and animal labor could not compete with machines.

What does this cylinder of gas have to do with stellar structure? A stream of radiation flowing from the deep interior of a star to its surface passes through many "cylinders" of stellar gas, one on top of the other. Why then, does not each of these "cylinders" of gas get increasingly hotter and expand explosively? We know that this does not happen because stars would blow up if it did happen. This means that the gas in a star must continually adjust itself properly in accordance with the gas law to allow the hot radiation from the interior to pass through and cool off by just the right amount to keep the star in equilibrium. The total amount of radiation per second that leaves the surface of any star (e.g., the sun) must equal exactly the amount of radiation (energy) that is produced per second in the core of the star. The radiation produced in the core of the star is quite different from the radiation that leaves the star's surface and accounts for the star's luminosity. We describe later how the interaction of the very hot core radiation with the cooler stellar gas surrounding the core produces this miraculous process.

We point out a few more gaseous phenomena stemming from the first law that play important roles in astrophysics. First, if we allow a gas to expand freely into a larger volume, the temperature of the gas remains constant. However, if the gas expands, thus pushing a piston with weights on the piston and doing work, the gas cools and thus loses energy. On the other hand, if we compress a gas in a cylinder by pushing the piston into the cylinder, thus decreasing the volume, the gas becomes hotter. These phenomena are clearly important in astrophysics because the gases in a star are constantly changing as they respond to the star's gravity and radiation.

We now come to the second law of thermodynamics which stemmed from the study of heat by Carnot, Rudolph Clausius, and Kelvin, and which led to the concept of entropy. The hint of the second law is contained in our description of how a gas in a cylinder with a movable piston responds when it is heated. Part of the heat is transformed into work (the piston rises) and part remains in the gas, increasing its temperature. This indicates that heat cannot always be changed completely into work: some of the heat goes into raising the temperature. In other words the process from work to heat is not always reversible. This irreversibility (the transformation of heat to work) was first clearly stated by Carnot, who studied the behavior of a heat engine in great detail. The engine receives heat (energy) from a hot source and then does its work (driving the engine), but at the end of this process it (the engine) gives up some of the heat (at a lower temperature). The difference between the amount of heat the engine receives initially and the amount of heat it gives up after it does its work, equals the amount of work done. In this investigation, Carnot proved conclusively that going from work to heat is a one-way process. We know that any amount of work (via friction) can always be changed to heat, but the transformation of heat into work is strictly limited. This is the essence of the second law of thermodynamics.

The next great step in this remarkable story was taken by Clausius, who was the first to recognize that all natural processes are irreversible and who introduced the concept of entropy as a measure of this irreversibility. Entropy is related to energy but, unlike energy, it cannot be observed or measured directly. Clausius expressed the second law of thermodynamics by relating it to the flow of heat. This is contained in his statement that heat can flow spontaneously only from a hotter to a colder body, and this is always accompanied by an increase of entropy which is defined as the quantity of heat flowing into or out of a body divided by the temperature of the body. In terms of entropy, the second law was stated by Clausius as follows: Heat always flows spontaneously from a high temperature source (e.g., a furnace) to a low temperature receiver (e.g., a cold water pipe from the furnace). This is equivalent to the statement that the entropy of a system can never decrease; it may remain constant but it can only increase if it changes at all.

Because the entropy concept has played an important role in philosophy, science (particularly astronomy), and even theology, it is important to have a clear understanding of its significance and not fall into the trap of applying it to support a particular philosophical or theological ideology, as has been done by the creationists. As entropy is also a measure of the disorder in a system in equilibrium, the creationists have argued that the second law is contrary to evolution and that we must therefore replace evolution with an act of creation. This is an incorrect interpretation of the second law and entropy which do not prohibit order. Indeed, the second law demands that disorder can occur only if order occurs at the same time. To illustrate this point we consider a cylinder containing two kinds of gases at a definite temperature. If we label the atoms of one gas $A$ atoms and those of the other gas $B$ atoms, and look at the mixture sometime later, we discover two changes in the gas: the temperature has increased and combinations $AB$ of atoms are present. Because these molecules are more highly ordered structures than the individual atoms *A* and *B*, the matter is more highly organized than previously. Indeed, the entropy of the matter has decreased. Does this change violate the second law of thermodynamics? No. In calculating entropy we must take into account not only the entropy of the matter but also the entropy of the energy (actually the radiation released when the molecule *AB* is formed) which has increased as evidenced by the temperature increase of the gas. The increase in the entropy of the released radiation exceeds the decrease in the entropy of the matter so that the total entropy has increased.

To understand the importance of this idea in astronomy (in particular for the formation of stars and galaxies), we consider the gas with its atoms *A* and *B* in more detail. Initially the individual atoms *A* and *B*, and hence the gas, are in equilibrium at a definite temperature and the entropy is at its maximum. But as the atoms move about and collide some are moving fast enough to combine to form molecules *AB*. When this happens, energy is released (the binding energy of *AB*) which increases the temperature and entropy of the gas. Order thus cannot occur without disorder. As long as the system's entropy can increase, order can occur. How does this apply to the universe which began in maximum disorder (the "big bang")? As we have pointed out, order occurs when particles (atoms) form bound systems. But this can occur only if these particles attract one another owing to the attractive forces they exert on each other. In the process, these particles radiate energy so that entropy increases. The understanding of the second law of thermodynamics is thus absolutely essential to the understanding of the formation of structures (creation of order) in the universe.

What was Lord Kelvin's contribution to the formulation of the second law? We have seen that Clausius expressed the second law by stating that heat always flows spontaneously from a higher to a lower temperature. Kelvin formulated the law in terms of heat and work. As we have seen, work can always be transformed into heat. Indeed, heat always accompanies work and limits the efficiency of an engine. Lord Kelvin stated the second law as follows: heat cannot be changed completely into work without leaving some change in the environment. With this we leave our present discussion of thermodynamics to return to it later when we go into great details about stellar structure (astrophysics).

We consider now the consequences of the laws of gases which have important applications to the understanding of stellar structure. As far back as Newton's time, physicists (including Newton himself) speculated about the structure of matter. But we can go back further in time and note that the Greek philosopher Democritus (c. 400 BC) proposed a primitive atomic theory of matter, based on the eternal existence of small solid spheres, which he called "atoms." His great Latin disciple Lucretius (c. 92 BC) expounded Democritus' atomic theory with his remarkable poem "On the Nature of Things." But all these early hypotheses about the nature and structure of matter were purely speculative since there was no way in those days to make any observations that would validate any particular hypothesis. This was altered with the study of gases. From the discovery of the three distinct states of matter, solid, liquid, and gaseous, physicists and chemists began to surmise that all matter consists of molecules, which, in turn, consist of atoms. Matter passes from one phase to another when heat alters the forces and relationships among the molecules.

We now know that when any substance is heated to a high temperature, the substance becomes a gas—a state in which the molecules move around freely—with no forces acting on them. Their bonds have all been cut and they interact with each other only occasionally, when they collide and bounce off each other. Collisions, of course, go on all the time among the molecules in a container, but any one molecule spends most of its time just moving freely in a straight line until it hits another molecule or the wall of the container. Clearly, a gas is the simplest state of a substance, and, all gases, regardless of their chemical nature behave, physically, the same way. The same laws apply to all gases and mixtures of gases. This is very heartening to astrophysicists because the temperatures inside all stars are so high (millions of degrees) that stars are just spheres (globes) of very hot gases. Astrophysics thus began their task with the study of the thermodynamics of gases.

To make clear just how intimately astrophysics is related to gas dynamics and to the thermodynamics of gases (e.g., kinetic theory) we explain the basic features or fundamentals of the kinetic theory of gases in somewhat greater detail. We begin with Amedeo Avogadro's important discovery of what is now called Avogadro's number or constant, which applies not only to gases but to all phases of a substance. To this end we introduce the concept of the gram-molecular weight (if we are dealing with molecules) or gram-atomic weight (if we are dealing with atoms). This concept stems from the concept of the atomic weight or the molecular weight which physicists, chemists, and astrophysicists use extensively. These weights are numbers assigned to atoms and molecules on a scale on which the hydrogen atom (the lightest atom) is assigned the number 1 and all other atoms are assigned the numbers which tell us how much more they weigh than the hydrogen atom. On that scale, the helium atom has the atomic weight 4, the carbon atom has the atomic weight 12, and the oxygen atom has the atomic weight 16.

The molecular weight of a molecule equals the sum of the atomic weights of its atoms. Thus the molecular weight of water ($H_2O$) is 2 + 16 = 18 and the atomic weight of carbon dioxide ($CO_2$) is 12 + 32 = 44. Consider now an amount of a given substance which, in grams, equals its atomic or molecular weight. This is called one gram-molecular or gram-atomic weight (or a mole). Thus 1 mole of hydrogen is 1 gram, 1 mole of helium is 4 grams, 1 mole of water is 18 grams, 1 mole of carbon dioxide is 44 grams, etc. Avogadro discovered that 1 mole of any substance contains the same number of atoms or molecules. This number, called Avogadro's number or constant, is 6.022 followed by 23 zeros (602 billion trillion). Moreover, Avogadro showed that any gas confined to the same volume at the same temperature and pressure contains exactly the same number of atoms or molecules. These investigations showed that a mixture of different gases behaves exactly the same as a single type of gas confined to the same volume at the same temperature.

All of these concepts are important for astrophysics; indeed, the behavior of stars and their various properties are proof on an enormous scale of the laws of thermodynamics and the kinetic theory. To pursue this discussion a bit further, we can see in a very elementary way how one can relate such things as temperature and pressure to the kinetic theory. We consider first the relation of the pressure in a gas to the motions of the molecules in the gas. We assign an average speed $v$ to the molecules in a gas even though some molecules are moving faster than this average and others are moving more slowly. We arrive at the correct kinetic picture by assuming an average speed. We may not do this for a complete star, but we may do it for a small volume in a definite region within the star. Clearly, the larger the speed $v$, the greater is the pressure of the gas in this small volume because if a molecule or atom is moving rapidly, it hits the imaginary walls of the volume harder and hence exerts a greater pressure than if it were moving more slowly. Thus the pressure increases with $v$. But we must multiply this quantity by $v$ itself to obtain the correct pressure because the faster a molecule moves, the more often in a unit time (e.g., a second) does it hit the walls. The pressure is thus proportional to $v^2$. But we must also multiply this quantity by the mass $m$ of the molecule because a more massive molecule hits harder than a less massive one. Thus the pressure is proportional to the product $mv^2$, which is twice the kinetic energy of a molecule. But the pressure depends on the temperature so we discover that the temperature in any region of a star depends on the average kinetic energy of the molecules or atoms in that region of a star.

From this discussion we can see how astrophysics became increasingly more dependent on physics and physicists. Most important in this development were the contributions of the British physicist James Clerk Maxwell, whom we have already discussed, as the founder of the electromagnetic theory of light, and the Austrian theoretical physicist Ludwig Boltzmann. Boltzmann contributed greatly to the development of what is now called "statistical mechanics," which is the application of probability theory and statistical methods to the analysis and the understanding of phenomena involving extremely large numbers of particles (atoms and molecules), as in the interiors of stars. Maxwell developed mathematical methods for finding the average values (such as the average velocity) of many particles moving about free of each other but colliding with each other every now and then as they move about. Maxwell not only developed the formula for the average velocity of a particle but a general formula from which one can calculate the numbers of particles in each velocity range (e.g., the numbers moving twice as fast as the average or one third as fast, etc.). This formula is known as the "Maxwell distribution" of velocities. The important parameter in this formula, as one expects, is the temperature.

The importance of this formula for stellar structure (astrophysics) is fairly obvious, for only if we know how fast atoms in a star are moving about at any point can we determine the amount of stellar mass that must be present to keep the atoms from dispersing into space.

Boltzmann is known not only for his basic work in statistical mechanics but also for his discovery of a famous formula called the "Stefan–Boltzmann law" which relates the intensity of radiation from a star (the star's luminosity) to the surface temperature of the star, which is really a misnomer because a star has no solid surface but looks as though it had one. In any case, the Stefan–Boltzmann formula enables us to assign a temperature to what looks like a star's surface.

Maxwell contributed not only indirectly to astronomy through his electromagnetic theory of light but directly through his work on the nature of the rings of Saturn. For a long-time astronomers had considered the rings around Saturn to be solid because they reflected light like a solid and they seemed to be opaque. But some doubt was cast on this assumption by the knowledge that comets that come too close to the sun are fragmented into many small pieces by the tidal action of the sun. These small pieces are then dispersed to form a ring of particles around the sun. The problem of this kind of tidal fracturing of a solid body was first analyzed by the French astronomer E. Roche in 1850, who proved mathematically that there is a distance from the center of each massive sphere like the sun (called the "Roche limit") below which a body approaching the sphere would be torn apart by the sphere's gravitational tidal action. Each sphere has its own Roche limit that depends on the mass of the sphere. The larger the sphere's mass, the larger is its Roche limit. Thus the earth's Roche limit is 1000 miles.

If our own moon ever gets as close as 1000 miles to the earth, it will be ripped apart by the earth's tidal action. Led by his own analysis, Roche believed correctly that each of Saturn's rings had started out as a moon within Saturn's Roche limit but had been fragmented into a ring by Saturn's tidal action.

That the rings of Saturn cannot be solid was proved from basic gravitational theory by Maxwell who, in 1858, showed, using Newton's law of gravity, that a solid ring would oscillate so violently that it would be shattered into fragments in a very short time. Maxwell did not pursue astronomy as a professional astronomer, but rather as a theoretical physicist who saw in astronomy certain challenging problems that he felt demanded the attention of a competent mathematical physicist. Among these was a precise mathematical treatment of Laplace's nebular hypothesis for the origin of the solar system. In his analysis, Maxwell encountered what appeared to him to be an insurmountable difficulty. He showed that if the forerunner of the solar system had been a Laplace-type of rotating nebula, its mass must have been 100 times as large as that of the present sun to have contracted down gravitationally to its present size. This immediately raised a question that could not and still has not been answered: where did all the extra mass go and how did it escape the sun's gravitational field?

Who else played a pivotal role in the story of astronomy during this era? Among the most important was the outstanding Irish mathematical physicist Sir William Rowan Hamilton, who contributed greatly to theoretical astronomy. He was productive in every branch of physics, many aspects of which were and are applicable to astronomy. His greatest contributions were in the area of dynamics which he reformulated in such a way that complex gravitational dynamical problems became manageable. Hamilton's great genius was quickly recognized and his mathematical achievements (particularly in the realm of optics) were quickly rewarded. In 1827 while still an undergraduate he was appointed professor of astronomy at Trinity College and, a year later, he became Astronomer Royal for Ireland and director of the Dunsink Observatory near Dublin.

The final years of the nineteenth century saw the coming of age of astronomy as a scientific discipline in its own right and not merely as a branch of physics. This recognition by the science community and the academic community has been marked by the introduction of independent astronomy departments and astronomy faculties throughout the world. Indeed, the status of a university and of a college increased in the eyes of most people if the institution had a separate astronomy department. This status was further advanced with the construction of great observatories. In the United States most of the observatories were named after the universities with which they were associated. In addition to the academic observatories, most countries constructed and financed national observatories. Thus in the United States we have the United States Naval Observatory in Washington, DC, which is not a research observatory but instead a service observatory which issues *The American Ephemeris and Nautical Almanac* each year, which provides important and useful information not only for the general public and the military but also for astronomers. Thus among its various data it lists the mean positions of 1078 permanent stars; the heliocentric and geocentric ephemeris (places of celestial bodies at regular intervals) of major and minor planets, universal and sidereal (stellar) time, etc. In England, the Royal Greenwich Observatory performs the same functions. In recent years, the almanacs published by both observatories have been merged so that one annual almanac is now published jointly by Washington and Greenwich.

During this period astronomy was also greatly advanced by the introduction of astronomical journals open to all astronomers and physicists who could submit original papers for publication. Thus the Royal Astronomical Society founded the Monthly Notices, B.A. Gould founded the Astronomical Journal now published by the American Astronomical Society; George E. Hale and James E.K. Keeler founded the Astrophysical Journal in 1895, now published by the University of Chicago Press for the American Astronomical Society; and, more recently, a European astronomical journal, Astronomy and Astrophysics appeared, which merged six astronomical journals that had been published separately in France, Sweden and Germany. In 1919, the International Astronomical Union (IAU) was founded to coordinate astronomical research in all countries. The IAU issues an information bulletin once a month to keep its membership informed of astronomical events and discoveries throughout the world. It also sponsors general assemblies of its membership and periodic symposia dealing with various astronomical problems.

Each astronomical society conducts its own annual conference during which papers (invited and unsolicited) are presented orally to unrestricted audiences for comments, questions, and criticisms from the floor. Thus the beginning of the twentieth century saw astronomy as a very energetic, expanding scientific and philosophic enterprise, but in a sense it had accomplished little in the way of answering basic questions. Its successes up to that point had been primarily in the areas of positional astronomy, solar system dynamics, and stellar statistics (distances, luminosities, motions). But such basic questions as the physical nature of stars, the clustering of stars, and the structure of the universe itself remained to be answered. As we shall see in the next chapter, these questions came at a time when two great discoveries occurred which changed physics and astronomy forever.

CHAPTER 13

# The New Physics and Its Impact on Astronomy

*There is something fascinating about science. One gets such wholesale returns of conjecture out of such a trifling investment of fact.*

—MARK TWAIN

As the nineteenth century drew to a close, physicists faced the twentieth century with a sense of great confidence, for it appeared to them that all that was left for physicists and astronomers to do was a kind of mopping up operation which involved nothing more than carrying out calculations to successive decimal points, that is, to more precise degrees of approximation. With Newtonian dynamics, including Newtonian gravity, and the Maxwellian electromagnetic theory of light, everything in nature that one could think of seemed to be covered. And yet at this very time new experimental discoveries and two new theories were about to burst upon the science community. They would produce a profound revolution not only in the basic principles of physics but also in the concepts of space and time which are the basis of all the principles of science and crucial to the understanding of the universe itself. The experimental work that was to revolutionize the science of matter dealt with the discovery of the basic electric charges that constitute matter. Because matter, under ordinary circumstances, is electrically neutral, physicists of the late nineteenth century reasoned that all matter contains equal quantities of negative and positive electric charges. That electric currents (the flow of electric charges) exist, which is conclusive evidence that electric charges can be separated, convinced these physicists that one should be able to isolate streams of electric charges of both kinds and study them. These experiments were designed to determine the magnitude of the unit (the elementary) electric charge (positive or negative) and the mass of the elementary charged particle. The magnitude of this experimentally measured individual charge was then to be accepted as the unit electric charge in nature (in that no fractional charge smaller than this exists).

These experiments were performed with what is called a Crookes tube named after Sir William Crookes, a British chemist and physicist, who directed the meteorological work at the Radcliffe Observatory at Oxford. The Crookes tube is a device for producing what we now call cathode rays so that the Crookes tube is what we now call a cathode ray tube. Such tubes are the image-producing devices in all television sets. The X-ray tubes (which produce X rays) are an extension of the Crookes tube, which are made of glass and contain air or some other gas at very low pressure (about one ten thousandth of normal atmospheric pressure). This is called a Crookes vacuum. An electrode is inserted at each end of the tube and a very high electric voltage (about 10,000 volts) is placed across these two electrodes, one electrically positive and the other one negative. This produces a stream of very rapidly moving particles which move away from the negative electrode (the cathode) to the positive one. Sir Joseph John Thomson (1856–1940), the British experimental physicist and probably the outstanding scientist of his day, was the leader in this research. He correctly interpreted the stream of cathode ray particles as a stream of the basic (elementary) negatively electrically charged particles that constitute one of the electrically charged particles in matter. He called these particles "electrons" and in 1906 he was awarded the Nobel Prize in Physics for "his discovery of the electron."

By subjecting the beam of electrons (the cathode rays) in his Crookes tube to electric and magnetic fields, cleverly arranged, he measured the ratio of the electric charge $e$ on the electron to its mass $m$, that is, the fraction $e/m$, but he could not measure $e$ and $m$ separately. Reasoning that positively charged elementary particles should be moving in a stream in a direction opposite to that of the electrons, he looked for such a stream and found it. The positively charged particles in this stream were (and still are) called protons. Thomson found that the charge to mass ratio, $e/m$, for a proton is smaller by a factor of about 2000 than that of the electron. He concluded from this that the mass of the proton is about 2000 times as large as that of the electron since the magnitude of the charge $e$ is the same for both.

These two tremendous discoveries were the beginning of the study of the structure of the atom which was to affect astronomy immeasurably. In 1910 the American physicist, Robert Millikan, measured the charge $e$ on the electron in his famous oil drop experiment. From this experimental measurement of $e$ and Thomson's measured value of $e/m$, the mass, $m$, of the electron was calculated and found to be very nearly one-thousandth of a trillionth of a trillionth of a gram. From this result the mass of the proton was calculated to be one-trillionth of a trillionth of a gram. Physicists and astronomers now pictured the simplest (the least massive) atom (the hydrogen atom) as consisting of one proton and one electron, held together by the electrostatic force of attraction between them. All that was needed to construct a correct model of an atom at that point was a correct theory of how the electron and proton cooperated to form a viable atom. It was not enough just to know that the electron and proton attract each other to construct a model of an atom. One had to construct a model that is not in conflict with the basic laws of electricity and magnetism (Maxwell's electromagnetic theory of light). But this knowledge was not available at that time. Not until 1913 was it clear as to how to perform such a feat.

The investigation into the electrical structure of matter had been carried out by Michael Faraday some years before the work with the Crookes tube was done. Faraday showed that when a molecule such as sodium chloride (NaCl) is dissolved in water, the molecule breaks up into positively charged ions of sodium and negatively charged ions of chlorine. He argued that the sodium atom of each molecule gives up a single negative charge to the chlorine atom of the molecule. He demonstrated this by showing that the salt solution conducts electricity with the sodium ions moving toward the negative electrode inserted in the solution and the chlorine ions moving toward the positive electrode.

All this experimental work, culminating with the discoveries of Thomson and Millikan, pointed to an entirely new approach to the study of stars. After all, if solid matter consists of atoms of various kinds, then this must also be true of stars. The question that arose, of course, was whether observers on the earth could determine the chemical (atomic) nature of stars from an analysis of the light coming from the stars. Fraunhofer had already answered this question in the affirmative by his discovery and analysis of the dark lines in the stellar spectra. Insisting that these lines are produced by the interactions of the material in the atmospheres of stars with the radiation passing from the stellar interiors through the stellar atmospheres, he was convinced that a correct interpretation of these dark lines can reveal the chemical nature of the atoms that produce these lines. With the discovery of the electron and proton, physicists now pointed the way to unraveling the mystery of the spectral lines by constructing a model of the atom with equal numbers of electrons and protons arranged in each atom to obtain an electrically neutral structure. If this modular atom was made to oscillate or vibrate, it would emit electromagnetic radiation. Only the electrons would vibrate because the electrons are some 2000 times lighter than the protons, which have too much inertia (mass) to vibrate. The protons would remain in place.

As we shall see, constructing a theoretical model of the atom was much easier said than done; indeed, two new revolutionary physical theories had to be discovered before the theory of the structure of the atom could be properly understood and used to account for spectral lines. We note, however, that as early as 1803, the British chemist John Dalton, having introduced the weight of hydrogen atoms as the unit atomic weight, proposed the revolutionary idea that the atomic weights of all the different kinds of atoms (e.g., carbon, sodium, iron, etc.) are very nearly simple numerical multiples of the atomic weight of hydrogen; he then proposed the bold idea that heavy atoms are built up from combinations of hydrogen atoms. This proposal was, of course, later verified.

While this exciting experimental work on the structure of matter was being pursued, experimental and theoretical work (theories) dealing with the behavior of light and of radiation, in general, was at the point of revealing two amazing facets of nature. These theories were, at that time, beyond the wildest speculations of the boldest physicists and, even today, they arouse skepticism in the minds of many. One of these discoveries is associated with Max Planck and the other with Albert Einstein. Planck's discovery introduced the constant $h$ (Planck's constant of action) into the laws of nature and Einstein's theory introduced the speed of light $c$ into the natural laws. These two constants together with $G$, the universal constant of gravity, are basic to an understanding of physics and astronomy, particularly astrophysics and cosmology.

Planck introduced his constant of action $h$ to explain the basic properties of the radiation (all forms of electromagnetic energy) emitted from a 1 square centimeter hole in the wall of a furnace at some definite temperature $T$. Such radiation is called "thermal" or "blackbody" radiation. This curious name arose from the observation that white bodies and blackbodies absorb radiation quite differently. To explain the relevance of this to the properties of radiation, we note that three things can happen to radiation that strikes the surface of a body: (1) it can pass right through the surface; (2) it can be reflected; and (3) it can be absorbed. We are concerned here only with the second and third processes. If a surface reflects all the radiation striking it, we call it a perfect reflector; the temperature of such a surface does not change when radiation is reflected from it. If a surface absorbs all the radiation (all colors or wavelengths) that strike it, it is called a perfect absorber or "blackbody" because it, in principle, cannot be seen. The temperature of such a surface rises as the radiation is absorbed so that the surface begins to reradiate the energy or radiation that it absorbs, but in doing this it redistributes the radiation into all colors in varying intensities from color to color. This reemitted radiation is called "blackbody" radiation. Regardless of the color of the radiation that is absorbed by the black body, the nature of the radiation emitted by the black body depends only on its temperature. One may wonder what all this has to do with astronomy; knowing the properties of blackbody radiation is essential for studying stars because the radiation from stars is quite similar to blackbody radiation.

The importance to astronomy of knowledge about radiation had been emphasized by Fraunhofer but it remained for the Heidelberg physicist Gustav Kirchhoff to put the final touches to what

Max Planck (1858–1947) ca. 1935 (Courtesy AIP Emilion Segrè Visual Archives; gift of Jost Lemmerich)

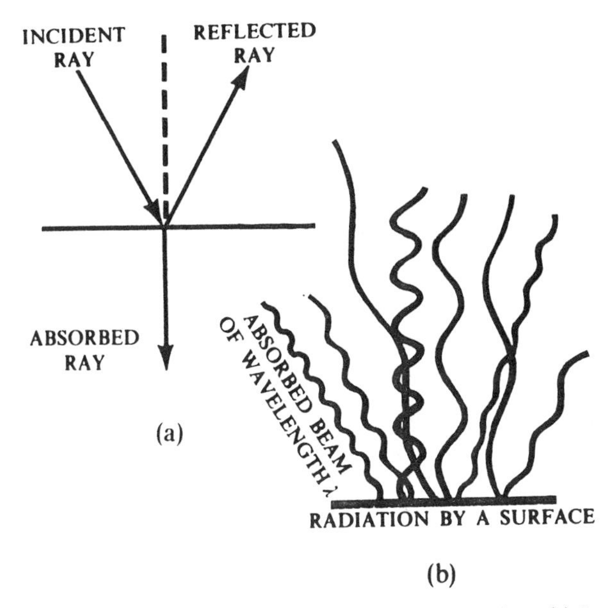


Figure 13.1. The behavior of a beam of light on striking a surface. (a) In addition to the incident beam there are a reflected and an absorbed beam. (b) The surface begins to radiate energy as soon as it absorbs energy. The radiated energy is spread over the entire spectrum.

we know of the properties of radiation in our understanding of stellar properties. Kirchhoff carried on the work of Fraunhofer but to a much greater depth, showing that the merest trace of a molecule such as sodium chloride or an atom such as calcium can easily be identified spectroscopically. But Kirchhoff went beyond spectroscopy in his study of radiation, discovering an important law that relates the rate at which a body at a given temperature emits radiation to the rate at which it absorbs radiation. This law, known as Kirchhoff's law of radiation, states that the rate at which a body radiates energy at a given temperature and frequency (color) equals the rate at which it absorbs radiation at that frequency multiplied by the rate at which a perfect blackbody at the same temperature absorbs such radiation. The importance of this law in studying the

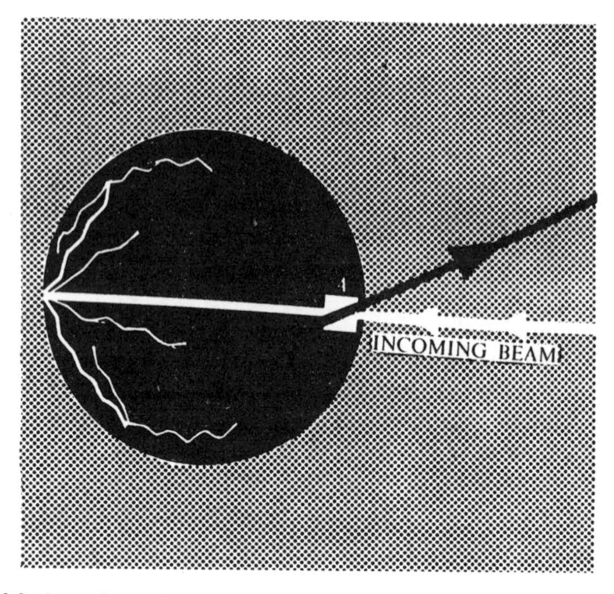


Figure 13.2. A small opening $A$ in the wall surrounding a hollow region (hohlraum) behaves like a blackbody.

properties of radiation is that we can calculate the rate any body radiates from the rate at which it absorbs and the rate of emission of a black body. For that reason physicists were greatly interested, near the end of the nineteenth century, in discovering the mathematical formula for blackbody radiation (radiation emitted by a perfect blackbody). As we shall see later, this kind of radiation (now called cosmic background radiation) fills all of space.

This problem was tackled by experimental and theoretical physicists at the same time. To the experimentalist, the principal difficulty associated with this problem was finding a perfect black body. It did not take them long to see that a small opening in the wall of a cavity (e.g., of a furnace) behaves exactly like a black body: every ray (regardless of its color) that strikes, that is, enters the hole, is completely absorbed; none of it ever gets back out. They reasoned, by the basic law of reversibility, that if the furnace walls (the walls of the cavity) are heated, the radiation coming out of the small hole is blackbody radiation. These studies showed that the radiation emitted from the hole (blackbody) in a furnace consists of all possible colors and that the "intensities" (amounts) of the various colors emitted depend only on the frequencies (colors) and the temperature of the furnace. Turning this idea around we see that the temperature of the furnace can then be determined from the distributions of the colors in the emitted radiation. Here we see the relationship of the study of blackbody radiation to understanding stellar structure; the analysis of stellar radiation leads us to the temperature of stellar surfaces, a basic stellar parameter.

The spectral analysis of radiation from a hole in a furnace is best represented by a graph on which the vertical axis (the ordinate) represents the intensity of the radiation (in appropriate units such as lumens) and the horizontal axis (the abscissa) represents the color, the frequency, or the wavelength (in Ångstroms). For a given temperature of the furnace (or cavity), the points on the graph, each point representing a given temperature and color, lie on a curve which is called the "blackbody" curve (later called the Planck radiation curve), which changes as the temperature changes. This blackbody radiation curve has very distinct, characteristic features that mark it as a "Planck radiation curve." The curve starts at very small values (low intensities, i.e., points close to the horizontal axis), at the blue end of the spectrum (for high frequency radiation), rises fairly steeply to a maximum value for some intermediate color, and then drops gradually toward the red end of the spectrum, becoming almost parallel to the horizontal axis (low frequency radiation).

The experimentalists also discovered that as the temperature of the furnace is increased, the general shape of the radiation curve does not change, but the entire curve lies higher on the graph, and the maximum point (the top of the curve) is shifted more and more toward the blue end of the spectrum. This, of course, corresponds to our experience. If we could look into a furnace as the temperature of the furnace rises, we would first detect an increasing outward flow of radiation (represented by the upward shift of the entire radiation curve in our graph) and then find that the color

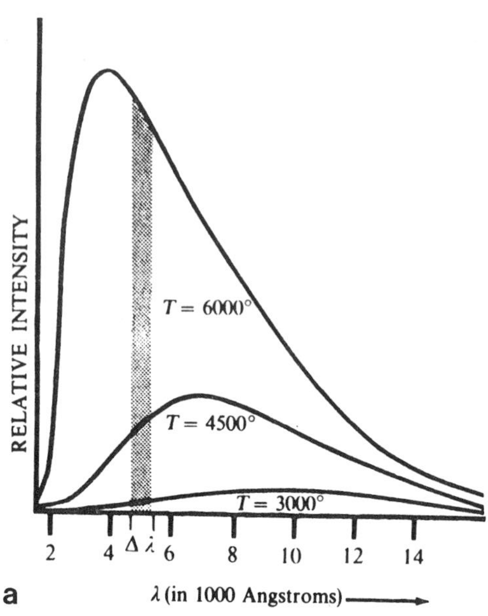


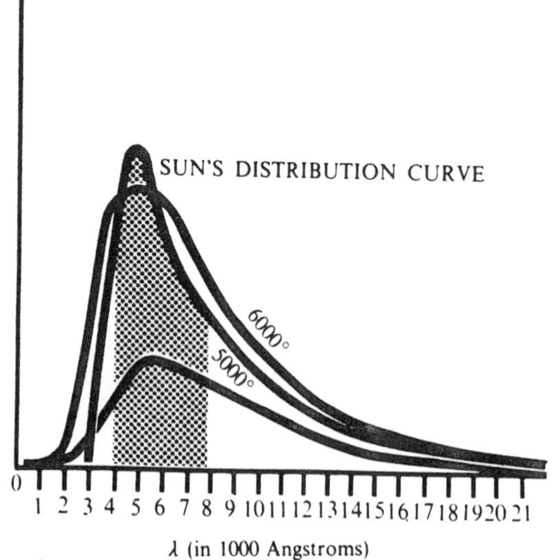


Figure 13.3. (a) A graph of the intensity of the radiation emitted by a black body as a function of the wavelength. The shift of the wavelength of maximum intensity with increasing temperature is shown. The small shaded segment represents the energy emitted per second in the wavelength interval $\Delta\lambda$. (b) The radiation emitted by the sun compared with the radiation emitted by a black body of the same temperature.

of the radiation changes from a cherry red to a yellowish glow and finally, if the furnace became hot enough, to a bluish white.

The experiments were clear enough but theoretical physicists were unhappy because they could not derive an algebraic formula for the blackbody radiation curve from first principles, that is, from classical Newtonian physics and Maxwellian electromagnetic theory. They did, however, deduce two formulas very useful both for physicists and astronomers. The first of these, known as the Stefan–Boltzmann law, states the total radiation emitted per second by the furnace is proportional to the fourth power of the absolute temperature of the furnace. This means that if the absolute temperature is doubled, the energy emitted per second increases by a factor of 16, and so on. This is why the radiation curve moves vertically up (parallel to the ordinate) as the temperature increases. This immediately permits us to determine the temperature of the surface of a star (the outer layer from which the stellar radiation appears to come).

The second important and astronomically useful blackbody radiation formula was deduced theoretically by the German physicist Wilhelm Wien in the first decade of the twentieth century. Known as the "Wien displacement law," it states that the wavelength (color) of the most intense radiation emitted per second by a furnace varies inversely with the absolute temperature $T$. This is given by the formula $\lambda_m = 0.290/T$ where $\lambda_m$ is the wavelength of the most intense color (the peak of the radiation curve) and $T$ is the kelvin (absolute) temperature of the furnace or the surface of the star. If we invert this formula, we can express this formula in terms of $\lambda_m$; thus $T = 0.290/\lambda_m$. All we need to do, then, to find the surface temperature of a star is to plot its radiation curve (intensity against wavelength), find the wavelength ($\lambda_m$) for the peak of the curve, and divide that number, expressed in centimeters, into 0.290; this quotient is the surface temperature expressed in degrees kelvin. Thus the solar radiation is the most intense in the yellow part of the spectrum, corresponding to a wavelength of about 6000 Angstroms or about 6/100,000 centimeters, indicating a surface temperature of about 5800 degrees kelvin. This is also the temperature obtained from the Stefan–Boltzmann law which takes into account the total radiation (radiation of all colors) emitted in 1 second by 1 squared centimeter of the sun's surface.

This "surface" temperature does not mean that the sun's surface is really a black body. Indeed, the radiation coming from the sun gives us the impression that we are looking at a "solar surface" which is reinforced by what appears to us to be a "sharp" edge when we see the sun faintly visible through a thin cloud. Actually, what appears to be a thin edge is really a layer about 300 miles thick which is how far we can "peer" into the sun, so that the radiation we receive is the accumulation from several layers extending 300 miles into the interior. If we could look more deeply into the sun we would find that the temperature increases steadily, reaching millions of degrees in the deep interior of the sun.

That Stefan, Boltzmann, and Wien had deduced simple formulas for two different features of the blackbody radiation curve left theoretical physics with much to be desired. A complete sense of victory in this remarkable struggle of physicists to understand the mysteries of radiation would be achieved only if a formula could be deduced which reproduces algebraically the complete blackbody radiation curve or spectrum. This would have to be a formula which relates the intensity of every color of the radiation for a given temperature of the black body (furnace) to the color (wavelength) of the radiation and the temperature of the furnace. Max Planck was most active and, ultimately successful, in this search for the mathematical "holy grail" of radiation physics. His formula, based on a remarkable and revolutionary hypothesis, permits one to calculate the intensity of the radiation from the knowledge of its wavelength and the temperature of the furnace alone. His discovery was so revolutionary that all of physics had then to be changed. Indeed, physics changed from Newtonian–Maxwellian physics to quantum physics almost overnight.

To understand the importance and significance of Planck's discovery, not only for physics but for astronomy, chemistry, and even for biology, we go back for a moment to Newtonian and Maxwellian physics. We recall that Newtonian physics stemmed from Newton's laws of motion, particularly from his famous second law, $F = ma$, which was accepted as the universal truth that was supposed to apply equally to galaxies, stars, planets, molecules, and atoms. That this law, combined with Newton's law of gravity, leads, by a simple mathematical procedure, to Kepler's laws of planetary motion, was accepted as clear evidence that Newtonian dynamics holds for the motions of objects interacting gravitationally. But people went beyond this point, arguing that the law holds for atoms and the constituents of atoms that interact electromagnetically as do electrons and protons. This was observed to be so for the electrons and protons moving in Crooke's tubes. But no one, up to that point, had any evidence that Newton's second law of motion is valid for electrons and protons held together in atoms by the electromagnetic force.

Newtonian dynamics led to three other concepts that play important roles in astronomy: momentum, energy, and action. The first two are governed by what are called "conservation" principles and the third one by what is called a "minimal" principle. To explain these ideas in the simplest possible way, we picture a particle of mass $m$ moving with speed $v$ in some direction. Newtonian physics assigns three quantities to this particle: its momentum $mv$ (its mass times its speed) in the given direction; its kinetic energy $0.5mv^2$ (one-half its mass times the square of its speed), and its action $(mv)d$ (its momentum multiplied by the small distance $d$ it moves with no change in its speed). If the particle is moving in a field of force (e.g., a planet moving in the sun's gravitational field) it also has potential energy which must be added to the kinetic energy (actually the potential energy is subtracted) to obtain the total energy.

The principle of conservation of energy means, then, that if any isolated body is moving freely in some kind of field (e.g., a gravitational field) its total energy (kinetic plus potential) cannot change. If it gains kinetic energy it must lose exactly the same amount of potential energy and vice versa. So since the earth (like the other planets) is moving in an ellipse, its speed, and, hence, its kinetic energy (like its potential energy) changes continuously. As it approaches the sun (like a rollercoaster going downhill), it loses potential energy and gains kinetic energy and vice versa. The potential energy of a planet (and, indeed, of an artificial satellite) increases as it recedes from the sun. In a system of many bodies (like the stars in a cluster), the total energy of the system remains unaltered; if one body loses energy owing to its interactions with other bodies, these other bodies must gain the energy lost by the one body.

The conservation of momentum is a somewhat more subtle idea because it involves the product $mv$, and $v$, velocity, has direction so that in considering the principle of conservation of momentum, we must take into account the direction of the motion of the body (the direction of the momentum). Suppose you are running at a certain speed toward a cart that is standing still. Considering you and the cart as a single physical system, we note that the total momentum of the system initially is just your own momentum (your speed times your mass) in the direction of the cart. When you jump into the cart, the total momentum of the system (you and the cart together) must remain the same. This can be true only if you and the cart move together at a slower speed in the direction in which you were running initially.

The concept of action, however, is quite different from that of either energy or momentum, although it involves both of these concepts. Action was introduced into physics in the eighteenth century as already noted by the French mathematician, physicist and astronomer Pierre de Maupertuis. As a metaphysicist and something of a mystic, Maupertuis was convinced of the existence in nature of a physical law that guides bodies in their motions in a field of force such as a gravitational field. He could not accept the idea of action at a distance. Even though he accepted Newton's law of gravity, he was puzzled how a planet moving around the sun, "knew" along which orbit it should move. He introduced the remarkable hypothesis previously discussed that each planet is aware of and possesses something intrinsic that guides it in choosing its orbit from among all other possible orbits. He called this intrinsic property "action," defining it as the momentum of the body ($mv$) multiplied by a small piece of the path $d$ along which it moves. He then proposed one of the most remarkable principles in science, the principle of least action: that the planet (or any particle) chooses that path along which its total change of action is a minimum. In the nineteenth century Hamilton extended Maupertuis's principle. He enlarged the action concept by subtracting from Maupertuis's action the product of the energy of the particle (or planet) and the time during which it is moving. This is what is now called Hamilton's principle of least action, where the action is defined as momentum $p$ times distance $d$ minus energy $E$ times time $t$ ($pd - Et$). Although the action concept and the principle of least action were introduced merely as mathematical devices to aid physicists, they have taken on remarkable lives of their own as guides to scientists in their choice of models to represent physical phenomena. We shall see later the role that the concept of action played in Planck's discovery of the quantum theory, which we introduce by a brief resumè of Maxwell's electromagnetic theory of light, which we discussed previously in some detail. We have expanded on our previous discussion of action to enlarge the reader's concept and understanding of it.

The essence of Maxwell's electromagnetic theory of light (conclusively verified by Hertz's experiments) is that oscillating electric charges generate electromagnetic waves which are transported through empty space at a constant speed (the same for all wavelengths). Maxwell suggested, at the time he proposed his theory, that the propagation of such waves requires the existence of a universal ethereal medium—the ether—as the light-propagating medium, similar to the terrestrial atmosphere that transports sound waves. To physicists and astronomers, Maxwell's proposal implied, at the time, that the ether is a kind of material medium which is endowed with contradictory, unacceptable properties that, following Einstein's rejection of it, was abandoned. The important feature of Maxwell's theory that had to be accepted at Planck's time, as the basis for describing blackbody radiation, was that it seemed to have banished particles (corpuscles) in favor of waves as the basic elements of radiation.

Planck's discovery was revolutionary precisely because it is based on corpuscles of radiation rather than on waves, and thus appears to contradict Maxwell's hypothesis. To account for the precipitous drop in the intensity of the blackbody (furnace) radiation curve at its blue-violet (short wavelength) end, Planck assumed that the radiation is emitted from the furnace in the form of tiny pellets which he called "quanta"—hence Planck's quantum theory of blackbody radiation. Planck described each quantum as a pellet of energy, vibrating with a definite frequency and, therefore, associated with a definite wavelength. Planck knew, from the classical dynamics of vibrating systems (e.g., springs, tight strings) that the energy of such a system equals the frequency of its vibrations multiplied by its action. Planck therefore proposed that the energy of each quantum or pellet of radiation equals its frequency multiplied by a universal constant of action, $h$, the famous Planck constant of nature, which stands with the speed of light $c$, and Newton's universal constant of gravity G, at the very peak of the fundamental constants of nature. We note that if we divide the product $hc$ by $G$, the square root of this quantity $(hc/G)^{1/2}$ is a universal unit of mass, known as the "Planck mass." This mass plays a basic role in cosmology which we shall explore later. If $E$ is the energy of a pellet or quantum of radiation of frequency $\nu$, we have Planck's remarkable formula $E = h\nu$ which has revolutionized physics.

A brief consideration of this simple energy–frequency (color) equation tells us at once why it led Planck directly to the correct formula for blackbody radiation. Because the blue-violet quanta have much higher frequencies than the red and yellow quanta, they are more energetic than the reds and yellows. The furnace thus finds it more taxing to produce and emit the blue and violet quanta than to produce the yellows and reds. The blackbody radiation curve therefore falls off rapidly at the blue–violet end of the spectrum. Planck's hypothesis and his equation introduce a unit of action $h$; this means that no phenomena can occur in nature associated with an action less than $h$. We may interpret this as a limit to continuity in nature; if we go down to very tiny dimensions, phenomena become discontinuous. We now draw another useful conclusion from the existence of a quantum of action $h$, which, by the way, is extremely tiny: 6 divided by 1000 trillion trillion (one followed by 27 zeros). We picture a particle with momentum $p$ moving a short distance $q$. The product $pq$ must always be larger than $h$. This means that if we try to determine the exact position $q$ of the moving particle we must make $q$ as small as possible.

This means that $p$, the momentum of the particle, must become extremely large to keep $pq$ larger than $h$ so that we lose precise knowledge of its value (the larger the entity, the greater the error in our knowledge of it). This led the German physicist Werner Heisenberg to his famous uncertainty principle which, in turn, led to what we now call quantum mechanics: the exact motion (momentum) of any particle and its exact position cannot both be known simultaneously.

Planck was so appalled by what appeared to him to be the demolition of Maxwell's beautiful electromagnetic wave theory of radiation that he introduced the compromising theory that the radiation emitted by the furnace comes out in the form of pellets (quanta) but then changes into waves. This compromise was rejected by Einstein, who showed in a famous paper written in 1905 that even before the radiation is emitted from the furnace it behaves as though it consists of particles. Indeed, Einstein showed that radiation, confined in a cavity, behaves just like a gas, consisting of Planck corpuscles (quanta), which themselves have wave properties. This was the beginning of the modern wave–corpuscle dualism concept of nature: all things are simultaneously both corpuscles and waves. Einstein's analysis of the behavior of confined radiation interacting with the walls of the cavity, which showed that the radiation consists of corpuscles with wave properties, led to the concept of the photon, a spinning, vibrating, wavelike corpuscle which has energy and momentum and can exert pressure on a wall just like a molecule. Einstein showed further that photons can produce the photoelectric effect (the knocking of electrons out of metal surfaces) in complete agreement with observations.

Einstein went on to make another important contribution to the development of the quantum theory by insisting that the quantum concept must be applied to all physical phenomena, and, in particular, to the various atomic and molecular processes in the universe. This philosophy led the Danish physicist Niels Bohr to his quantum model of the atom which brought all of physics, chemistry, and astronomy to their present advanced states. We recall that Fraunhofer's discovery of the dark absorption lines in the solar spectrum led him to the belief that these dark lines are produced when the solar continuous radiation interacts with the atoms in the solar atmosphere, and each type of atom leaves its distinct and unique imprint (characteristic to its type, e.g., hydrogen, helium, etc., like a fingerprint) on the outgoing radiation. These discrete Fraunhofer lines indicate that a kind of quantum process for each line occurs in atoms. He was fortified in this belief when he discovered that heated atoms, like sodium, when salt is thrown into a flame, emit bright spectral lines, which are the bright counterparts of the dark absorption lines.

Before we discuss Bohr's great contribution to the structure of the atom and thus to all phases of spectral theory, we review briefly the state of knowledge about spectra in general in the early years of the twentieth century. Physicists and astronomers knew that all radiation coming from matter is produced in some way by the cooperative interactions of the atoms in the matter. But how do these atoms produce three different kinds of the known spectra: bright emission lines, absorption lines, and continuous spectra? We can understand all of this if we note first that emission lines are produced by excited atoms that are not close to each other, which is so if a gas is confined to a Crookes tube and the individual atoms are excited by an electrical discharge. As the atoms and electrons move at fairly high speeds, collisions occur which excite the atoms. As the atoms are deexcited, they emit radiation of discrete frequencies which is then observed as discrete spectral lines: hence we observe the discrete spectrum, with each type of gas emitting its own kind or set of spectral lines.

One now observes that these spectral lines begin to broaden as more and more atoms of the given gas are injected into the tube until, at a certain concentration (or number density) of the atoms, the lines become so broad that they practically touch and we have a continuous spectrum, which is no longer characteristic of the type of gas in the tube. The continuous spectrum is thus produced when atoms are so closely packed that they interfere with each other.

If the tube is now surrounded by a cool gas of any kind, the atoms in the cool gas are excited by that part of the continuous radiation that the atoms of the cool gas would emit if they were first excited and then deexcited. This accounts for the dark lines of the absorption spectrum. We see then that the three types of spectra are produced by essentially the same atomic processes. The next step in this story was to develop a model of the atom from which, by pure reasoning, one can predict all the properties of the radiation emitted or absorbed by the atom. At that time—about 1905 to 1906—it seemed that all the "raw material," electrons and protons, needed to construct atoms were available. Only the right ideas were missing. To see why the ideas or the physical concepts available at that time did not lead to a correct atomic model we note that Maxwell's electromagnetic theory leads to the conclusion that an accelerated electric charge, a charge moving with a varying velocity (changing speed, changing direction, or both) must radiate and therefore lose energy. This meant to the early nineteenth century atomic model builder that a planetary model of the atom (negative charges—electrons—moving in closed orbits around a positively charged nucleus containing protons) was forbidden by Maxwell's theory.

We can best describe these early difficulties by considering the planetary model of the simplest atom (the hydrogen atom) consisting of a single electron revolving around a single proton. As the proton's mass is roughly 2000 times that of the electron, the proton is practically fixed in this model. The electron at a distance of about one one hundred millionth of a centimeter from the proton in this model would revolve about two thousand trillion times per second. Because this means that the electron would experience an enormously large acceleration, it would, according to the then accepted electromagnetic theory, radiate energy very rapidly, and therefore fall into the proton within 100 millionths of a second. It was precisely at this time, in the 3 years from 1910 to 1913, that experimental physics, under the direction of Ernest Rutherford in England, and theoretical physics (the quantum theory) in the hands of the Danish physicist Niels Bohr saved the nuclear model (the planetary model) of the atom.

Rutherford, through his study and analysis of natural radioactivity (the radioactive decay of heavy nuclei such as uranium and thorium) had established himself as the leading experimentalist of the day (he won the Nobel Prize in chemistry in 1908); he also initiated the experimental study of atomic nuclei (nuclear physics). To determine the structure of the atom experimentally he devised an ingenious experiment to study the electric charge distribution within the atom; he used positively charged particles, called alpha particles, emitted from the uranium nucleus, as charged probes. Alpha particles, emitted spontaneously from the decaying uranium nuclei, and later discovered to be the nuclei of helium atoms, were ideal missiles for Rutherford to use in his experiments. Bombarding very thin gold foil with rapidly moving alpha particles (traveling at about 10% of the speed of light), he discovered that most of the alpha particles passed right through the gold foil, hardly deviating from their straight line paths, but that, occasionally, an alpha particle was repelled violently and deviated by 90 degrees or more from its straight line path. Rutherford correctly interpreted this phenomenon as proving that the gold atom (and, by inference, any heavy atom) consists mostly of empty space in which electrons are moving about a massive, positively charged nucleus. Most of the alpha particles pass right through the empty space of the atom, but every now and then, an alpha particle meets a nucleus head-on and is, therefore, repelled violently. This is conclusive, incontrovertible, experimental evidence that the atom consists of a massive, positively charged nucleus, surrounded by a cloud of encircling electrons.

The next step in this atom-model building enterprise was taken by Niels Bohr who proposed what is now called the "Bohr atom," which is now universally accepted. Starting from Rutherford's experiments with "solid gold," he sought some way of circumventing the proscription imposed by electromagnetic theory against the motions of electrons in an atom around the atom's nucleus; Bohr found, in Planck's quantum theory, just what he wanted. To see this in its simplest form we turn again to the hydrogen atom (the simplest atom). We picture the electron circling the proton as close as it is permitted by the quantum theory, and follow Bohr's reasoning. He saw that the introduction of the concept of the quantum of action $h$ (Planck's constant) immediately restricts the electron to a set of possible orbits, with a definite lowest orbit (closest to the proton and called the Bohr orbit) which is defined by a single unit of action, *h*, so that the electron cannot by radiating energy fall into a smaller orbit without violating the quantum theory. In other words, as Bohr understood it, the quantum theory permits the electron to move in a permitted orbit without obeying Maxwell's demand that it radiate. Quantum theory is thus the "master theory," superseding the Newtonian and Maxwellian theories. Bohr thus looked upon the permitted orbits as "stationary states" of the electron.

Bohr developed the simplest form of his quantum model of the atom, working with hydrogen, by treating the permitted orbits of the electrons as circles (a quite unrealistic but a very heuristic assumption). Bohr numbered each orbit by integers 1, 2, 3, . . . starting with the lowest orbit. Each of these integers expresses the action, in units *h*, associated with that orbit. Thus the action of the electron in the first orbit is *h*, in the second orbit, 2*h*, in the third orbit, 3*h*, etc. These integers, called the principle quantum numbers, are also a measure of the electron's energy (negative) in that orbit. Bohr went on to argue that if an electron is in some orbit above the lowest (the atom is in an excited state), it remains in that orbit for a very short time (about one hundred millionths of a second) and then jumps down to the lowest state directly or in a series of steps (from a higher orbit to a lower orbit) emitting a single quantum of energy (photon) with each jump. The frequency of the photon emitted in any downward jump is large if the energy difference between the two levels is large and vice versa. Using these simple ideas and no more than elementary algebra, Bohr deduced the frequencies (or wavelengths) of the hydrogen spectral lines. Because blackbody radiation is produced by the electrons in the excited atoms in the walls of a furnace, we see how the Bohr theory explains the Planck formula and the radiation curve for blackbody radiation. Some years later (1917) Einstein deduced the Planck radiation formula from the Bohr model of the atom. To do this he had to introduce a process called "the stimulated emission of radiation by an excited atom," which in the 1950s led to the development of the laser by Charles Townes.

Bohr completed his picture of atomic processes according to his model by describing the absorption of radiation by an atom as

Neils Henrik Bohr (1885–1962) (Courtesy AIP Emilio Segrè Visual Archives; Margrethe Bohr Collection)

the inverse process of emission. Not only can an electron emit photons by jumping spontaneously from higher energy states to lower ones, but it can also be hoisted from lower to higher states by absorbing photons of just the right frequencies. The electron is thus very selective when it absorbs photons. Before discussing how the Bohr model of the atom explains the formation of the dark (absorption) Fraunhofer lines in stellar spectra, we note that from 1913 to about 1920 improvements were made in the primitive Bohr model to account for certain detailed features of atomic spectra (e.g., line doublets and triplets, or, in general, multiplets) for which the simple Bohr model could not account. In searching for an atomic mechanism for such multiplets, physicists traced the difficulty back to the circular orbits of the primitive Bohr model and argued that the orbits should be ellipses.

Because ellipses come in all shapes, physicists had to introduce some restrictive rule to limit the number of shapes; they then discovered the correct rule that the number of shapes must equal the principal quantum number of the overall orbit. Thus three elliptical shapes are assigned to the orbit of principal quantum number 3, and so on. These shapes vary from a complete circle to a very thin ellipse (essentially a straight line). These shapes (the second set of quantum numbers) represent or express another property of the motion of the electron: the angular momentum of the electron (the rate of its revolutionary motion around the proton). These orbital shape quantum numbers thus, in a sense, are, for the electrons, like Kepler's second law of motion for planets (equal areas are swept out in equal times). The change of the spectrum of an atom placed in a magnetic field required a third set of quantum numbers called "magnetic quantum numbers" which were introduced shortly after the second set of quantum numbers.

The need for the set of magnetic quantum numbers can be understood from the following brief description of how a magnetic field affects the dynamical behavior of an atom. One finds that each of the spectral lines of the atom is split into a group or multiplet (doublets, triplets, etc.) of lines. The reason for the splitting of the lines is that the motion of the electron in the atom is equivalent to an electric current so that, like all such currents, it is accompanied by a magnetic field. The atom thus behaves like a spinning magnet. We know, however, that when a bar magnet is placed in a magnetic field, it aligns itself along the field. But if the magnet is spinning, it precesses around the magnetic field, like a tilted top spinning on a horizontal surface precessing around the vertical (around the gravitational field). Because the atom is a spinning magnet, it precesses around the magnetic field; this precessional motion, requiring a quantum number and adding energy to the electron, alters the energy levels (stationary states) of the electron and thus alters the spectral lines. This effect of a magnetic field on the spectral lines of an atom is called the "Zeeman effect," named after the Dutch physicist P. Zeeman, who discovered this effect in 1896. Because stars, like the sun, generate magnetic fields (e.g., sunspots) we must use the Zeeman effect to study these fields. We note that, as the precession of the atom in a magnetic field is "action," it is governed by the quantum theory. This is manifested by the restriction the quantum theory places on the alignment of the spinning axis of the atom. The quantum theory allows only a discrete number of orientations (called space quantizations), each of which is given a number. The number of such orientations, called the magnetic quantum number, equals twice the orbital quantum number plus 1.

The introduction of the magnetic quantum number did not end the quantum number story; a fourth quantum number, the spin quantum number (having only two values which we may write as +1 or −1) was introduced to take account of the spin of the electron itself in 1925 by two young Dutch physicists, Samuel Goudsmit and George Uhlenbeck. This fourth quantum number is required by the appearance of doublet spectral lines (e.g., the sodium *D* line) even when the atom is not in a magnetic state. These four quantum numbers which we write as $n$ (principal quantum number), $l$ (azimuthal or orbital quantum number), $m$ (magnetic quantum number), and $s$ (spin quantum number) describe all the dynamics of the electrons in atoms. The four quantum numbers cannot be assigned arbitrarily but are related to each other in a very definite way. For a given $n$, $l$ can have any one of the $n$ integer values from 0 to $(n - 1)$, and $m$ can have any one of the negative or positive integer values from –1 to +1. As we show later, these quantum numbers determine the chemical properties of atoms, but to understand just how this process occurs, we describe another principle, the famous Pauli exclusion principle. This principle was announced in a paper in 1924 by the Swiss physicist Wolfgang Pauli, who received the Nobel Prize for his work.

Physicists and chemists alike had become convinced that the electrons in an atom containing many electrons cannot all be in the same ground state (lowest orbit) because one could not then explain the valence and the general chemical properties of atoms. Only if the electrons are arranged in layers with some outer electrons (called the "valence electrons") can we explain the different kinds of atoms and their different chemical properties. But this layering (or shell arrangement) of electrons defies our intuitive understanding of dynamics. If we compare the electrons arranging themselves around the nucleus of an atom with perfectly round balls falling into a smooth hole, we are at once puzzled. We know that all the balls settle into the bottom of the hole: they do not arrange themselves along various circular recesses along the walls of the hole. Why then do the electrons arrange themselves in concentric orbits instead of all falling into the lowest (ground state) orbit? To explain this problem, Pauli proposed his famous exclusion principle which states that no two electrons in the same atom can have the same set of four quantum numbers.

We can best illustrate this principle by describing according to this principle the arrangement of the electrons in a few light atoms. The hydrogen atom, consisting of one proton and one electron, is quite straightforward. Its one electron is in the lowest orbit (or shell) with quantum numbers $n$, $l$, $m$, $s$ equal to 1, 0, 0, +1 or 1, 0, 0, –1, respectively. It presents no problem. Next comes helium with a two-proton nucleus and two electrons. Both of these can occupy the lowest orbit or shell with one electron with the set of quantum numbers 1, 0, 0, +1 and with the other electron with the set of quantum numbers 1, 0, 0, –1. This difference simply means that the two electrons move in the same way, but with one spinning clockwise and the other counterclockwise.

With lithium we see a very important change. It has three electrons because the lithium nucleus contains three protons, but now we must place the third electron in the second orbit (shell) with $n = 2$; only two electrons can be placed in the first orbit since the four quantum numbers of a third electron would have to be identical to one of the two sets of the four quantum numbers assigned to the other two electrons. The first shell of electrons—called the K shell in an atom—can accommodate only two electrons, in which case, as in helium, we say that the shell is closed and the atom, like helium, is inert. Because the lone electron in the second orbit of the lithium atom is much further from the nucleus than the two innermost electrons, and is partially electrically shielded from the nucleus by the two inner electrons, this lone electron is fairly easily torn away from the nucleus so that the lithium atom becomes ionized (a positive ion). Thus lithium is chemically quite active because its third electron jumps around quite easily; this electron, called the "valence electron," gives lithium its chemical properties. In general, then, the chemical valence of an atom is determined by its electrons in its outermost shell.

With this in mind we now consider the number of electrons that can be accommodated by the second shell, with principal quantum number $n = 2$. For $n = 2$, the atom can have two different azimuthal, or orbital, quantum numbers $l$: 0 and 1. If $l = 1$, $m = 0$ but for $l = 1$, $m$ can have the values $-1$, 0, 1. Considering only the quantum numbers $n$ and $l$, four distinct sets of quantum numbers, $n$, $l$, $m$ are permitted (2, 0, 0), (2, 1, $-1$), (2, 1, 0), (2, 1, 1). Taking the spin quantum number $s$ into account, which may be $+1$ and $-1$, we then have the eight distinct sets of quantum numbers (2, 0, 0, +1), (2, 0, 0, $-1$), (2, 1, $-1$, +1), (2, 1, $-1$, $-1$), (2, 1, 0, +1), (2, 1, 0, $-1$), (2, 1, 1, +1), (2, 1, 1, $-1$). These sets of eight quantum numbers mean that the second shell (called the L shell) in an atom can accommodate only eight electrons. We illustrate this scheme by describing the carbon and oxygen atoms which are important not only for stars but for life processes as well. The carbon atom has six protons in its nucleus so that it must have six electrons arranged in the atomic shells. Two of these electrons are in the K shell (lowest state), leaving four electrons for the L shell, which means that the carbon L shell can accommodate four more electrons. Coming now to the oxygen atom, with a nucleus containing eight protons, we see that it has eight electrons, two of which are in the K shell and six of which are in the L shell. Thus the oxygen L shell has room for two more electrons. We now go on to the atoms neon and sodium to illustrate a very important point. Because neon's nucleus has 10 protons, its outer shells contain 10 electrons; two of these electrons are in its ground state (K shell) and 8 are in its L shell, thus closing that shell so that neon, like helium, is a chemically inert gas. Sodium, with 11 protons in its nucleus and 11 external electrons has two electrons in its closed K shell, 8 electrons in its closed L shell, and 1 electron in its third shell (principal quantum number 3). Chemically speaking, then, sodium behaves like lithium since the single electron in each case (the valence electron) determines the chemistry.

We now point out the relationship of this information to the chemical valence of atoms and to the explanation of the periodic table of the chemical elements introduced to the scientific world in 1871 by the Russian chemist Dmitri Ivanovich Mendeleev. The chemical valence of an atom is a measure of the chemical activity of the outermost shell of electrons in the atom; the electrons in the inner shells, being more tightly bound to the nucleus than those in the outer shell, play only a minor role in the chemistry of the atom and therefore do not contribute to the valence of an atom. If an atom tends to give up an electron in any chemical reaction, we say that its valence is positive. Thus hydrogen, with only one electron, tends to give up that electron to another atom when it binds to other atoms to form a molecule. This is also true of lithium and sodium, which are therefore members of a group of elements with similar chemical properties. Carbon with four electrons in its outer shell and four open slots in that shell can give up electrons or gain electrons equally easily. Thus carbon, with both positive and negative valence is, chemically speaking, very versatile. Oxygen, with six electrons in its outer shell and two open slots in that shell tends to accept electrons so that its valence is negative. All of chemistry is built on these ideas, so that, in a sense, chemistry is the physics of molecules.

One may wonder what all of this has to do with astronomy. As we shall see in our discussion of the classification of stars according to the spectral features of their radiation, the dynamics of the electrons in the atomic shells determine these spectral features. A brief description of this phenomenon before we pursue it in greater detail in the following chapter will help us understand the logic of stellar spectra classifications. The spectral features (the Fraunhofer absorption lines) of stellar radiation are determined by the states of the electrons of the various atoms in the stellar atmospheres. But these states, in turn, are determined by the temperatures of the atmosphere (essentially determined by the temperature of the thermal, blackbody, radiation passing through the atmosphere). Thus the distribution of the atomic electrons in stellar atmospheres are quite different from what they are in our cool terrestrial atmosphere; the stellar spectral classification is thus a temperature classification. In the next chapter we show how these new ideas led to a remarkable classification of the stars.

CHAPTER 14

# Relativity and Astronomy

*In this life we want nothing but facts, Sir; nothing but facts.*

—CHARLES DICKENS

We introduced the preceding chapter by mentioning the two great physical theories that changed not only the Newtonian and Maxwellian laws of nature, but also the Galilean–Newtonian concepts of space and time. We described the first of these, Planck's quantum theory, which introduced into physics the corpuscle–wave concept and led Bohr to the correct nuclear–planetary model of the atom. We now present the second of these two revolutionary theories, the theory of relativity, without which modern astrophysics and cosmology could not have been developed. Albert Einstein presented the theory of relativity in two steps: the special theory of relativity first appeared in 1905 in one of Einstein's three famous papers that were published that year in *Annalen der physik.* The great importance of these three papers may be gauged by their evaluation, many years later, by the Nobel laureate Max Born, who stated that each of these papers opened up a "whole new branch of physics." The second step in the presentation of the theory of relativity appeared in 1915 as the general theory of relativity.

Before presenting a detailed but nonmathematical discussion of Einstein's special theory of relativity we consider his interpretation of the word "relativity" which does not appear in the title of his first paper in which he develops the basic features of his special theory of relativity. The title of that paper is "The Propagation of Electromagnetic Phenomena through Rapidly Moving Media." The word "relativity," assigned to Einstein's work later, and having become very popular since then, is now used, in a loosely defined way, in many different disciplines, particularly in connection with knowledge and the meaning—or lack of meaning—of "absolute truth." In common usage, "relativism" implies the introduction of a standard which may change from society to society and even from individual to individual. In saying that this or that person is "relatively good" at some activity or other, we imply that we measure his or her skill against some higher standard set by someone else (an outstanding performer). Such comparisons are very loosely defined since there are no quantitative measurements on which one can base such comparisons.

In our daily lives we deal with relative quantities which our experience immediately translates into "real" quantities. Thus the apparent size of a distant object (e.g., a tree) is related to our distance from the object. If we draw a straight line from our feet to the top of a tree, its relative size is determined (or measured) by the angle this line makes with the horizontal (the smaller this angle, the smaller the apparent size). As an extreme example, we note that the diameter of the moon, as seen from the earth, subtends an angle of half a degree so that its apparent size to an observer on the earth is that of a penny held at arm's length; owing to the closeness of the moon compared to the distance of the sun from the earth, the apparent size of the moon is very nearly the same as that of the sun. This is why total solar eclipses of the sun can occur. Stars are so far away that they appear to have no size at all beyond that of mere points of light. We note that relativism plays a very important role in our perception. Again, our experience has taught us how to draw conclusions about the true shapes of objects from appearance in which our depth perception plays a very important role. To the artist the difference between apparent and real is delineated in "perspective," which the artist uses to represent the "actual aspect" of an object from a given point of view, which may differ dramatically from artist to artist.

Returning now to the theory of relativity, we emphasize that the concept of relativity introduced in Einstein's theory of relativity has nothing to do with the various relativity concepts we have just discussed. The theory of relativity deals with the question of the invariance (the constancy) of the laws of nature as these laws are stated by observers in different frames of reference. The word "different" here refers to the state of motion of the frame of reference.

To present this concept of relativity as clearly and simply as possible, we first discuss the concept of "the frame of reference" of the observer, which stems from the need of observers, going back to the early Greeks, to be able to describe their observations (positions, times, and motions) of particles (or events) in a meaningful, quantitative way from which important conclusions about nature might be drawn. If the observations were limited to objects or events on the earth, then the frame of reference was a system of two sets of intersecting lines which emanated from the observer's position (some point on the earth's surface). Because the earth was thought to be flat, these sets of intersecting lines were "straight lines" in the Euclidean sense of "straight." If these early observers were limited to a particular ancient city, these lines became streets and avenues which were absolutely essential to the multitudinous activities of a civilized metropolis. It was only natural then to label these streets and avenues numerically starting from some important landmark (e.g., the ruler's residence or the capitol). If, instead of being concerned with where in the city the events the observer was studying occurred, but rather where in his immediate neighborhood the events were happening, he laid down his own set of intersecting lines starting from some point which he labeled *O* and called the origin.

This idea of laying down two sets of intersecting lines to locate events or objects became very important, indeed, imperative, with the rapid development of shipping, which led to terrestrial mapping and to the geographic system of latitude and longitude. The city of Greenwich, England, was adopted as the "origin" of this system. The whole city of Greenwich thus became a mere point in the geographic system of "streets and avenues" (latitudes and longitudes).

In going from the Euclidean straight lines of streets and avenues in a city stretched out over a "flat" earth to the curved lines of latitude and longitude on the spherical earth, we now consider a very important question: Does anything intrinsic in the phenomena we are observing change? The answer is no! We only change our way of describing the phenomena. What does this mean? To elucidate this question in a way which makes more understandable our discussion of relativity theory we consider the distance and the time interval between two events (e.g., the explosions of two bombs at different points in our frame of reference) as intrinsic. We know then that no matter how we lay off the two sets of intersecting lines (streets and avenues) that we use to specify the positions of the exploding bombs, and no matter what kind of timing device we use to measure the time interval between the two explosions, we must always obtain the same distance and the same time interval between the two events (explosions). These, then, are the intrinsic features of the events that are not altered by changing our frame of reference.

To quantify these ideas and present them in an elementary algebraic and geometric way, we define the concept of the frame of reference by starting from the concept of the coordinate system introduced by the seventeenth century French philosopher and mathematician Rene Descartes, who, in his book *Geometrie* showed that all geometry can be represented algebraically and all algebra geometrically by means of what he called a "coordinate system" which is constructed by laying off two intersecting sets of parallel lines (streets and avenues). We obtain a Cartesian coordinate system by laying off one set of parallel horizontal lines in any direction and another such set perpendicular to this first set of lines so that the position of any point on the plane defined by these two sets of mutually perpendicular lines can be located by specifying it as the point of intersection of one particular line from set 1 and one particular line from set 2. To do this we must label each horizontal and each vertical line starting from the intersection of one particular set 1 line and one particular set 2 line. This starting point of intersection is called the origin $O$ of the coordinate system. Any point on the plane may, of course, be taken as the origin of the coordinate system.

Descartes went on to show that if one set of intersecting parallel lines (e.g., the set 1) is labeled $x$ lines and the other set is labeled $y$ lines, then the position of any point on the $x$–$y$ plane can be expressed by giving its $x,y$ position in this $x,y$ coordinate system. The numbers $x$ and $y$, called the coordinates of the point, are the perpendicular distances of the point from a particular vertical ($y$

line) and a particular horizontal ($x$ line) respectively. The point of intersection of the particular $y$ and particular $x$ reference lines is called the origin of the coordinate system. The $x$-coordinate of the point is then the $x$ distance of the point from the origin $O$, and the $y$ coordinate of the point is the $y$ distance of the point from $O$.

Descartes went on to show that curves (e.g., straight lines, circles, ellipses, etc.) can be expressed or described by algebraic formulas. As examples we note that a straight line passing through the origin and making an angle of 45 degrees with the $x$ lines and the $y$ lines is expressed or defined by the algebraic equation $y = x$. This simply means that every point on the line is equidistant from the $x$ line (above the $x$ line) passing through $O$ and from the $y$ line passing through $O$. Also the algebraic formula for a circle of radius $r$ around the origin $O$ is $x^2 + y^2 = r^2$, which is just the theorem of Pythagoras for every point on the circumference of the circle. This mathematical merging of geometry and algebra, called "analytic geometry," was very important, not only for the advance of mathematics (it was particularly important for the rapid development of the calculus) but also for physics and astronomy. Using his laws of motion and his law of gravity, Newton deduced Kepler's three laws of planetary motion from the equation of the ellipse.

Returning now to the concept of the distance between two events on our plane which are separated by the distance $d$ on the plane, we can express this distance algebraically in our coordinate system just using the coordinates (the $x,y$ specifications) of the events on the plane. Thus if the $x,y$ coordinates of event 1 are $x_1,y_1$ and the $x,y$ coordinates of event 2 are $x_2,y_2$ (the subscripts 1 and 2 of the $x$'s and $y$'s are merely identification symbols), then the distance $d$ equals the square root of $(x_2 - x_1)^2 + (y_2 - y_1)^2$, which is just the theorem of Pythagoras.

Suppose now we change our coordinate system, which we can do in three ways: we may shift the coordinate system in the plane to another origin; we may rotate the coordinate system in the plane; we may use curved lines (e.g., circles, ellipses, parabolas) as our $x,y$ lines instead of straight lines. If we shift or rotate our coordinate system the basic algebraic formula (expressed as the Pythagorean theorem) remains unaltered. But if we replace the straight parallel lines by curves, such as concentric circles around the origin and straight lines radiating from the origin (known as a "polar coordinate system,"), the algebraic distance formula becomes more complex, but still, essentially a Pythagorean formula. If, now, we replace both sets of straight lines by curves (e.g., circles and ellipses or parabolas and hyperbolas, two different conic sections) the distance formula becomes very complex and no longer recognizable as the theorem of Pythagoras. In all of these transformations, however, the distance between the two events remains the same. Thus the distance is an "invariant": the same for all observers, regardless of the kinds of coordinate systems (frames of reference) they use.

As a noninvariant we cite the direction from event 1 to event 2 which may change from one coordinate system to another. If we rotate our coordinate system in its plane, all directions change so that direction has no absolute meaning; it is relative to the frame of reference we use to define distance. However, we can introduce an entity associated with direction which is an invariant (the same in all coordinate systems) if we only rotate our frame of reference; this produces the change in direction to any point but it does not change the direction from event 1 to event 2. To define this change we draw two straight lines from our origin *O*: one to event 1 and the other to event 2. The change in direction from event 1 to event 2 is defined by the angle between these two straight lines from our *O*. Indeed, the angle is a measure of the amount we must turn our eye to look first in one direction and then in the other. If we just change our coordinate system from straight lines to curves, we find no change in direction. The change in coordinate does not alter the direction from 1 to 2.

Up to this point we have been discussing distances and the distance formulas for coordinate systems in which we describe events on a flat surface, which is somewhat unrealistic because the surface of the earth, the surface on which we live, is spherical. We cannot lay off straight line coordinate systems on the earth; indeed, as we know, the coordinate system on the earth consists of circles of latitude (small circles) and circles of longitude (great circles), each of which corresponds to a straight line on a plane; that is, the shortest distance between two points. This shortest distance concept plays a very important role in Einstein's theory of gravity (the general theory of relativity). We note that if we limit ourselves to a small enough surface area on the earth, we may treat it as a plane and lay off on it a straight line coordinate system, exactly illustrated by the streets and avenues in a city. We may then use the Pythagorean formula to calculate the distance between two points on this restricted surface. But if we consider the distance between two cities (e.g., New York and Philadelphia) we cannot express this distance as a simple formula in which only the latitudes and longitudes of the two cities appear. In dealing with distances between widely separated points on a spherical surface such as the earth's surface, we must measure that distance along the shortest path connecting these two points. On the earth (or any sphere) this shortest distance is the arc of the great circle (one and only one such circle for a given arc exists and its radius equals that of the earth) that connects these two points.

We have devoted these few pages to a discussion of this purely geometrical aspect of distances and coordinate systems to emphasize the purely mathematical aspects of the relativity concept. As our discussion here reveals, the relativity concept carries with it the concept of "absolute" or "invariant." As we have seen in pure geometry, certain features (e.g., mathematical distance formulas) of events appear to change as we change our frames of reference. But certain features such as distances and time intervals remain unaltered. These, then, may be called the "absolutes" or "invariants" of the geometry.

Einstein's theory of relativity goes beyond the invariants of pure mathematics and deals with the invariants and the relativity of the laws of nature and how these laws are to be formulated so that they are the same in all frames of reference. But in the theory of relativity the concept of invariance refers not to different frames of reference that are at rest with respect to the phenomena being studied but instead frames of reference that are moving with respect to such phenomena and with respect to each other.

Because we may have two kinds of motion and two kinds of frames of reference—constant velocity or accelerated motion—we must study separately the questions of relativity and invariance with respect to these two kinds of frames of reference. If we limit ourselves to frames moving with constant velocity (called inertial frames) we have what Einstein called the "special theory of relativity." The special theory rests on the basic assumption that the laws of nature (of physics) must be "invariant" to a transformation from one inertial frame of reference to any other inertial frame. We can best explain the significance of this concept by presenting it in a somewhat different way. We consider any number of physicists, each in his own inertial frame, that is, each moving with a definite speed along some particular straight line. Because no absolute frame of reference exists in the universe, only relative motion has any physical meaning. This means that any one of these physicists may assume his frame to be at rest so that the velocity he assigns to any one of the other physicists is not absolute velocity but relative velocity (relative to his frame). The basic task of each of these physicists is to find or deduce those measurable statements about the universe that are the same for all these inertial observers. These are the laws of nature. This principle of invariance provides us with a very powerful analytical tool which can separate the false statements from the truth about the universe. We shall see later that this "sieve" allows us to sort out and isolate not only the true laws of motion and the true law of gravity but also the correct space–time concepts.

We have spoken here about comparing the "laws" discovered by any one observer with those discovered by any of the other observers, and accepting only those "laws" as true laws if they do not change as expressed by these various observers. Now, offhand, it seems that this way of gleaning the "seed" from the "chaff" in science is a hopeless task, for it seems to require examining and evaluating the observations and conclusions of every possible observer whose velocity differs from that of any other observer. But this is not so because all we need to do is formulate the laws so that they do not depend on the velocity of the observer. If we can do this, then the laws will be the same for all observers and we shall have achieved our goal.

But what is to guide us in this task? The goal of finding the universal laws goes back to Galileo and Newton, who were well aware of the need to express laws that are independent of the influence of the observer. Newton was convinced that he had discovered such laws in the laws of motion and the law of gravity. Moreover, he was also convinced that these laws are in complete accordance with the concepts of absolute space and absolute time. This is true for all observers who are in frames of reference that are at rest with respect to the phenomena being studied. Newton and Galileo went beyond that point, however, and accepted the idea that this statement is true even if the observers are moving with respect to these phenomena. They did not believe that the laws had to be velocity-independent. Why, then, did Einstein depart from this "reasonable, commonsense" point of view, and how did he discover how the laws were to be made velocity-independent? He arrived at his discovery by studying Maxwell's electromagnetic theory of light, noting that in Maxwell's wave equation for the propagation of a beam of light in a vacuum, the speed $c$ of the wave appears simply as a mathematical consequence of the properties of the electromagnetic field and nothing else. To Einstein, who was then a very young man (about 20 years of age), this meant that $c$, the speed of light, is a universal constant of nature—like Newton's universal constant of gravity $G$ and Planck's constant of action $h$. Einstein then interpreted this idea as meaning that the speed of light is the same for all observers, regardless of their relative motions.

This is the only conclusion Einstein could have drawn from Maxwell's theory, for Maxwell's electromagnetic wave equation contains nothing about the motion of the observer who is measuring the speed of a beam of light passing through his frame of reference. Maxwell, himself, was not entirely happy about this point and, therefore, introduced the idea that all of space is permeated by a universal stationary medium which propagates the electromagnetic waves and that the speed $c$ is relative to this medium. Maxwell then suggested that one might test this ether hypothesis by measuring the speed of light on the earth at different times of the year. As the direction of motion of the earth around the sun varies from moment to moment, its direction of motion with respect to the postulated "ether" would also change, and this would then show up as differences in the speed of light moving in a definite direction. At certain times of the year the earth would be moving in the same direction as that of the light and 6 months later the earth would be moving in a direction opposite to that of the light. In the first case the measured speed of the light beam should be $c - v$ where $v$ is the speed of the earth relative to the "ether" and 6 months later the speed of light should be $c + v$.

In 1887, the American physicist Albert Michelson, the first American to win the Nobel Prize in physics, devised an ingenious experiment to test this theory. Using a very accurate device called the "Michelson interferometer," Michelson and a colleague, Edward Morley, measured the speed of light in many different positions of the earth in its orbit (different directions of motion) and found that the measured speed of light was the same. This, one of the most surprising experimental discoveries ever made, confronted physicists with a most disturbing conundrum, for not only did it defy "common sense," but also the most elementary law of arithmetic, that of the addition of physical quantities (speeds). One might ordinarily have dismissed this "aberrant" behavior of light as stemming from an experimental error, but this criticism had to be discarded very quickly owing to the fantastic accuracy of the Michelson interferometer.

Accepting the accuracy of the Michelson–Morley experiment, the physicists at the beginning of the twentieth century tried desperately to find an acceptable explanation for the "negative result" of the Michelson–Morley experiment. "Acceptable" here means an explanation that was not based on the rejection of any of the accepted laws of physics or the accepted concepts of space and time, which had been set by Galileo and Newton. The most ingenious explanation was presented by the Dutch physicist H.A. Lorentz, one of the great physicists of the early twentieth century, and the Irish physicist George Fitzgerald, who proposed the hypothesis that the dimension of the Michelson interferometer in the direction of its motion was diminished (contracted) owing to its own motion, by just the right amount to give the Michelson–Morley observations, that is, to give the same speed of light in all directions, regardless of the direction of the earth's motion.

Einstein criticized this explanation as ad hoc and artificial, arguing that the correct conclusion to be drawn from the Michelson–Morley experiment is that the speed of light is the same for all observers and that the concept of the ether as an absolute frame of rest in the universe must be rejected. Einstein went further to point out that the constancy of the speed of light means that we must also discard the Newtonian–Galilean concept of absolute space and absolute time. To demonstrate that the concept of absolute time is untenable Einstein presented the first of his famous thought experiments; this experiment, if performed, would prove that the simultaneity of two events separated in space has no absolute meaning. The experiment involves two observers, one fixed next to railroad tracks, and the other in a train moving past the fixed observer. We have the observer on the railroad siding exactly midway between two train windows which are separated from each other by exactly the distance between two fixed trees on the siding. We now have two bolts of lightning strike the two trees simultaneously as seen by the fixed observer. He concluded that these two bolts struck the two trees simultaneously because he received the light from the two strikes at exactly the same time.

What about the moving observer on the train, which is moving so smoothly that this observer does not know he is moving? To answer this question, we impose two other conditions on this experiment: (1) the lightning bolts strike the two trees at the exact moment when the moving observer and the fixed observer are opposite each other (each one at that moment is halfway between the two trees). (2) each tree, at the moment it is struck, falls against the window next to it in the train and smashes it so that the fixed observer concludes that the two windows were smashed simultaneously. With these conditions, the man in the train concludes that the two windows were not smashed simultaneously. He receives the light from the smashed forward window (the one further down the track in the direction of the train's motion) earlier than from the smashed back window because, without knowing it, he is moving toward the light coming to him from the forward window and away from the light coming from the back window. He therefore concludes that the front window was smashed before the back window. This experiment destroys the concept of absolute simultaneity and therefore of absolute time.

We now consider another thought experiment which shows that the constancy of the speed of light for all observers destroys the concept of absolute space, that is, the concept of absolute distance between events. We again consider two observers, one fixed in a railroad siding and the other in a moving train moving to the right as seen by the fixed observer and ask each one to measure the speed of a beam of light moving parallel to the railroad tracks, that is, parallel to the motion of the train. We give each observer an imaginary rod 186,000 miles long, held parallel to the tracks, and give them identical clocks. We now picture the beam of light striking the left ends of the two rods simultaneously, when these two ends exactly coincide. The fixed observer notes that the beam of light reaches the right end of his rod in exactly one second and concludes that the speed of light, as he measures it, is exactly 186,000 miles per second.

What about the moving observer in the train? Suppose that the train and his 186,000 miles long rod are moving at 185,000 miles per second so that the right end of his rod, as estimated by the fixed observer should have advanced 185,000 miles in one second. If the fixed observer, not knowing that the speed of light is constant, were now asked to estimate what speed of light the moving observer would obtain (i.e., how long the moving observer would find that it took the beam to reach the right end of his rod) he would say "more than 1 second." Imagine his great surprise, then, on being told by the moving observer that the beam traversed his 186,000 mile long rod in exactly 1 second. The fixed observer, refusing to accept this statement offhand, decides to repeat the experiment and does so, keeping very careful track of the events in the moving train. He is then amazed to discover that the moving rod shrinks and the moving clock slows down by just the right amount (no less, no more) to give the same speed of light as measured by the moving observer as he, the fixed observer, found. In other words, distances and time intervals as measured in every frame of reference adjust themselves to give the same speed of light. This is the essence of the "special theory of relativity." This theory, now universally accepted as correct in every detail, has replaced the three-dimensional Newtonian dynamics by the four-dimensional Einsteinian dynamics which merges space and time into a single space-time manifold.

This was so revolutionary and remarkable a development, with important applications to our understanding of every phase of nature from the structure of the nucleus of an atom to the structures of stars, galaxies, and the universe itself, that we discuss it in somewhat greater detail to show that the "theory of relativity" is really a theory of "absolutes," that is, of those aspects of natural phenomena that remain the same (are invariant) when we pass from one moving frame to another moving frame. This change is called a transformation of coordinates, a concept which plays a very important role in the theory of relativity. To compare the Newtonian dynamics operating in absolute three-dimensional space and in the "absolute" flow of time with the Einstein four-dimensional space–time dynamics, we first consider a single event in its simplest form: a moving particle at a definite position along its path at a definite time. In Newtonian dynamics this "event" is described in any particular frame (a three-dimensional coordinate system) by specifying its the position coordinates $x$, $y$, $z$ in that frame (coordinate system) and noting the time $t$ when it is observed; the three position coordinates $x$, $y$, $z$ and the time $t$ are specified independently of each other. This appeared to be correct to Newton and Newtonians, in general, as it does, even today, to most people. The reason for this is quite simple: we can shift the position of a particle from point to point in any direction, but we cannot "shift" it in time; we have no control of the flow of time. But to Einstein this did not mean that the "flow" of time is the same in all frames of reference. Einstein argued that space and time cannot be treated independently of each other and that the time $t$ of an event must appear "on an equal footing" with the position in space.

This point of view was developed in its all-embracing mathematical form by the early twentieth century mathematician Hermann Minkowski, and is now the standard four-dimensional geometry of relativity theory. Whereas the geometry of Newtonian dynamics is three-dimensional in which dynamic quantities such as position, velocity, force, etc., are presented as triplets (so-called three-vectors), in Einsteinian geometry all dynamical quantities are presented as four-vectors, each consisting of three spatial components and one time component. We illustrate this distinction by going back to the consideration of the simple event of a particle at a definite point in space at a definite time $t$. In Minkowskian four-dimensional geometry, this is specified by three spatial dimensions (coordinates) and one "dimension" in time. To do this properly so that the time $t$ appears as a fourth dimension, similar to the three spatial dimensions $x$, $y$, $z$, we multiply $t$ by $c$—the speed of light—the universal or absolute speed. We also multiply it by $i$, the square root of –1, an imaginary number that can be represented as $i$. The time $t$ now assumes its role in Minkowskian geometry as the imaginary "distance" $ict$. Thus the four space-time coordinates of a particle are $x$, $y$, $z$, $ict$. We pass from Newtonian three-dimensional vector dynamics to Einsteinian four-dimensional geometry by appending to every spatial Newtonian triplet a fourth time component to obtain what is known as a four-vector. As mentioned in the previous sentence, the three-vector $x$, $y$, $z$ of Newtonian physics becomes the four-vector $x$, $y$, $z$, $ict$ of Einsteinian physics. We see from this change that the "assumed to be absolute" spatial distances that define Newtonian dynamics become the true "absolute" four-dimensional space–time intervals that define Einsteinian dynamics.

We consider now the transformation of one other Newtonian three-vector into an Einsteinian four-vector by the addition of a "time" component to its three spatial dimensions because this four-vector—called the "momentum-energy" four-vector—is of extreme importance in our understanding of the structure of a star. To this end we return to the definition of the momentum and the kinetic energy of a moving particle which will lead us to one of the most important four-vectors in science. The kinetic energy $E$ of a particle of mass $m$ moving with the velocity $v$ is defined in Newtonian dynamics as 0.5 of $mv^2$, which is what is called a "scalar" in dynamics since it has no direction. This means that the kinetic energy of a particle is the same regardless of the direction of its motion. The momentum $p$ of this particle is now defined as $mv$, which is not a scalar, but a vector because $v$ can have three different directions $x$, $y$, and $z$ in a Cartesian coordinate system: $v_x$, $v_y$, $v_z$. We thus have the momentum three-vector $mv_x$, $mv_y$, $mv_z$, which we label as $p_x$, $p_y$, and $p_z$, respectively. We now obtain the momentum-energy four-vector of relativity dynamics by combining the three space components $p_x$, $p_y$, $p_z$ of $p$ with $E/c$, the kinetic energy divided by the speed of light $c$, which we may call the time component of the momentum: $p_x$, $p_y$, $p_z$, $iE/c$. As we now show, we obtain from this four-vector Einstein's mass–energy formula $E = mc^2$, the most celebrated and, certainly, the most important discovery in the history of science.

The most remarkable feature of this striking relationship between energy and mass is that it inevitably follows from the constancy of the speed of light for all observers and the expression of absolute dynamical entities as four-vectors instead of the three-vectors of Newtonian dynamics. To see how this comes about we note first that the distance $r$ of an event from an observer fixed at the origin of a Cartesian coordinate system is given by the extended theorem of Pythagoras as $r^2 = x^2 + y^2 + z^2$, which, in Newtonian mechanics, is the same for every observer, however he may be moving, if he is at the origin of the fixed observer's coordinate system at the time $t$ of the event. But this is not so in relativity dynamics, in which one must apply the extended Pythagorean theorem to the four-vector $x$, $y$, $z$, $ict$ to obtain the square of the space–time interval $s^2 = x^2 + y^2 + z^2 + (ict)^2 = x^2 + y^2 + z^2 - c^2t^2$ or $s^2 = r^2 - c^2t^2$. This measured quantity is the same for every observer coincident at the time of the event with the fixed observer. Each such observer measures a different $r$ and $t$ (depending on his relative velocity) but finds $s$ to be the same, so that $s$—the space–time interval—is an invariant.

We must apply this same analysis to the relationship between every Newtonian three-vector and its Einsteinian four-vector counterpart. Here we are interested primarily in the Einsteinian momentum–energy four-vector obtained from the separate Newtonian momentum and energy. The four-vector is $p_x$, $p_y$, $p_z$, $iE/c$. Applying the extended Pythagorean theorem to this formula we obtain the invariant $p_x^2 + p_y^2 + p_z^2 - E^2/c^2$ or $p^2 - E^2/c^2$. This is the square of an invariant four-vector momentum which is the same for all observers, and therefore must be the product of an invariant mass and an invariant velocity. We apply this to a particle of rest mass $m_0$ (the mass as measured by the observer when he is at rest with respect to the particle) which is the same for all observers by definition. Thus the momentum $m_0c$ is an invariant for all observers. Because this is the only "invariant" we can construct, the negative of its square invariant is $p^2 - E^2/c^2$; this enables us to derive several equations which allow us to see that even if a particle of rest mass $m_0$ is at rest (if its momentum $p$ is zero, i.e., no kinetic energy) its energy $E$ is not zero. Indeed, we then obtain $E = m_0c^2$: Einstein's famous relationship between mass and energy. We now note that another famous deduction from this equation was made by the British physicist Paul Dirac in 1928. To obtain the energy $E$ from the equation for $E^2$ we must take the square root of both sides of the equation, but in so doing, as emphasized by Dirac, we must take both the positive and negative square roots. We have no trouble understanding the positive square root for it is the energy of a particle of rest mass $m_0$ moving with momentum $p$ in empty space. But what of the negative square root? What does it mean? The answer requires a brief explanation.

When Dirac first proposed that the positive and negative square roots be used in the study of the relativistic motions of an electron, physicists were inclined to disregard the negative square root as a physically meaningless entity for they wondered how a particle can have negative—less than zero—kinetic energy. In 1929 Dirac suggested that these negative energy states of an electron represent real particles, which are the exact counterparts, with opposite electron charge, of positive energy particles. Dirac thus proposed the revolutionary concept that positively charged particles corresponding to electrons, later called "positrons," each with the exact mass of the electron, exist. In 1932, the American experimental physicist Carl Anderson announced his discovery of positrons as components of cosmic rays. Anderson and other physicists later discovered that positrons have only a fleeting existence, disappearing suddenly on combining with an electron. This we now describe as an "annihilation process" in which the positron and electron annihilate each other. Dirac pictured this process as though the positron were a "hole" in the vacuum into which the electron falls, canceling the mass $m$ and the electric charge of the positron. The electron fills the "hole" so that the "hole" (the positron) disappears, as does the electron. This means that an amount of mass $2m_0c^2$ (the sum of the mass of the positron and that of the electron) disappears, and is replaced by two photons (gamma rays) whose total energy equals $2m_0c^2$, in complete agreement with Einstein's mass–energy relationship. All these relativistic phenomena play a most important role in our understanding of astronomical phenomena, in particular, in our analysis of stellar interiors (astrophysics).

We complete our discussion of special relativity by describing how such entities as mass, distance, and time intervals (the slowing down of clocks) change from one observer to another observer owing to the difference in the velocities of these observers. We can see immediately how the mass changes from the mass–energy momentum equation we discussed above. If $m$ is the moving mass of a particle (moving relative to an observer, which means that it does not matter whether we say that the particle is at rest and the observer is moving or vice versa), $m_0$ is the rest mass of the particle and $v$ is the relative velocity of the particle and observer, so that $E = mc^2$, $p = mv$, then the equation $E^2 = c^2p^2 + m_0^2c^4$ becomes $m^2c^4 = c^2m^2v^2 + m_0^2c^4$ or $m = m_0/\sqrt{1 - v^2/c^2}$. The moving mass $m$ of a particle increases without limit as its relative velocity $v$ approaches the speed of light $c$. This, of course, means that no body can move at the speed of light because its mass and its energy would then become infinite, which is a physical impossibility. As we shall see, this is of great importance in the theory of the expansion of the universe and in the theory and structure of "black holes."

We now turn to the variation of distances and time intervals as we transform from one inertial frame of reference to another moving with a different speed. We consider two different events $A$ and $B$ separated in space by a distance $r$ and in time by the interval $t$, which, as we have already seen, gives us the four-vector $(r, ict)$. As the square of the this four-vector is $r^2 - c^2t^2$, this, as we previously noted, is invariant (the same for all observers) but $r$ and $t$

separately change from observer to observer. Taking $r$ first we find that the change in $r$, for any given observer, depends on the direction of the velocity of the observer; an observer moving at right angles to $r$ finds no change in its magnitude, but an observer moving at the speed $v$ parallel to $r$ find that it is reduced by the factor $\sqrt{1 - v^2/c^2}$, so that $r$ becomes $r\sqrt{1 - v^2/c^2}$, which reduces to zero as $v$ approaches $c$.

Considering the time interval $t$, as measured by a clock at rest with respect to the two events $A$ and $B$, we find that all observers moving at speed $v$ with respect to these events (with respect to the fixed clock) measure a longer time interval. Put differently, the clocks moving along with the moving observers run faster than the fixed clock which to all the moving observers runs slower than their clocks. If $t$ is a time interval measured by a moving observer's clock moving with a speed $v$ with respect to the fixed clock, this moving observer finds that the fixed clock shows that an interval $t\sqrt{1 - v^2/c^2}$ has elapsed. As the speed of the moving observer approaches the speed of light, the fixed clock slows down more and more until, at the speed of light, it shows no passage of time at all so that it is as though it had stopped altogether. All of this does not depend on whether we picture the observer as moving and the clock as fixed or vice versa. All that matters is the relative velocity of clock and observer. Considering the order (in time) in which the two events $A$ and $B$ occur we cannot define this order in any absolute way. To one observer $A$ and $B$ may appear to be simultaneous, to another observer $A$ may appear to precede $B$, and to a third observer $B$ may appear to precede $A$.

Finally, we describe what has come to be known as the "twin paradox." One of two 18-year-old brothers (twins) moves off into space at very nearly the speed of light toward the star Sirius which is about 9 light years distant from the earth. The brother left behind keeps track of his brother's journey, aging 18 years from the moment his brother left the earth to the moment he returns. Now 36 years old, this "left behind" twin is amazed to find his space-traveling twin as young as when he left. From the point of view of the traveling brother, his youthfulness is easy to explain. Traveling at his very high speed he finds that the distance to Sirius is no more than a few hundred miles so that his round trip measured by his own clock (let us say by the number of his heartbeats during his trip, which, of course, are normal) he is as young as when he left. The apparent paradox in this set of events stems from the symmetry requirement imposed by the theory of relativity on the observations of the two brothers: all that matters in these events should be the relative velocity of the two brothers. Just as the earth-fixed brother sees the Sirius-bound brother recede at very nearly the speed of light so the Sirius-bound brother sees his earth-fixed brother recede from him at the same speed. Why, then, does the Sirius-bound brother remain young while the other one ages? The reason is that the symmetry between the two brothers is not a total but instead a broken symmetry; the brother traveling to Sirius must turn around after reaching Sirius and this breaks the symmetry so that no paradox is present if the various accelerations are taken into account to which the traveling twin is subject. These accelerations violate the constraint of the special theory of relativity that the frames of reference (the coordinate systems) of the two twins be inertial frames (no acceleration). This constraint is violated by the noninertial frame of the travelling twin.

Einstein, knowing that the special theory of relativity is not the final chapter in the story of space, time, and matter, spent a decade from 1905 to 1915 completing this story in his construction of the general theory of relativity, which is probably the single greatest and boldest creation of the human mind, for it merged physics and geometry. Since the special theory is "special" precisely because it deals only with inertial observers, leaving out accelerated observers, and, hence, the dynamics of bodies in gravitational fields, Einstein understood that the "principle of invariance" must apply to noninertial observers and accelerated systems as well as to inertial observers. This means that all laws of nature must be the same in all frames of reference—inertial and accelerated frames.

No law can be used to prove absolute motion. Offhand, we are tempted to deny this statement because we know from our daily experience that we can discover the difference between uniform motion in a straight line and accelerated motion. If we are in a train moving uniformly in a straight line, we can make no

Albert Einstein (1879–1955) (Courtesy AIP Emilio Segrè Visual Archives)

observation in the train (carry out any experiment) that can tell us that we are moving. But this is not so if the uniform motion changes. We can immediately detect any such changes. If the train suddenly slows down, we are immediately thrown forward, and if the train suddenly speeds up, we are thrown backwards. This seems to deny the possibility of extending the "special theory" to a "general theory" that applies to all moving frames of reference.

But Einstein's vision again came into play here; he showed, in a very ingenious way, that our senses may mislead us in our conclusion about the state of motion of our frame of reference when we are "thrown" forward or "pulled" backward; he pointed out that our physical responses to accelerations cannot be distinguished from our responses to the pull of gravity. This statement by Einstein, known as the principle of equivalence, is nothing more nor less than the generalization of Galileo's discovery of the equivalence of inertial and gravitational mass. Before Galileo's discovery that all bodies in a vacuum (no atmospheric friction) at the same point in a gravitational field (from the same height above a given point on the earth) fall at the same speed, physicists did not know that the gravitational mass and inertial mass of a body are identical.

When we consider this in somewhat more detail we are struck by the similarity between our reaction to gravity and to acceleration, the first resulting from what we may call our "gravitational mass" and the other from our "inertial mass." Note that we speak of the "gravitational mass" of a body as the mass that generates the gravitational field surrounding the body and as the mass that responds to any other gravitational field. In the same way we refer to the inertial mass of a body as the mass which responds to accelerations. Such responses are often referred to as inertial forces, which means that Galileo's discovery proves that gravitational forces cannot be distinguished from inertial forces and vice versa. Einstein stated this as a general principle now called the "principle of equivalence." This is the basis of Einstein's general theory of relativity, also referred to as Einstein's law of gravity.

This principle of equivalence led Einstein to his principle of general invariance: the laws of physics are the same for all frames of reference, regardless of how they are moving. No observation that an observer can make in any frame of reference can lead him to deduce anything about the state of motion of his frame. If an observer, in a frame moving in any arbitrary way, objects to this, stating that he "knows" his frame is accelerating because of his body's response to this "acceleration," Einstein would state that the "principle of equivalence" as proved by Galileo's observations denies this observer's inference about the frame's acceleration. Einstein would point out that an observer would experience the same sensation if his frame were at rest or moving uniformly in a gravitational field. This impossibility of distinguishing between inertial and gravitational forces led Einstein to his geometric theory of gravity.

We consider Einstein's famous elevator thought experiment in which an observer makes any observations in an accelerated elevator far from any body but with the same vertical acceleration away from the earth that a freely falling body has on the surface of the earth. The observer in this elevator, too far from the earth to feel the earth's acceleration, nevertheless experiences every phenomenon that he would if he were standing on the earth.

Imagine now a body thrown horizontally across the elevator (at right angles to the acceleration of the elevator). To the observer in the elevator, the body falls to the floor of the elevator just as it would if thrown horizontally on the surface of the earth. To an observer in an inertial frame of reference (constant velocity) outside the elevator the body is moving in a straight line. Thus the concept of a straight line depends on one's frame of reference or equivalently whether one is outside a gravitational field or is in such a field. Einstein therefore argued that gravity is equivalent to a distortion (bending) of space–time so that non-Euclidean geometry must replace Euclidean geometry in describing gravity. This has had an enormous influence on our understanding of cosmology. If no masses were present in the universe, the geometry of the universe would be Euclidean. But the presence of stars, galaxies, and huge dust clouds causes a curvature of space–time, which we explore further in the chapter on cosmology.

Because the presence of mass in space distorts space-time, the Minkowskian expression $r^2 - c^2t^2$ for the invariant special relativistic interval between two events is no longer valid; it must be replaced by a general relativistic space–time interval which takes into account the non-Euclidean character of the geometry of space–time produced by mass. This is equivalent to the non-Euclidean formula we use for the shortest distance between two points on the earth's surface owing to the curvature of this surface. To find the correct formula for the space–time interval in a region of space in which a measurable gravitational field is present, we must be able to compare rods and clocks in a gravitational field with these same rods and clocks in field–free space. Considering a point at a distance $r$ from the center of the sun, we reformulate our question: how can an observer, pictured as fixed at this distance from the center of the sun determine how the length of a rod at his point changes compared with its length when the observer sees it in field-free space? The same question applies to a clock. Offhand, it may appear that such comparisons are impossible; how, it may be asked, can an observer, at a fixed point in a gravitational field, simultaneously compare the length of a rod at that point with its length when it is not in a gravitational field? The answer to this question is simpler than it appears for all we need do is apply the principle of equivalence to the observer. We picture him as falling freely toward the sun (in a freely falling elevator). His speed squared at the distance $r$ is just $2GM/r$ where $M$ is the sun's mass. To this observer a rod held fixed and parallel to his direction of fall is moving past him at the speed $\sqrt{2GM/r}$ and therefore appears contracted by the amount $\sqrt{1 - 2GM/c^2r}$ by the Einstein–Lorentz law of the contraction of moving bodies. By the same law clocks slow down in a gravitational field. These results were first deduced theoretically in 1917 by the famous German theoretical astronomer Karl Schwarzschild from Einstein's basic equations. This was about 2 years after Einstein had announced his general theory of relativity.

Some of the most important advances in modern-day astronomy spring from the general theory of relativity. The radiation emitted from an atom in a strong gravitational field as on the surface of a "white dwarf," a very dense, small, but massive sphere, sends out light that is redder than the light it sends out when it is not in a strong gravitational field. The reason for this is that a vibrating atom is essentially a clock and therefore vibrates more slowly in a strong gravitational field than in a region of low gravity. This reddening effect, called the "Einstein red shift," is important in finding the radii of white dwarfs and other very dense stars. Because the dimensions of a measuring rod are altered in a gravitational field if the rod is placed parallel to the field (e.g., vertically on the sun) but is not altered if it is held at right angles to the field (e.g., horizontally on the surface of the sun), Kepler's third law of planetary motion is altered. This produces what is known as the advance of the perihelion (the point closest to the sun) of the planet's orbit. These two effects (the Einstein red shift and the advance of the perihelion) have been fully confirmed observationally.

A third effect predicted by Einstein's general theory of relativity is the bending of the path of a ray of light passing close to a massive body, such as the sun. Einstein announced his general theory of relativity in 1915 and Karl Schwarzschild presented the first solution of Einstein's equations for a point source of gravity like the sun or for any other star. But not until 1919, when a British expedition led by the British astrophysicist Sir Arthur Eddington, observed a total eclipse of the sun in West Africa, was actual proof of this predicted effect obtained. Einstein's theory predicted that during such an eclipse the paths of light from stars on either side of the sun (as seen from the earth) would be deviated and, therefore, that the positions of such stars would appear to be different from their true positions as viewed at nighttime. This prediction was fully confirmed and the angle of deviation of the rays of light agreed almost exactly with that calculated from Einstein's theory.

The announcement of this success of Einstein's general theory of relativity struck the world like a thunderbolt for it placed Einstein with Newton as the preeminent scientists of all time. Suddenly the name Einstein became a household word, while his theory of relativity remained a great mystery, mentioned with reverence and awe but much too esoteric to be understood except by a few chosen ones. Since then the cloak of mystery that surrounded this theory has been swept away and, if properly taught, it can be understood by high school students. In the chapter that follows we shall see that most of the observed phenomena in the universe cannot be understood without the application of this theory.

CHAPTER 15

# The Origin and Development of Astrophysics

*Silently one by one, in the infinite meadows of heaven*
*Blossomed the lovely stars, the forget-me-nots of the angels.*

—HENRY WADSWORTH LONGFELLOW

The two great nonclassical physical theories, the quantum theory and the theory of relativity, that led to the nuclear model of the atom, also led to astrophysics, and to a rational cosmology. Astronomy began to change from a discipline concerned with the study of the positions, the motions, and the clustering of stars to the study of the structure of stars, to "astrophysics," which is the application of the laws of physics to the analysis of stellar structure. The aim here was, and still is, to lay down the blueprint for the theoretical structure of a model of a star. Because stars differ considerably in their physical features, a single stellar model cannot describe all kinds of stars. Astronomers, clearly recognizing this, understood the need to find those common physical characteristics that can be represented by a single stellar model. To this end they began to classify stars into groups with the idea that all the stars in a single group can be described by a single stellar model. Thus they reasoned they could construct a correct model of the sun and that that model would describe all stars like the sun. The classification of stars had already been begun on a simple scale with the introduction of the magnitude (apparent brightness) scale by Hipparchus and Ptolemy. This classification of the apparent brightnesses of stars made precise by the British astronomer Pogson in 1850 led, with the measurements of stellar parallaxes (distances), to the classification of stars according to their luminosities (the luminosity of a star is the actual radiation emitted in all directions per second by the star). This is one of the most important classifications of the intrinsic properties of stars, and ranks with the spectral classification introduced by Fraunhofer. With the luminosity and spectral classifications of stars introduced first as separate, unrelated classifications, it occurred to the astronomers interested in astrophysics that these classifications might be correlated to each other in some way, and a search of such correlations was begun. We discuss these activities later in this chapter and turn our attention now to the rapid progress then being made in improving the accuracy of observational astronomy with improved optical and photographic procedures.

As we mentioned in a previous chapter, this emphasis on accuracy and precisely organized procedures in the observatory was instituted as an inviolable principle by the royal astronomer Sir George Biddell Airy, who dominated astronomical research in England from 1845 until his retirement in 1881. During this period he organized the Greenwich Observatory and set up six new science and technological institutes in England. The research activities of these institutes ranged from chronometry to astrophotography. Concerned as he was with precision in measuring stellar positions, Airy discovered as previously noted an optical limit to the precision of the telescope itself which is now known as the Airy disk. This disk, the image formed on a photographic plate by the objective (front lens) of a telescope, not a true image of the disk of the star, which is far smaller than the Airy disk. Because the star, at its great distance from the earth, appears like a point of light, its image in the telescope should also appear as a point on a photographic plate. But its image is a disk. The reason for this difference is the diffraction of light passing through an aperture. Because the light is a wave, it cannot be concentrated in a point, regardless of the accuracy of the telescope. When a wave passes through an opening, different parts of the wave-front are deflected differently (diffraction) so that these different parts cannot be focused at the same point. This produces a disk image (the Airy disk).

Airy, recognizing this limitation which the wave structure of light imposes on the accuracy of any telescope, discovered that the size of this disk is reduced if the diameter of the objective of the telescope is increased and the wavelength of the light coming from the sun is decreased. Because astronomers have no control over the wavelengths of the various rays of light from a star, they emphasized the importance of using very large aperature telescopes in studying stars to increase the precision of their measurements. This marked the beginning of the age of large telescopes with the American observatories assuming the major roles in these developments.

Although astronomers were beginning to branch out into the new, inviting branches of astronomy, the emphasis was still on the time-honored activity of the measurements of stellar positions. This had one very important consequence: it led to the rapid internationalization of astronomy, for it was obvious to all working astronomers that the task of cataloging the positions of even a small fraction of stars in our galaxy would require the work of generations of astronomers and is well beyond the capacity of any single observatory. Friedrich Argelander, who became director of the Bonn Observatory in Germany in 1837, the last of the individual workers in this phase of astronomy, completed the Argelander catalog in which the positions and magnitudes of 324,000 stars are listed. Fully aware of the futility of one man's attempt to proceed very far in this important activity, Argelander founded the Astronomische Gesellschaft specifically for extending this work in collaboration with different observatories in many countries. Working without photography (naked-eye astronomy) this association enlarged Argelander's work considerably.

Naked-eye astronomy gave way to photography and more than 22,000 photographic plates were exposed to the entire sky during this early period of astronomical photography. This was important not only for positional astronomy but also for measuring the motions (relative to the sun) of nearby stars. This was done by comparing photographs of the same groups of star plates exposed to the same part of the sky taken at different times. This led to the very important discovery that certain stars in a given region of the sky are moving together (they are all moving in the same direction). Such a collection of stars, now called a "local" or "galactic" cluster, is very important in the study of the evolution of stars. The Hyades cluster in Taurus is an excellent example of a local stellar cluster. The comparison of photographs of the stars in the Ursa Major constellation shows that all the stars in the Big Dipper are not moving together and that this constellation will lose its present shape 10,000 years from now. The study of the photographs of stars taken at different times generated a whole new branch of astronomy called "stellar motions," which is of great importance in the study of galactic structure.

The improved and enlarged telescopes had their greatest impact and influence, however, on the rapid growth of astrophysics. In 1859 the German physicist Gustav Kirchhoff, an expert in spectroscopy, began to study the solar Fraunhofer (absorption) lines, listing 2000 such lines. Studying the solar spectrum became very popular among astronomers, who were turning their attention from positional astronomy to astrophysics; in Sweden the physicist Ångstrom listed 6000 solar Fraunhofer lines and the American physicist Rowland listed 21,000 solar lines. They found that these lines continue into the infrared and the ultraviolet parts of the spectrum. The infrared lines were measured very carefully and accurately by the American astronomer Samuel P. Langley, who had invented a very sensitive thermometer with which he measured the heat associated with each of these lines.

In all of these remarkable advances in astronomy, photography played a decisive role with glass photographic plates becoming the repositories of all astronomical knowledge. The many advantages of such repositories over the naked eye are immediately obvious regardless of the area of astronomy being considered. Photographic plates are permanent records, which can always be called upon to refresh our knowledge. Moreover, in measuring the brightnesses of celestial objects, we can expose the photographic plate to the radiation from an object as long as may be necessary to obtain a photographic impression on the plate. The retina of the eye cannot accumulate optical information and store it. This means that the photographic plate can register images of very distant, faint objects. Photography thus opened up a whole new branch of astronomy, which, in a relatively short time, led to galactic astronomy, and, ultimately, to cosmology, perhaps the most exciting and challenging branch of astronomy today.

Returning now to the classification of stars, we see that spectral photography revolutionized the study of stellar spectra. Comparing the photographs of the spectra of different stars led quite naturally to stellar spectral classification with the Fraunhofer dark (absorption) lines constituting the base of this classification. By arranging the spectral photographs of properly chosen stars in a vertical linear sequence, astronomers easily demonstrated the gradual variations from the sharp lines of the well-known spectra of different atoms in the periodic table of the chemical elements to the spectral bands of molecules. To understand the stellar significance of these spectral variations in the radiation from one type of star to a neighboring type, we must keep in mind that the Fraunhofer lines (absorption lines) result (as noted by Fraunhofer and Kirchhoff) from the interactions of the continuous radiation from the hot photosphere (surface) of a star with the atoms of its much cooler atmosphere. These lines give us a picture of the chemistry of and the physical conditions in the star's atmosphere. One might then be inclined to interpret the spectral variations as indications or proof of the differences in the chemical compositions of different stellar atmospheres. But this is a wrong interpretation, for, as we shall see later, all stars are chemically the same. The spectral variations arise, then, from the different physical conditions (primarily the temperatures of the various stellar photospheres).

To accept this idea, we point out that the absorption lines of a given type of atom (e.g., hydrogen or helium) arise only if the radiation passing through an atmosphere consisting of these atoms is energetic enough to cause the valence electrons in this atom to jump from orbit to orbit, and thus to absorb and reemit this radiation. An absorption line is thus actually produced by what physicists call a "scattering process." We can perhaps best understand this process by drawing an analogy between the scattering of sound and the scattering of light. We therefore consider a collection of tuning forks, which, when struck, emit the note A and which surround in all directions a central source of sound that emits all notes. If the tuning forks were not present, an observer in any direction from the central source would find that all notes in the sound are equally intense. But with the tuning forks distributed uniformly in a sphere around the source, the observer would find that the note A is weaker than the other notes. The reason is that a tuning fork at any point vibrates in response to the A vibrations in the sound and then reemits these vibrations in all directions so that the note A in any one direction is weakened because most of the sound that hits the tuning fork leaves the fork in all directions. This is called scattering. The atoms in the atmosphere of stars scatter light of different colors and this scattering process produces the observed absorption lines.

To see how the temperatures of the photospheres come into play here we note that this temperature determines the intensities of the various wavelengths (or frequencies) of the light in the continuous radiation emitted from the stellar photospheres into the stellar atmospheres. Each atom in the atmosphere scatters the radiation of a set of discrete wavelengths that correspond exactly to the set of discrete wavelengths that this atom would emit if it were excited energetically in some way. Thus each atom behaves like a collection of tuning forks in the analogy with sound we drew previously.

Owing to this atomic scattering of radiation, the spectrum of this radiation produced by a spectroscope in the astronomical observatory is a continuous spectrum threaded by dark lines (the Fraunhofer absorption lines). These lines are used by astronomers to classify stars or arrange stars into "spectral classes" or "types." The atoms associated with these lines (the atoms in the stellar atmospheres that scatter the radiation to produce a particular set of absorption lines) can easily be determined by placing the bright line emission spectrum of this atom next to the stellar absorption spectrum. If a one-to-one correspondence exists between the bright emission lines in the atom's spectrum and a definite set of the dark lines in the star's Fraunhofer's spectrum, we conclude that this species of atom is present in the star's atmosphere. In this way the chemical composition of stellar atmospheres were (and still are being) determined.

In establishing a stellar spectral classification scheme, which was begun by Fraunhofer in 1823 (who classified about 600 solar absorption lines) and carried on by Sir William Higgins in 1864 (who first actually identified some of the solar absorption lines with those of known terrestrial chemical elements), the astronomers who did so used only a tiny portion of the thousands of absorption lines in the spectra of normal stars. The absorption lines they chose are the most prominent lines that can be easily identified as the lines of well-known and definite atoms. This type of classification was begun in 1863 by the Jesuit astronomer Angelo Secchi, who arranged the stars in four general groups. This scheme was modified in time and extended so that today astronomers work with seven principal spectral classes.

Because the most abundant atoms in the universe are those of hydrogen and helium, we find, as expected, that the lines of these atoms are most important in establishing spectral classification schemes. Because we find that the intensities of these lines vary among different stars, we might be misled into believing that these differences mean that the chemical compositions of the atmospheres of these stars differ. But this is not so; everything we know about the chemistry of the universe teaches us that all stars can be divided into two broad, chemically different groups which astronomers have named population I and population II stars. All stars like the sun, called population I stars, are chemically similar. Stars much older and redder than the sun are called population II stars; all population II stars are chemically similar. Because the spectra of population I stars may differ from one such star to another, we conclude that this difference is produced by a physical property (physical parameter) not by a chemical variation. We now know that this physical parameter is the surface (photosphere) temperature of the star.

To see how surface temperature plays this dominant role in the production of stellar spectra we recall that the valence electrons in atoms (the outer electrons) radiate and absorb (scatter) radiation to produce the bright line emission and dark line absorption spectra; we can therefore describe how these spectra arise by using the simple Bohr model of the atom, according to which the valence electrons circle the nucleus of the atom in elliptical orbits. Here we limit our discussion to the hydrogen atom and the helium atom. Because the hydrogen atom, consisting of a single proton as its nucleus, and a single electron revolving around this proton, is the simplest atom, and because its spectrum has been studied in greater detail than that of any other atom, we use its stellar absorption spectrum as the best introduction to stellar spectroscopy.

As the hydrogen absorption spectrum can be produced in stellar atmospheres only if the electron in the hydrogen atom responds to the photosphere radiation passing through the atmosphere, this radiation must be energetic enough to affect the lone hydrogen electron. To see what is involved we note that the electron in an undisturbed hydrogen atom revolves in an orbit about one hundred millionths of a centimeter from the proton. This is called the ground state of the atom and the atom is then said to be "unexcited." The absorption of radiation occurs if the electron absorbs a photon from the sun's photosphere which is energetic enough (high enough frequency) to lift this electron to a more distant, permissible orbit. The more distant this orbit is from the proton, the more energetic the photon must be (the higher its frequency). These frequencies of the photons correspond to ultraviolet photons and the array of absorption lines produced by these photons constitute what physicists call the Lyman series of absorption lines, which lie in the usually unobservable ultraviolet part of the hydrogen spectrum. Obviously, if the temperature of the star's photosphere is not high enough (at least 11,000 degrees kelvin) to produce these energetic photons, the Lyman hydrogen lines are not present in such a star's spectrum.

But other sets of lines (other series) may be produced by electrons that are not in their ground states in the hydrogen atoms but are in the orbits just above the ground states. To see how such series of absorption lines can arise we consider the atmosphere of a typical star like the sun. In such an atmosphere the collisions among the hydrogen atoms produce a distribution of the hydrogen atoms with their electrons in various orbits. If the number of hydrogen atoms per unit volume (the number density) is $n$, a certain fraction of these atoms, $n_0$, have their electrons in the ground state, a certain fraction $n_1$ have their electrons in the next higher orbit (called the K shell), and a certain fraction, $n_2$, have their electrons in the next higher orbit (called the L shell), and so on. The numbers $n_0$, $n_1$, $n_2$, etc., in this distribution depend on the temperature in the star's atmosphere.

Consider now the $n_1$ atoms with their electrons in the K shells. These electrons are more easily dislodged by the photosphere photons (less energetic photons, i.e., lower frequencies) than those in the ground state. The absorption of these photons produces a series of absorption lines—the Balmer series—which lies in the visible part of the hydrogen spectrum. This clearly illustrates the very important role that the photospheric and atmospheric temperatures play in the formation of various stellar spectra we observe. This analysis applies to every kind of atom with differences arising as we go from one kind of atom to another because the energies required to displace electrons depend on how close these electrons are to the atomic nucleus. We illustrate this by comparing the helium atom with the hydrogen atom. Each of the two helium electrons is about twice as tightly bound to the helium nucleus as the single electron in the hydrogen atom is bound to its proton. We conclude from this that the absorption spectra of the helium atoms in stellar atmospheres are more difficult to produce (require a higher photospheric and atmospheric temperature). This is precisely verified by the observed stellar spectra.

We are now prepared to describe the various spectral classes that are used by astronomers; here we are guided by our previous discussion of the hydrogen and helium absorption (Fraunhofer lines) spectra and their dependence on the temperature of the atmosphere and of the radiation streaming through the atmosphere (the temperature of the photosphere). We introduce the temperature by correlating the temperature to the color of the stars. Astronomers had been aware of the color differences among the visible stars (Sirius is called a white star whereas Betelgeuse, in Orion, is a red star, and the sun itself is a yellow star). Of course, in designating the color of a star as definite, such as red or blue, we do not mean that the star emits light of only that color; we mean that it is the dominant color in its emitted radiation. Although the sun is said to be a "yellow" star, it emits all colors, including the infrared and ultraviolet. The importance of the dominant color is that it is directly related to the absolute (kelvin) temperature of the star's surface (photosphere). This enables us, then, to correlate the spectra of stars to their colors.

Returning now to the spectra of hydrogen and helium, we first note an important difference between the responses of these two kinds of atoms to temperature, which arises because the hydrogen atom has one valence electron and the helium atom has two. If the hydrogen atom is subjected to very hot (high temperature) radiation, it loses its single electron (it is then ionized), and it no longer exhibits a spectrum. Helium, however, can lose one of its electrons (becoming ionized) and still show a spectrum. Put differently, we can say that only neutral (un-ionized) hydrogen has a spectrum, whereas both neutral helium and singly ionized (loss of one electron) helium exhibit spectra.

With these facts in mind we now trace the spectra of neutral hydrogen, neutral helium, and singly ionized helium as we go from red and orange colored stars (Arcturus) to blue-white stars (Rigel). The surface (photosphere) temperatures of red stars are of the order of 3500 degrees kelvin, far too low to excite the hydrogen or the helium atoms, so the spectral lines of these atoms are not found in the Fraunhofer spectra of these red stars, which are classified as M-type stars. The temperatures of the atmospheres of M-type stars are low enough for molecules such as titanium oxide (TiO) and OH (hydroxyl ion) to exist. This is verified by the presence of such molecular bands in the spectra of M-type stars.

As we go to orange stars called K-type stars we observe the absorption spectra of neutral metals such as iron. The surface temperatures (now called their effective temperatures) are of the order of 5000 degrees Kelvin. This is hot enough to excite the valence electrons of the metals since these electrons are far enough away from their nuclei so that they are rather loosely bound by the nuclear–electron electrostatic force and, therefore, easily subject to the disruptive force of the radiation. The trend here as we go from red to blue-white stars is in the direction of the excitation of the electrons in the more tightly bound states of ionized metals, hydrogen, neutral helium, and ionized helium. Thus in the hotter atmospheres of yellow stars (called G-type stars), like the stars with temperatures of the order of 6000 degrees kelvin, the lines of the ionized metals dominate. The lines of hydrogen are also faintly visible, primarily because hydrogen is so abundant that it shows up even under unfavorable conditions. Passing on to the creamy colored F-type stars such as Canopus and Procyon with temperatures of the order of 7500 degrees Kelvin, the hydrogen spectrum becomes quite pronounced, becoming dominant for white stars like Sirius, with temperatures about 10,000 degrees Kelvin. As we go to still hotter stars such as those found in Orion (e.g., the blue-white star Rigel), the hydrogen spectrum fades because these stars are so hot that most of the hydrogen atoms have been stripped of their electrons (i.e., most of the hydrogen has been ionized). The spectra of the blue-white stars (which have photospheric temperatures above 15,000 degrees Kelvin) are dominated by the lines of neutral helium. At still higher stellar temperatures the spectra of singly ionized helium atoms begin to come into play.

This dependence of stellar spectra on the temperatures of stars makes spectral classification an extremely important phase of astronomy. This task was begun in 1885 by E.C. Pickering at the Harvard College Observatory and was pursued with great success by Annie J. Cannon. This was summarized in the production of the *Henry Draper Catalogue* which, consisting of nine volumes and completed in 1924, gives the spectral classification of 255,000 stars, divided into seven main classes. Starting with the bluest (hottest) stars, these classes are labeled O, B, A, F, G, K, M, each of which is divided into 10 subgroups (e.g., B0, B1, . . . , B9). The sun is a typical G3 star.

The discovery of the spectral classification of stars was the first step in the study of astrophysics (the physics of stellar interiors) and ultimately, in the study of the birth, evolution, and death of stars. The second step after stellar spectral classification was the discovery of important physical differences among stars belonging to the same spectral class. This discovery required the development of an optical technology that enabled astronomers to assign large numbers of stars to their appropriate spectral classes almost at a glance. This was achieved as previously described by using what is now called the "objective prism"; a large prism (large enough to place in front of and cover the objective of a telescope). This arrangement of a prism in front of the telescope objective produces images of stars that are stretched out into spectra. The well-trained observer can then arrange these images into their spectral classifications.

With such large numbers of spectra, astronomers quickly discovered differences among stars in the same spectral class. These differences were first discovered in 1911 by the Danish astronomer E. Hertzsprung; he discovered that the cooler yellow and red stars can themselves be divided into distinct groups: those that are intrinsically faint (low luminosity) and those that are intrinsically bright (large luminosity or small absolute magnitudes). Hertzsprung called the low luminosity stars "dwarfs" and large luminosity stars "giants." Thus the sun and Capella belong to the same spectral class (they are both G-type stars) but Capella, called a giant, is about 100 times as luminous as the sun. This work begun by Hertzsprung was later extended by the Princeton University astronomer Henry Norris Russell, who extended these ideas to the hotter stars such as the A- and B-type stars. These discoveries of Hertzsprung and Russell were finally summarized in what today is called the Hertzsprung–Russell (H–R) diagram.

This diagram or graph can be described as follows: we draw two perpendicular lines, one horizontal (east–west) and the other vertical (north–south). These lines divide the paper into four quadrants but we limit ourselves to the upper right-hand quadrant. Starting from the point of intersection of the two perpendicular lines we mark off points, going to the right, giving the change in the color, the temperature or the spectral class. Starting from the same point of intersection, but going to the north, we lay off points, giving the absolute magnitudes (decreasing) or luminosities (increasing) as we move along upward. Because every star belongs to a particular spectral class (definite surface temperature or color) and has a definite luminosity, it can be represented by a point in this quadrant of the H–R diagram or graph. Not every point in this quadrant represents a star since most of the points are in regions of the quadrant which represent impossible (physically unreal) luminosities and temperatures. If we take a thousand stars at random and determine their representative H–R points we expect these points to lie within reasonable physical bounds.

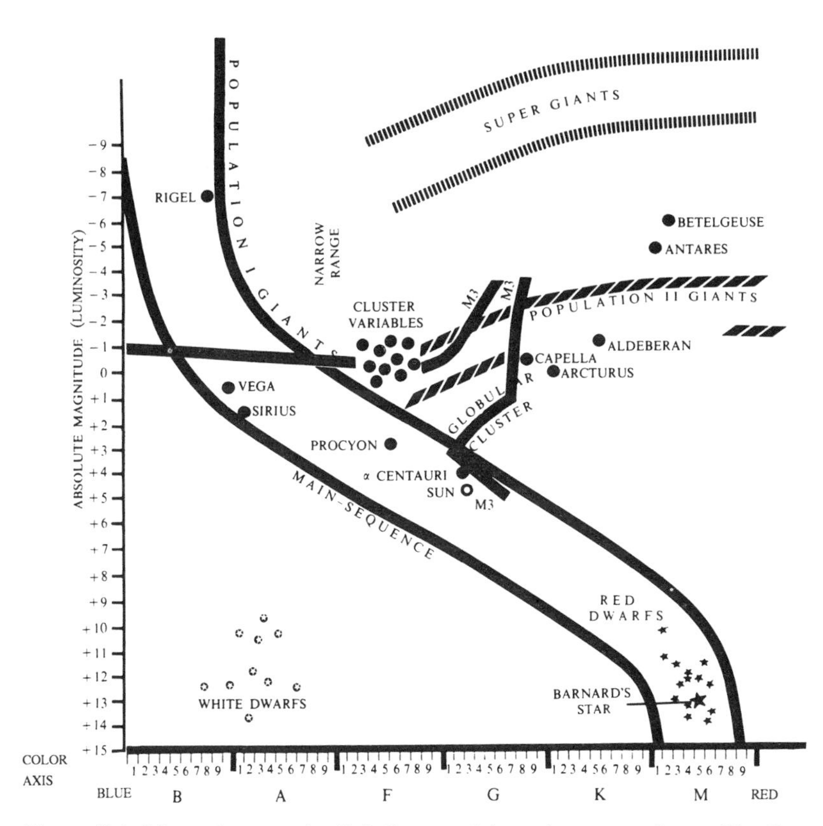


Figure 15.1. Schematic composite *H–R* diagram of the various types of stars. The diagram of the globular cluster M3 is also shown.

This was, indeed, found to be so but the points are not distributed randomly within these bounds but instead lie along definite, narrow bands or branches, the most populated of which (containing about 90 percent of the thousand stars) runs diagonally from the upper left-hand region of the quadrant (very luminous blue-white hot O-type stars) down to the lower right hand region (cool, low luminosity red M-type stars). This branch is called the "main sequence." The upper left-hand stretch of the main sequence is dominated by stars like Rigel, which is 62,000 times as luminous as our sun. The star Sirius (a white A-type star about 30 times as luminous as the sun) lies about one-third of the way down the main sequence and the sun—a G-type star—is halfway down. Another branch of stars called the giant branch, containing about 10 percent of the stars, lies above the main sequence and runs parallel to the horizontal color (temperature) axis. These stars are all about 100 times as luminous as the sun. In recent years stars have been assigned to two other regions of the H–R quadrants. One of these is a branch parallel to but above the giant branch consisting of mostly yellow and red stars some 10,000 times more luminous than the sun (100 times as luminous as the "giants"). These rare stars are called supergiants. A good example is the red supergiant Betelgeuse in the upper left-hand corner of Orion; another is the red supergiant Antares in the constellation Scorpio. The other region, in the lower left hand corner of the quadrant, consists of white, very hot low luminosity stars, called "white dwarfs."

A brief description of the physical properties of the stars in these various regions of the H–R diagram, using the sun as a kind of standard, reveals to us how the visible characteristics of a star are determined by its internal physical structure. A detailed analysis of this structure, using the basic laws of physics (essentially optics, thermodynamics, atomic physics, and nuclear physics) brings astrophysics into play. But we need not employ the rigorous discipline of astrophysics here to present a clear, acceptable description of how a star operates and why stars lie in different branches of the H–R diagram.

Accepting the sun (a sphere) as our standard, we note that its surface temperature is about 5800 degrees kelvin, its radius is $7 \times 10^{10}$ centimeters (70 billion centimeters or 433,000 miles), its mass is about $2 \times 10^{33}$ grams (about 335,000 times that of the earth) and its luminosity (the rate at which it emits energy) is about $4 \times 10^{33}$ ergs per second. Its density is 1.4 grams per cubic centimeter. Its surface gravity—the acceleration of a freely falling body at its surface—is 2700 centimeters per second per second. The earth's acceleration of gravity at its surface, by contrast, is 980 centimeters per second per second. The speed of escape from the sun's surface is 618 kilometers per second, which is to be compared to the 11.2 kilometers per second speed of escape from the earth's surface.

Here we pause for a moment to explain the erg, the unit of energy used in physics and astronomy. First, we recall that energy is measured by the amount of work we do on a body when we set it in motion by pushing it. If we push a mass until it acquires a speed $v$, we say the body has an amount of energy (kinetic energy) $\frac{1}{2}mv^2$. If the mass of the body is 2 grams and we push it until it acquires a speed of one centimeter per second ($v = 1$), we say that we have done 1 erg of work on the body and that its kinetic energy equals 1 erg. To relate this unit of energy (1 erg) more closely to our everyday experience with energy we note that a 1 watt current (a watt is a unit of energy of electricity supplied per unit of time) supplies us with 10 million ergs per second or 1 joule of energy per second. This unit of electrical energy—the joule—was named after the nineteenth century British physicist James Joule, who did extensive research in electricity and thermodynamics.

In thermodynamics the unit of energy is the calorie, which equals 4.19 joules or 42 million ergs. In physics, the calorie is defined as the heat (energy) required to raise the temperature of one gram of water by 1 degree kelvin. Now, a 1 watt electric current is a current of 1 ampere of electricity flowing across an electrical potential difference of 1 volt. (An example is a conductor whose ends are connected to a 1 volt outlet and which has just the right electrical resistance (one ohm) to allow a 1 ampere current to flow through it.) This brings us to another unit of energy—called the electron volt, which is used extensively in astronomy, physics, and chemistry. The electron volt is related to the energy an electron in a vacuum (no friction) acquires when it moves freely from one negatively electrically charged plate to another positively charged plate. If the electrical potential difference between the two plates is exactly 1 volt, produced, for example, by connecting one of the plates to the negative pole of a 1 volt battery and the other plate to the positive pole—then the kinetic energy the electron acquires when it starts from rest on the negatively charged plate and moves to the positively charged plate is called "1 electron volt."

This unit of energy, the electron volt, equals 1.6 trillionths of an erg. The energies involved in moving electrons around in atoms (in chemistry) are of the order of a few electron volts. Thus the energy required to tear the single electron completely away from the proton (the ionization energy of the hydrogen atom) is 14.5 electron volts (ev). The variations in the spectra of the various spectral groups arise from changes of a few electron volts in the temperatures of stellar atmospheres. In high energy physics, in nuclear physics, in cosmology, and in the study of cosmic rays, scientists deal with millions and even billions and trillions of electron volts written as Mev, Gev, and Tev. The energies of the particles in the most energetic cosmic rays are of the order of 1 hundred million Tev.

Einstein expressed his very important discovery of the equivalence of mass and energy in what is perhaps the most famous equation in science, $E = mc^2$, where $E$ stands for energy, $m$ for mass, and $c$ for the speed of light. This equation, in time, led astrophysicists to the solution of the most important problem in stellar structure (astrophysics): what is the energy source that powers stars? We return to this question later and show how its answer enabled astrophysicists to construct models of stellar interiors. For the moment, however, we return to a more detailed exposition of Einstein's equation. In brief, it means that if we can destroy $m$ grams of mass, this mass is immediately replaced by (releases) an amount of energy equal to this mass multiplied by the square of the speed of light. If the mass $m$ is expressed in grams and $c$ is expressed in centimeters per second, the energy $E$ is then expressed in ergs. To illustrate this equation in a quantitative sense, we imagine that 1 gram of matter is completely destroyed which, according to the equation, generates 900 million trillion ergs, the amount of energy needed to lift a million tons of mass 30,000 feet in the air. Although mass is not directly destroyed in this way in stellar interiors, the nuclear reactions that occur in stellar interiors result in the transformation of mass into energy. We return to this point later and describe just what kind of nuclear transformations produce mass diminutions with the release of energy.

We return now to our discussion of the sun, pointing out that its most important property as a star is that it is a source of energy (of power). The sun's luminosity is 400 trillion trillion watts. If the sun's rays could be concentrated to move in one direction, they would in 1 second melt a column of ice extending from the sun to the earth (about 93 million miles long) with a cross-section diameter of 3.5 miles. Because the solar radiation leaves the sun in all directions, however, the earth's surface above the atmosphere receives only half of one billionth part of this enormous output. And the earth's surface receives only about 75 percent of this solar radiation; the earth's atmosphere absorbs and scatters the other 25 percent.

From our knowledge of the sun's luminosity, its mass, and its radius, astrophysicists, during the first four decades of the twentieth century, faced the problem of constructing theoretical models of the sun. Because stellar model building is essentially the same whether we consider solar models or any other stellar models, we outline the basic features of solar astrophysics. We can then pass to the general problem by appropriate generalizations. The task the solar model builder faces then is the following: Given a mass of material equal to the mass of the sun and dispersed over all space, compress this material and arrange it into a sphere of gas equal in size to the sun so that this sphere radiates energy at a rate equal to the luminosity of the sun. By "arrange" we mean deduce, from the basic laws of nature, the temperature, density, and pressure at each point within the sphere so that the sphere for some 5 billion years neither explodes nor collapses. In constructing such a model, the astrophysicist must include the basic laws of nature to insure that the physical conditions change uniformly from point to point as we move from the center of the model to allow a steady flow of radiation to be maintained to produce the observed conditions on the solar surface.

In constructing his solar model the astrophysicist has a very important choice: he has the freedom to choose the chemical composition of the raw material from which he builds his solar model. In this choice, he is guided by his knowledge of the chemical composition of the solar atmosphere which is available to him from the solar spectrum. The basic laws that guide him also tell him that the abundance of the two most common elements in the sun, namely hydrogen and helium, must be taken into account to enable him to construct the correct solar model he is seeking.

We now consider the physical laws that must be applied (must hold) at each point in the solar model's interior for the model to represent the sun correctly. In constructing this model, the astrophysicist makes two very important assumptions: first, that conditions at all points at the same distance $r$ from the center of the model are the same. This is called spherical symmetry—the conditions in the solar interior depend only on $r$, the distance from the center. This is certainly true if the sphere is not rotating or is rotating very slowly. Rapid rotation would cause the poles of the sphere to contract toward each other, producing a bulge at the equator.

The astrophysicist also makes a second assumption: that the solar material is in the form of a perfect gas, governed throughout by the thermodynamic "equation of state" of a gas. This simplest state of matter greatly reduces the complexity and difficulty of constructing a theoretical model of the sun. It is theoretical only in the sense that we have done it with pencil and paper, but it is real in the sense that if we could transport ourselves to any interior point of the sun with the appropriate measuring instruments, these instruments would record the numerical values of such measurable quantities as pressure and temperature exactly as predicted by the model.

To see how all this is done, we picture ourselves at some interior point a distance $r$ from the center of the sun, and consider the solar matter (gas) within a cube 1 centimeter on a side (a cubic centimeter in volume). Clearly the matter within this cube is in a state of violent agitation with its atoms, ions, electrons, protons (and molecules, if they exist at all), moving about randomly and very rapidly, owing to the high temperature of the gas. But the essential features of this cube of gas do not change over long periods of time, as we know from the sun's extreme stability; it has remained pretty much the same for billions of years (as it will continue to do for some 10 billion future years). This stability must, of course, stem from the laws of physics governing this unit cube of solar gas. The task of the astrophysicist is, then, to use these laws to deduce the physical conditions that keep every unit cube of gas in a state of equilibrium. As we move closer to the center of the sun, or move away from the center of the sun, the conditions must change in such a way as to keep the entire solar sphere in a steady state for billions of years.

The astrophysicist now divides the problem into two parts—the dynamic part and the radiative part. The dynamic part of the problem is expressed as a relationship between the force of gravity which tends to pull the cube of gas inward (like an object immersed in water that tends to sink) and the outward gas pressure that tends to push the cube of gas outward (causing an outward expansion). These two opposing forces on the gas cube must be equal, for if gravity dominated, all the gas cubes at all points would sink, and the sun would collapse. If the outward gas pressure dominated, by comparison, all the unit cubes of gas would rise and the sun would explode. Because the sun is neither collapsing nor exploding, the astrophysicist obtains, from the laws of physics, two equations which determine some of the physical conditions that must exist at every point in the sun to keep the sun steady.

But the requirement that dynamic equilibrium hold at each point inside the sun, is, in itself, not sufficient to permit the astrophysicist to determine all the physical conditions; in particular, the conditions of dynamical equilibrium cannot be used to determine the change of temperature from point to point inside the sun. That the surface of the sun radiates energy in all directions equally means that a constant flow of energy from the deep interior to the solar surface must exist. The astrophysicist thus concludes that the solar internal temperatures must decrease steadily from a maximum value at the solar center to its smallest value at the solar surface. Moreover, one of the physical conditions (e.g., the temperature) from point to point (moving outward) must decrease steadily in such a way as to allow the heat to flow out at just the right rate to account for the observed solar luminosity.

This problem was more difficult than the dynamical equilibrium problem because very little was known about the nature of the transfer of heat through the interior of the sun. We know from our own daily experiences that heat can be transferred in three different ways: radiation, convection, and conduction. In the late part of the nineteenth and early part of the twentieth century, astronomers were convinced that convection plays the dominant role in solar heat transfer (flow). This belief continued to hold sway until the British astronomer and astrophysicist Sir Arthur Eddington published a series of basic papers which initiated modern astrophysics. Eddington may properly be called the father of astrophysics. His most important contribution was his demonstration that in most stars (such as the sun) the heat is transported by radiation and that convection may play a role close to the center of the star and in its outer envelope.

By accepting radiation as the dominant mechanism for stellar heat transport, astrophysicists were well on their way to constructing acceptable stellar models and, in particular, solar models. But they could do this only if they could write down the proper equations to describe the transport of radiation through the stellar interior. This required some additional knowledge about the resistance of the stellar gaseous medium to the flow of radiation. Here, too, Eddington was the leader, pointing out that at the high temperatures in stellar interiors all the electrons are stripped away from their nuclei so that the stellar material is essentially what physicists call a "plasma"—a medium of freely moving positive and negative electrical charges. Eddington made one more important contribution, pointing out that most of the stellar plasma consists of electrons, protons, helium nuclei, and only traces of heavy nuclei—no more than about 3 percent. Eddington's great contributions are summarized in his book *The Internal Constitution of Stars*, the first modern book on astrophysics and still a standard text.

While Eddington was laying down the foundations of modern astrophysics, his British contemporary, Sir James Jeans, like Eddington an excellent theoretical physicist as well as an astronomer, was extending the domain in which physics and astrophysics overlap so that one could justifiably speak of "astrophysics" as a field of study in its own right with physics as important to it as astronomy. Jeans was best known as a physicist for his theoretical contribution to our understanding of the laws of radiation. These contributions and his astrophysical researches are described in his book *Astronomy and Cosmogony*, which goes beyond astrophysics and discusses such diverse astronomical fields as galaxies, the evolution of stars, stellar formations, stellar rotation, and the fission of stars.

Even with the seminal astrophysical work of Eddington and Jeans, a number of basic questions remained unanswered. The most important of these questions was the origin of the radiation in stars. How can stars continue to generate energy at the enormous rates they do for billions of years? The great British physicist Lord Kelvin had already shown that ordinary burning (ordinary chemical reactions) could have supplied such energy for only a few thousand years and that even the release of gravitational energy from the solar contractions of stars, could, at best, have supplied stellar energy for only a few million years. Some new, unheard of, and inexhaustible source of energy was required, which was found by Eddington. He proposed nuclear energy as such a source, arguing that in the cores of stars, where temperatures reach tens of millions of degrees, protons are constantly being fused to form helium nuclei and thus release vast amounts of energy. But even without any knowledge of the origin of the solar (stellar) energy, astrophysicists could deduce the primary physical conditions from the solar center outward by general arguments, applying the appropriate physical laws as needed. A careful analysis of this astrophysical problem showed that knowledge of only four interior parameters (physical variables) is required to construct a model of the sun (in fact, of any star).

To understand these physical parameters better, we picture ourselves in the interior of the sun at some distance from its center. At this distance, the weight of a thin shell of gas at this distance must be balanced by upward pressure on it. For the astrophysicist to formulate this condition he must know the inward pull of gravity on the shell, which is determined only by the amount of matter (mass) in the sphere of gas on the surface of which the shell rests. This mass, which increases as the distance $r$ increases, is one of the four physical parameters mentioned above. Two others are the temperature of the shell and the pressure. Finally, the fourth parameter is the rate of flow of heat through the shell.

Eddington, as well as Jeans, pointed to the way these four parameters can be related to each other by four equations which are the basis of all astrophysics. Without going into any of the mathematical details of these equations, we point out how we can understand these equations in a very general way and accept their theoretical consequences with confidence. Considering how the mass of a star (its internal gravity) alone plays the dominant role in producing and controlling the internal conditions of a star, we note that if the mass of the sun were increased, the internal gravity of the sun would increase and the sun would begin to collapse. To prevent this collapse, the temperature at each interior point would have to increase, which would result in an increased rate of flow of heat and an increased solar surface temperature. In other words, the sun would no longer be a G-type star; it would move up the main sequence with a change in color (whiter) and an increased luminosity. Astrophysics thus explains the H–R diagram, and, in particular, its "main sequence"; the more massive a star is at its birth, the higher up on the main sequence it must lie. This conclusion was one of the great triumphs of astrophysics.

In closing this chapter, we point to another important astrophysical deduction which became a very useful guide in our search for astronomical truths: the famous "mass luminosity" law. We see in a very general way that massive stars, on the average, are more luminous than stars of small mass. The precise relationship between stellar mass and luminosity can be deduced from the astrophysical equations. But this does not tell the full story of the power of astrophysics to reveal the mysteries of nature's deepest stellar secrets. The most important—the birth, growth, and death of stars, remains to be discussed. Eddington described astrophysics as an "intellectual delving machine," for it permits us to probe deep into stellar interiors and to explain their superficial appearances as the consequence of their internal structures. All of this by properly applying the basic physical laws to the behavior of atoms and radiation at temperatures of millions of degrees. This is one of science's great successes.

CHAPTER 16

# Stellar Evolution and the Beginning of Galactic Astronomy

*But he, with first a start and then a wink,*
*Said, "There's another star gone out, I think."*

—GEORGE GORDON, LORD BYRON

In the previous chapter we saw how astrophysics grew from the need to understand how stars are constructed and how we can use astrophysics to deduce the internal conditions of stars from their visible properties—including their mass, their size, and their luminosity (their surface temperature, or color in conjunction with their size) by applying the appropriate physical laws. The wonder of it all is that the results are very simple and that the laws of physics (only a few in number), as we know them, are sufficient to give us a complete picture of stellar interiors. No new laws are needed or even indicated, which is quite remarkable, for it tells us that our knowledge stemming from our limited experience on the earth and our own intellectual synthesis has projected us into the deepest interiors of stars and the boundless depths of space. But, as we shall see, these very disciplines unravel for us as well the mysteries of time, revealing the past, and predicting the future.

The basic theory of stellar models we have already discussed, taken together with the theory of the evolution of stars, which we now discuss, accounts for the giant and supergiant branches in the H–R diagram as well as for the white dwarf domain of the H–R diagram. Instead of going directly to the evolution of stars from their present state to their ultimate demise, we first consider the initial stages of star formation, again taking the sun as a standard.

Because the sun is about 5 1/2 billion years old, it was not born from the primordial material (hydrogen and helium) that filled the universe some 500,000 years after the big bang but from recycled primordial material. This material first spent some 5 billion years in a star that became a supergiant and then exploded as a supernova, ejecting mostly hydrogen and helium into space, but also about 3 percent of heavier elements such as carbon, oxygen, sulfur, chlorine, and iron which had been synthesized in the core of the supergiant before it became a supernova.

We now trace the history of the supernova ejecta from which our solar system emerged, beginning with the birth of the sun. We discuss this process from the point of view of the energetics of the transformation of the supernova ejecta from a widely dispersed gas to a highly concentrated, very hot gas (the sun) and a few cold bodies (the planets) circling the sun. The energetics tell us that no matter what transformations the ejecta undergoes, the total energy must remain the same. But this energy may change from one kind to any other; in this solar problem we are dealing with dynamical energy (kinetic and potential) and radiation (electromagnetic energy) into which some of the dynamic presolar energy must ultimately have changed if the sun was to be born as it was.

To make our discussion as clear and as simple as possible we consider the dynamical energy in more detail. If we push an object of mass $m$ until it acquires a speed $v$, the work we do is 0.5 of $mv^2$ as previously described and we define that quantity as its kinetic energy, noting that kinetic energy is always positive because $v^2$ is always positive. (Here we disregard certain esoteric particles such as positrons called antimatter particles.) If we now lift a particle a height $h$ above the ground, we do an amount of work $wh$ where $w$ is the weight of the body. We then say the body has potential energy. In dealing with gravitational problems, in general, we find it convenient to take the potential energy as negative. We may do this because we can choose $h$ in such a way as to make $wh$ negative. Thus if the body is in a hole of depth $h$, its potential energy in the hole is negative because we do work to lift it to the surface where $h$ is zero and where its potential energy (in this example) is taken as zero. We eliminate this kind of ambiguity in dealing with the gravitational interactions of celestial bodies by defining the configuration of infinite dispersion (all bodies are infinitely separated from each other) as the configuration of zero potential energy. This means that if the bodies are separated from each other by a finite distance, their mutual potential energies are negative (less than zero). All of this is just a convenient arithmetic way of treating the energetics of gravitationally interacting bodies.

Returning now to the raw material (the nova ejecta) from which our sun was formed, we picture it as a gas whose total energy initially was zero: the molecules and atoms of this material were infinitely dispersed (zero potential energy) and their kinetic energies were zero. Under the mutual gravitational attraction these molecules acquired positive kinetic energies (they started moving toward each other) and negative potential energies. But their total mutual potential energy became more and more negative as they approached each other, so that their total energy (kinetic plus potential) remained zero (conservation of energy). Clearly a condensed stable configuration (the sun) can be produced only if the total energy drops below zero, for if it remained zero, the gas would just as well expand out to an infinite dispersion again. With this energetic constraint imposed on a gas cloud—if it is to contract to the stable solar configuration—we must ask how the total energy is to sink below zero, to become negative. This can happen only if the gaseous presolar cloud loses energy, that is, if it radiates energy. How did this happen?

We can understand this better if we keep in mind that "radiation" here means electromagnetic radiation; the radiation mechanism then becomes obvious: the atoms and molecules in the gas must radiate energy. At first sight this mechanism may seem impossible because the atoms and molecules in the gaseous configuration are electrically neutral and such particles do not ordinarily radiate. But nature has a very effective mechanism up its sleeve to force the atoms to radiate: atomic and molecular collisions. As the atoms and molecules are crowded together by their gravitational free fall toward each other, they collide with each other, causing the atoms to become excited so that their electrons are thrown into higher orbits around their respective nuclei. Then, on falling back in a very short time to their ground state orbits, they radiate in accordance with the electromagnetic laws.

Here we see nature operating magnificently to produce the sun, which, of course, was the first step in the evolution of intelligent life. Note that gravity alone could not and cannot produce stable gravitational structures; the electromagnetic force must also be present. The principal burden of producing stars, however, falls on gravity; the electromagnetic forces play no role in their formation, but without the electromagnetic forces there would be no radiation to reduce the total energy below zero. This is a manifestation of what we call the second law of thermodynamics, which states that processes involving a net flow of heat (radiation) out of a system are irreversible (expressed in physics as an increase of entropy). In this instance of the gravitational collapse of a gas to form a star, the entropy is represented by the outflow of radiation. We see that this outflow is irreversible because the radiation can never return to bring back the infinitely dispersed gaseous configuration.

We now follow the flow of radiation as the gas collapses, slowly at first, and then more and more rapidly. As this collapse proceeds, the temperature of the collapsing gas rises (the mean velocity and, hence, the kinetic energy of the atoms and molecules increase), and this manifests itself in the increasing temperature on the surface of the forming sphere. This temperature then gives us the rate at which the collapsing sphere is losing energy. Because this loss must be compensated in some way, the sphere of gas continues to collapse, releasing gravitational energy, which is the only source of energy available to the star in its formative stage.

From thermodynamics and the physics of gravitational collapse, we can easily show that only one half of this gravitational energy escapes from the surface of this prestellar phase and the other half remains within the collapsing sphere, raising the temperature of the contracting gas continuously. One might conclude that the sphere must continue collapsing down to a point. But this is not so because when the internal core temperature reaches 10 million degrees kelvin, the contracting sphere taps another source of energy: nuclear energy.

The lack of knowledge about nuclear energy, which Eddington had suggested as a source of the sun's power, held up the development of astrophysics for some 20 years. Astrophysicists were puzzled then by the stability of stars, which, from all the observable evidence (e.g., earth dating using the uranium: lead ratio of uranium ores on the earth) are billions of years old. The door to solving the stellar energy problem was opened by the discovery of the neutron by the British physicist James Chadwick in 1932. Up to that time the structure of the atomic nucleus was a deep mystery. From the measured nuclear masses, physicists thought that the nucleus of an atom must contain enough protons to account for its mass. Thus the mass of the hydrogen atom is one atomic mass unit (the proton's mass is taken as the atomic unit of mass—one trillionth of a trillionth of a gram). Physicists therefore thought that atomic nuclei contain electrons in addition to protons to account for the electric charge on the nucleus which, for the light nuclei, is numerically about half the mass (in atomic units). Under this assumption, to account for the mass of the helium nucleus, for example, we must assume that the nucleus also contains two electrons to account for the electric charge (2 units of charge) on the helium nucleus (the unit of charge is taken as the charge on the electron—a negative charge). But placing electrons in atomic nuclei leads us to an irreconcilable conflict with the Heisenberg uncertainty principle. The reason for this conflict is that the diameter of an atomic nucleus is a tenth of a trillionth of a centimeter (called 1 fermi unit of length). This means that we would then know the position of any electron in a nucleus within an error of 1 fermi of length. But the uncertainty principle then tells us that our knowledge of the electron's momentum is highly uncertain; the kinetic energy of the electron would then be very large, indeed, so large that the electron would not be held within the nucleus by the pull of the nucleus.

Placing electrons in atomic nuclei would also upset (in fact, contradict) the measured spins of nuclei. Thus the known spin of the nucleus of the nitrogen 14 nucleus (14 units of mass) is zero, but if we place 7 electrons in this nucleus to account for its electric charge of 7, and 14 protons to account for its mass of 14 units, we cannot obtain its correct spin, because the total number of basic particles (protons and electrons) in the nucleus would then be 21, each with a half-unit of spin. But no way exists of arranging 21 half-units of spin to give a total spin of zero.

The discovery of the neutron eliminated these difficulties because the neutron has zero electric charge, has a mass only slightly larger than that of the proton, and has a half unit of spin. Because its mass is about 2000 times that of the electron, it can be confined in the tiny nucleus without being in conflict with the uncertainty principle; its large mass gives it the large momentum demanded by the uncertainty principle and, at the same time, allows it to move around so sluggishly within the nucleus that it cannot get out. Moreover, because the neutron has zero electric charge we can add to or remove neutrons from nuclei without altering the chemical properties of their atoms, which are determined by the electric charges on their nuclei (their external electrons). Thus neutrons account for atomic isotopes (atoms with the same chemical properties but with different atomic weights). Finally, neutrons, which have the same spin as protons (half-unit of spin) correctly account for the nuclear spins. Thus the nitrogen 14 nucleus with seven protons and seven neutrons (just 14 basic particles) can have a spin of zero because the proton and neutron spins are arranged in equal and opposite spin pairs, each pair with spin zero so that the total spin of the nucleus is zero.

The existence of the neutron leads us to the existence of another particle, called the neutrino, which plays an important role in physics and astronomy—particularly in the generation of nuclear energy. The neutrino was introduced in 1930 by the theoretical physicist Wolfgang Pauli to account for problems associated with the decay of certain radioactive nuclei that decay by emitting electrons, which are called beta rays. The British experimentalist Sir Ernest Rutherford, who had worked extensively with radioactive atoms (nuclei), was the first to distinguish among the different components of the radiation emitted by such nuclei, calling them alpha, beta, and gamma rays. He identified alpha rays as positively charged particles with the same mass as the helium nucleus and then identified them as helium nuclei. He identified the gamma rays as very high energy photons (electromagnetic particles of very high frequencies) and identified the beta particles as high energy electrons.

The gamma rays and alpha particles presented no problems at the time of their discovery but the beta rays (electrons) presented a very serious problem which Pauli solved by proposing the neutrino. We can understand the difficulty presented by beta rays if we consider the energetics of the emission of beta rays from beta radioactive nuclei. When a nucleus of mass $m$ emits a beta ray, it changes to a nucleus of a smaller mass, let us say $m'$. According to the theory of relativity, the difference in mass produced $m - m'$ must show up as the mass of the emitted electron plus its kinetic energy, divided by the square of the speed of light. In other words, the kinetic energy of the emitted electron (its speed) should always be the same, but that is not observed. In fact, the beta ray (electron) may be emitted over a continuous range of energies: sometimes with very little kinetic energy, sometimes with the maximum allowed energy, and at other times with energies anywhere in between these two extremes. We describe this strange property of beta ray emission by saying that the energy spectrum of the beta rays (electrons) is continuous. In other words, the beta ray (electron) is not emitted with the total energy available to it but a fraction of it which may be a small or a large fraction of it. This, of course, was not acceptable to physicists for, if no other particle were involved in the beta ray decay of a nucleus, the conservation of energy (one of the most inviolable principles in physics) would have to be discarded. Aware of this crisis and unwilling to discard the conservation of energy principle, Pauli proposed the concept of the emission of another kind of particle with the electron in the beta decay of any beta-radioactive nucleus such as cobalt 60. This companion particle of the electron was named the neutrino because it is electrically neutral and, therefore, difficult to detect. Indeed, it was not detected until 1957 when Cowan and Reines detected neutrinos emitted from a nuclear reactor.

We can infer the basic properties of neutrinos by considering the simplest example of beta decay—that of the neutron. The lifetime of the neutron, about 15 minutes, was measured and deduced theoretically. In this simplest beta-decay process, the neutron, slightly more massive than the proton, emits an electron and a neutrino, thus becoming a proton. We see that the emission of the electron is required by the principle of the conservation of electric charge (the neutron, before the decay, has zero electric charge and the proton and electron pair together after the decay has zero electric charge).

But another conservation principle must be obeyed in this neutron decay—the conservation of spin. As we know, the electron, the proton, and the neutron have spins of a half-unit each so that if we start out with a single neutron, the total spin of our system is half and this must be the total spin after the decay. But this could not be so if only an electron and a proton were the product of the decay because the total spin (combining 2 half-units of spin) would then be either 1 or 0. This means that a third spin half particle is involved in this decay. This is the neutrino. The three half-unit spins (electron spin, proton spin, and neutrino spin) can now be combined to give a half unit spin.

Finally, we must determine the mass of the neutrino, which is a more difficult task and has been pursued since the neutrino was first postulated. The preponderance of evidence is that its rest mass is zero, which means it travels at the speed of light.

From this discussion of the decay of the neutron we see that the beta decay of the heavy nuclei is produced by the decay of neutrons in these nuclei. In the light nuclei such as helium (2 protons, 2 neutrons), carbon (6 protons, 6 neutrons), and oxygen (8 protons, 8 neutrons), the numbers of protons and neutrons are equal, but as we go to heavier nuclei the neutrons begin to dominate in number. This sets an upper limit on how heavy a nucleus can be, for if we try to build up a very heavy nucleus, the excess of neutrons over protons becomes so large that the neutron decays prevent the build up of such heavy nuclei.

With this preliminary discussion of nuclear structure we can return to our description of the pre-stellar stage of the gravitational collapse of the gaseous cloud that ultimately became our sun. We saw that this collapse could not have continued during the entire lifetime (about 5 1/2 billion years) of the sun but must have stopped when the sun tapped another source of energy (nuclear energy) and that this happened when the temperature in the sun's core reached about 10 million degrees. This 10 million degrees was critical because it triggered the fusion of quartets of protons to produce helium nuclei.

From the nuclear masses involved in this fusion process we can easily calculate the amount of energy released in each such fusion and see that this is just the right amount to account for the solar luminosity. This calculation is not as difficult as it may appear for all we need to do is note that whereas the mass of a helium nucleus is very nearly 4 in atomic mass units, the mass of 4 protons is 4.032 atomic mass units. This means that 0.032 units of mass is transformed into energy in accordance with Einstein's famous equation $E = mc^2$ (energy released equals the mass diminution multiplied by the square of the speed of light). In terms of grams of mass and ergs of energy we find that when 4.032 grams of protons are fused to produce 4 grams of helium, $0.032 \times 9 \times 10^{20}$ ergs of energy are released.

Because the sun emits every second $4 \times 10^{33}$ (4 followed by 33 zeros) ergs in all directions, we can calculate the number of quartet proton fusions (number of helium nuclei) produced per second in the solar core to produce the observed solar luminosity—100 trillion trillion trillion. This rate of quartet proton fusion requires a higher temperature than 10 million degrees; this higher temperature is found in the core of the sun (at the sun's center) where the temperature is 15 million degrees. But even this high temperature by itself is not sufficient to keep the "nuclear fires" burning in the sun's core; a high density of protons is also required and, fortunately for us, the sun's core has this high proton density. In fact, each cubic centimeter of volume in the sun's core contains 150 grams of protons or 100 trillion trillion protons. All of these data have been provided to us by the astrophysicists who have constructed solar models.

But even with these high temperatures and high proton densities the fusion of protons to produce helium nuclei is not along a straight unhindered path; it proceeds in a series of distinct steps, each of which has a certain probability of occurring, which can be calculated by nuclear physicists from the ambient conditions and the nuclear physics that governs the nuclear reactions involved.

The first step in this chain of reactions that leads from protons to helium nuclei is the fusion of two protons to form the deuterium nucleus (the deuteron $H_2$—heavy hydrogen—which consists of one proton and one neutron). But this is a very slow process—so slow that a given free proton moving around in the solar core will continue doing so for billions of years on the average before it combines with another proton to produce a deuteron. The reason for this slow rate is that the process cannot proceed unless one of the two protons that are to merge (fuse) changes to a neutron by capturing an electron and an antineutrino (or emitting a positron and a neutrino). This is just the reverse of the process of the spontaneous decay of the neutron. Here we emphasize the emission of a neutrino, which plays a very important role in the direct detection of this first step in the proton–proton chain by neutrino detectors on the earth. In a sense, the neutrinos permit us to penetrate theoretically into the very center of the sun and thus to check the validity of our theoretically constructed solar models.

Once a deuteron is produced in this remarkable way, it quickly captures another proton (with the emission of energy) to become the stable He3 isotope of helium. Thus helium 3 nuclei will be built up until enough of these nuclei are all around to combine in pairs rapidly to form He4 (ordinary helium nuclei) and in the process emit protons. If we do our bookkeeping properly, we see that since two He3 nuclei (4 protons in all) are required in this chain of nuclear reactions, two protons are returned to the medium leaving over the other 2 protons and the 2 neutrons required by the He4 nucleus (alpha particle). Because the neutrinos interact only very weakly with protons, neutrons, and electrons, they leave the sun immediately and reach the earth in about 8 minutes after they are emitted, where they can be detected with special neutrino detectors. Solar neutrinos have been detected, but only about one-third of the number that should be detected if the picture of the hydrogen → helium chain of nuclear reactions described above is the correct chain of reactions to account for the solar luminosity. Simple calculations show that each square centimeter of the earth's surface should receive 100 billion neutrinos every second, but only about 30 billion neutrinos are detected every second.

This proton–proton chain of reactions for the generation of energy in the cores of main sequence stars like the sun was first investigated by Hans Bethe and C.F. von Weiszacker. Bethe, however, was the first to distinguish between the energy producing nuclear cycle in stars on the lower part of the main sequence and those, like Rigel, on the upper part of the main sequence which are hundreds of times more luminous than the sun. Bethe pointed out that to account for the luminosities of extremely luminous stars on the upper part of the main sequence, one must use a nuclear cycle that is much more sensitive to (dependent on) temperature than the proton–proton chain, which goes as the fourth power of the temperature so that the luminosity is proportional to $T^4$; the central temperatures of such stars would therefore have to be about 20 times as high as the sun's central temperature. Such stars could not be dynamically stable. Bethe therefore suggested another nuclear cycle (the carbon cycle) which involves the carbon nucleus as a catalyst. This cycle begins with a carbon nucleus in the core of the star capturing a proton and becoming a nitrogen 13 nucleus. Five more steps occur in this cycle with the capture of three more protons and the emission of two neutrinos and two positrons. In the last (sixth step), the original carbon nucleus reappears with an alpha particle (a helium 4 nucleus). About 25 percent more energy is released in each of these cycles than in the proton chain of reactions.

Because this cycle is much more highly temperature dependent (it goes as the temperature to the 20th power) than the proton–proton chain, it contributes little to the generation of energy in stars like the sun or stars lower than the sun on the main sequence (orange and red stars) because their central temperatures are too low. But it takes over in stars above the sun on the main sequence, which accounts for the way the main sequence turns upward above the sun's position on it. We can understand this very critical role of the star's core temperature in the generation of nuclear energy if we recall that the nucleus of an atom and a proton (the most abundant, and easiest to ignite nuclear material) repel each other.

The heavier the nucleus (the larger its positive electric charge), the greater the repulsion.

In spite of this repulsion, a proton in the stellar core, if it is moving fast enough, can penetrate this electrical barrier and merge with the repelling nucleus to form a new nucleus with the repelling nucleus. Theoretical nuclear physicists worked out the mathematical details of such nuclear processes and Bethe worked out the details of the carbon cycle, receiving the Nobel Prize for this important astrophysical work. The rate of release of energy via the carbon cycle goes as $T^{20}$, which means that an increase of the stellar core temperature by 1 degree increases the rate of the carbon cycle by a factor of 20. Of course the central temperature alone does not determine the rate of energy generation, but if we take all the contributing factors into account, as Bethe did, we see that if the central temperatures of stars like Rigel are about 40 million degrees kelvin, the vast luminosities of these blue-white supergiants on the upper part of the main sequence can be explained.

We devote the rest of this chapter to the evolution of main sequence stars away from the main sequence. This phase of astrophysics is one of the most beautiful and dramatic developments in astronomy in which theory and observation went hand in hand to establish beyond any doubt the evolutionary track of stars in the H–R diagram. As we shall see, the distribution of stars and the various branches of the H–R diagram pointed the key to understanding the evolution of stars.

That the various branches in the H–R diagram of the distribution of stars point to stellar evolution was first proposed by Russell himself. He argued that all stars begin as cool luminous red giants (owing to slow gravitational collapse) at the extreme right end of the giant branch in the H–R diagram (see Figure 15.1). To account for the constant luminosity of the giants in the giant branch of the H–R diagram (the cool red giants at the right end of the giant branch and the hot white giants at the left end, where the giant branch merges with the main sequence, are all equally luminous) Russell suggested that the young giants continue to contract gravitationally, becoming hotter and smaller, with these two changes in the structure and thermodynamics of the giants to help their luminosities remain constant.

To account for the main sequence stars, Russell continued his evolution theory in the same vein, with one very important change: he suggested that the stars, instead of remaining huge spheres of gas on reaching the main sequence, solidified and moved down the main sequence as they cooled off. However appealing this Russell scenario may be, it is fatally flawed and untenable because, as we have already stated, the luminosities of stars on the main sequence cannot, even fractionally, be accounted for by a cooling off process. But Russell proposed this evolutionary scheme before the era of nuclear physics. We now know that all stars are gaseous spheres, as we have already observed, and their interiors, down to their very centers, are governed by the law of gases.

The only correct part of Russell's hypothesis is that the various stellar branches of the H–R diagram are, indeed, irrefutable evidence of stellar evolution, but the direction of stellar evolution is just the opposite of Russell's picture; stars do not start their stellar lives on the giant H–R branch but on the main sequence (also called the zero age branch) and evolve from the main sequence to the giant and supergiant H–R diagram branches. The structure of the main sequence itself is not a thin line with stars strung along it like beads on a string but a rather broad band which is particularly pronounced at its cool red end (which contains many nearby stars whose distances are fairly accurately known). This means that the breadth of the main sequence cannot arise from errors in measurements of the distances of the main sequence stars. It must arise from the variation in some intrinsic property of these stars.

The position of a star along the main sequence (from right to left) is determined by its mass when it starts the stellar part of its life on the main sequence; the more massive it is, the higher up (to the left) and the whiter and hotter it is. The broadness of the main sequence is to be accounted for by the differences in the chemical compositions of stars having the same mass. We therefore look to differences in chemical compositions (the differences in helium as opposed to hydrogen abundances) of main sequence stars. How was this hypothesis confirmed?

Because we cannot watch individual stars evolve owing to their long lifespans, we must resort to observations of groups of stars which were born together but which evolved at different rates owing to their different masses so that each such group consists of stars in different stages of evolution. Where are such groups to be found? They are scattered throughout our galaxy and are called "local" or "galactic" clusters of stars, which we know must have been born together from the same primordial galactic cloud of gas and dust because they are all moving together in the same direction in the galaxy. These local (galactic) stellar clusters, consisting of population I (solar type) stars, are, therefore, ideal laboratories for testing the stellar evolutionary theories of the astrophysicists. Because all the stars in such a cluster started their stellar lives with the same chemical composition, they are all equally old now and any differences among them are a measure of the differences in their rates of evolution which in turn could have arisen only from their mass differences.

When a cloud of gas fragments gravitationally, the fragments produced have different masses and, therefore, evolve at different rates. We may illustrate this point by comparing the rate of evolution of a massive hot blue-white star like Rigel with that of the sun. The mass–luminosity law alone tells us that Rigel's mass is some 20 times the sun's mass. Because its luminosity is 62,000 times that of the sun, Rigel is using up its nuclear fuel (hydrogen) 62,000 times more rapidly than is the sun. Moreover, Rigel's chemical composition is changing just as rapidly, which means that it is evolving very rapidly.

The observational astronomers saw a simple, dramatic, and foolproof way to prove or disprove this evolutionary hypothesis. They constructed an H–R diagram for each of the local (galactic) star clusters and studied how the H–R diagrams of these clusters differ from or resemble the various branches in the H–R diagram of stars chosen at random. Various observational astronomers thought of this approach but most of the credit for the thorough analysis of the evolutionary data contained in these cluster H–R diagrams goes to Alan Sandage of the Carnegie Observatories in Pasadena, California. The H–R diagram of stars in any cluster is a

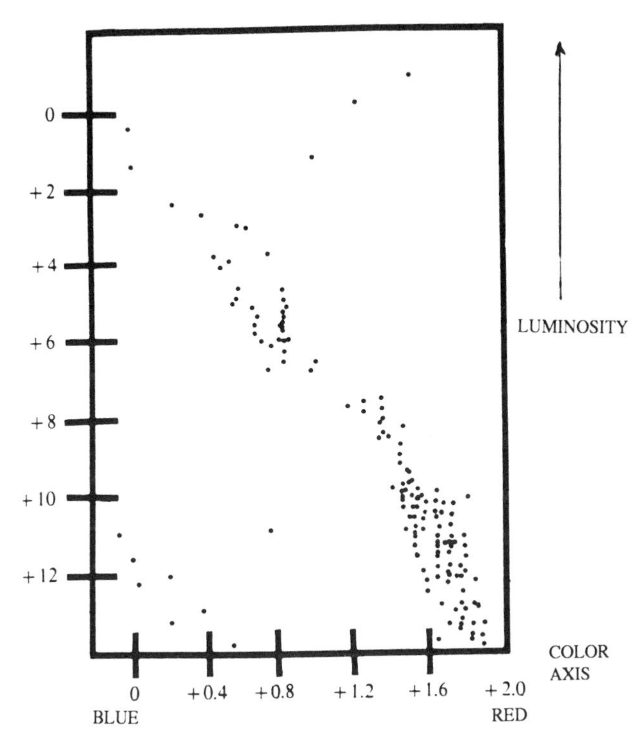


Figure 16.1. H–R diagram of stars lying within 10 parsecs. [Adapted from H. L. Johnson and W. W. Morgan, *Astrophysical Journal* CXVII (1953), 313. ©1953 by the University of Chicago.]

fairly thin line, the lower part of which lies along the main sequence, but the upper part of which (the white and the blue-white stars) is turned to the right away from the main sequence. In the H–R diagrams of most of these clusters, only a small piece of the upper track is off the main sequence, but in some, the upper track extends through the giant branch into the supergiant branch and then turns back toward the main sequence.

Sandage interpreted these stellar cluster H–R diagrams as evolutionary tracks, arguing that all the stars of a cluster along the

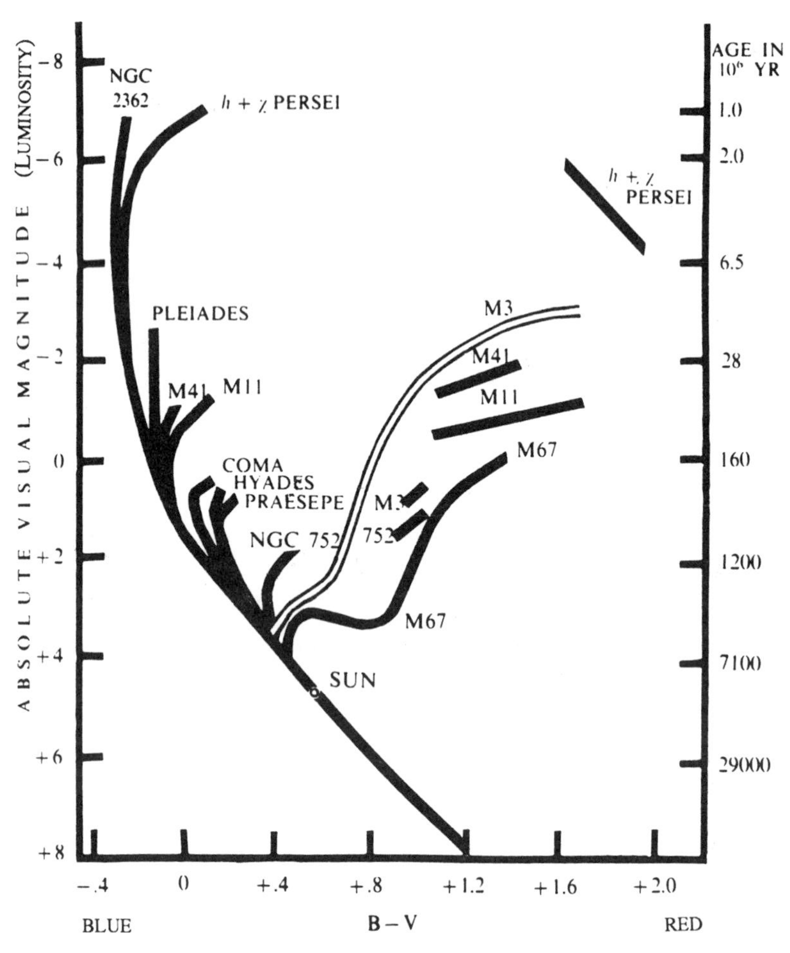


Figure 16.2. A composite H–R diagram of various open clusters, showing ages. (After Sandage)

line of the cluster's H–R diagram began their lives on the main sequence and that the more massive ones evolved away from the main sequence along tracks that took them to their present positions on the H–R line. Sandage went on further to point out that

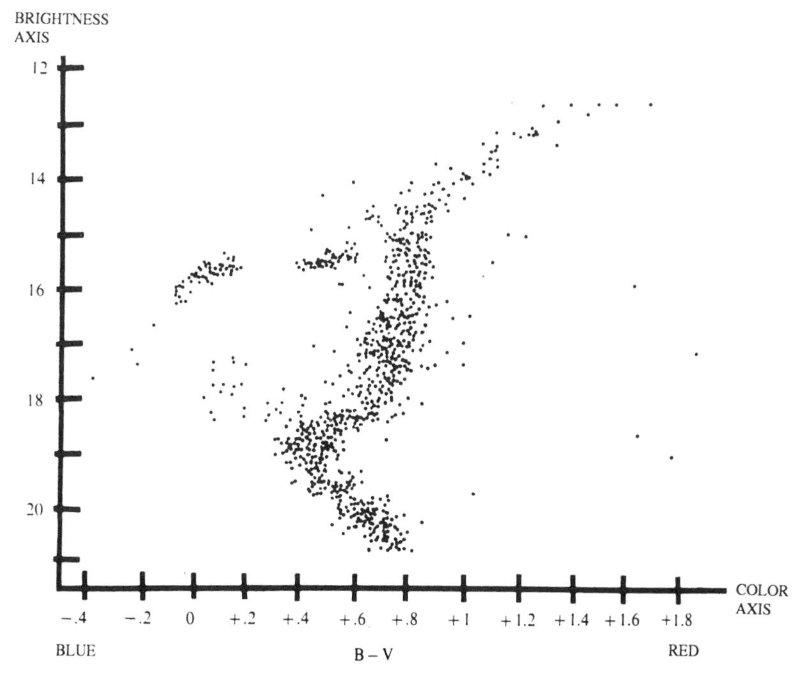


Figure 16.3. A detailed H–R diagram of globular cluster M3. [H. L. Johnson and A. R. Sandage, *Astrophysical Journal* CXXIV (1956), 379. ©1956 by the University of Chicago.]

the age of each cluster can be calculated from the length of the evolutionary track of any star from its initial position on the main sequence to its present position in the cluster. In this way, the ages of clusters were found to range from tens of millions of years to billions of years.

If one now places the H–R lines of all the known clusters on a single H–R diagram, one finds that all these lines coincide at the main sequence but the points along the main sequence where the lines of the H–R evolutionary tracks of the various clusters depart from the main sequence differ. The older the cluster, the further down along the main sequence (toward the red stars) does its departure point from the main sequence lie.

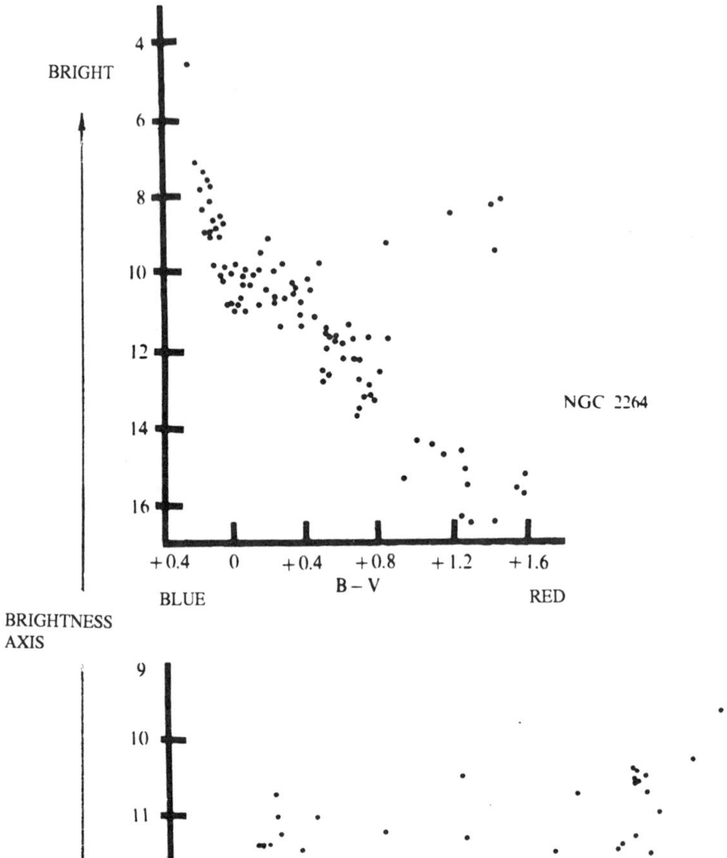
BRIGHT
4
6
8
10
12
14
16
NGC 2264
+0.4
0
+0.4
+0.8
+1.2
+1.6
B – V
BLUE
RED
BRIGHTNESS
AXIS

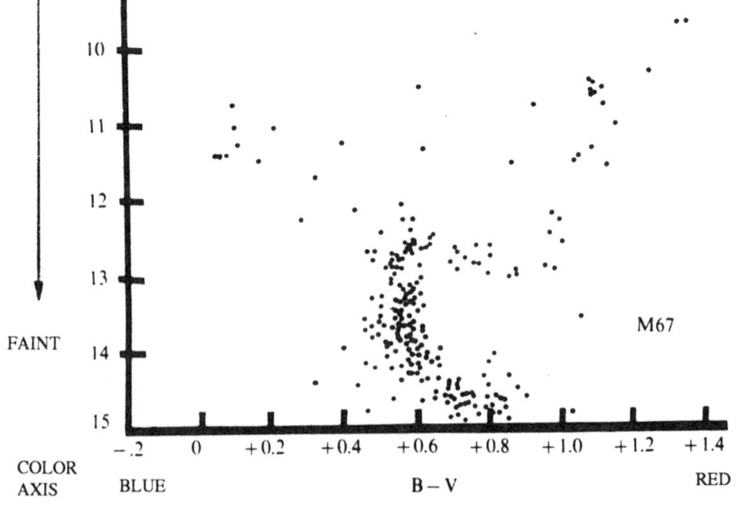
9
10
11
12
13
14
15
FAINT
M67
–.2
0
+0.2
+0.4
+0.6
+0.8
+1.0
+1.2
+1.4
COLOR
AXIS
BLUE
B – V
RED

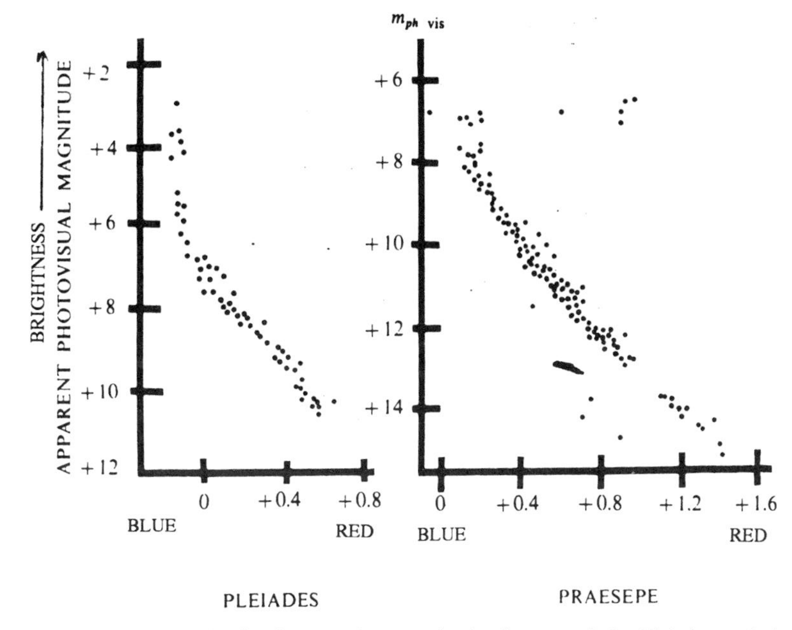


Figure 16.5. Color–luminosity or color–magnitude diagram of the Pleiades and the Praesepe clusters. [Pleiades: H. L. Johnson and W. W. Morgan, *Astrophysical Journal* CXVII (1953), 313; Praesepe: H. L. Johnson, *Astrophysical Journal* CXXVI (1952), 640. ©1952, 1953 by the University of Chicago.]

This brilliant analysis of the relationship of the age of a cluster to the shape and position of its H–R evolutionary diagram track was just the beginning of the observational feature of the theory of the evolution of stars from the zero age main sequence branch of the H–R diagram to the giant and supergiant H–R branches. The theoretical aspect of this phase of astronomy was initiated and developed by the Princeton astrophysicist Martin Schwarzschild, the

---

Figure 16.4. A detailed H–R diagram of two extreme galactic clusters, NGC 2264 (a very young cluster) and M67 (a very old cluster). [NGC 2264: M. Walker, *Astrophysical Journal,* Suppl. No. 23, 1956; M67: H. L. Johnson and A. R. Sandage, *Astrophysical Journal* CXXI (1955), 616. ©1955, 1956 by the University of Chicago.]

son of the German cosmologist and astrophysicist, Karl Schwarzschild. This work was developed to its fullest extent and is still being avidly pursued by Professor Icko Iben of the University of Illinois. Mainframe computers have played a critical role in this story, for without them it would have been impossible to carry out the long complex calculations that the pursuit of this discipline required.

Without going into any of these details and complexities we can see in a general way that stars must evolve away from the main sequence. Their chemistries change more or less rapidly because the ratio of their hydrogen to helium abundances change continuously as they generate and release nuclear energy. We give a brief discussion of the future history of the sun owing to the continual nuclear fusion of protons (hydrogen) and the resulting growth of the helium abundance in the solar core at the expense of the hydrogen abundance. This change in the mean molecular weight of the solar matter will change the sun's position on the main sequence, bringing it ever closer to the upper edge of the main sequence.

Stars more massive than the sun start their lives higher up on the main sequence than the sun's position and stars less massive than the sun start their lives on the main sequence lower than the sun. Stars with a higher helium/hydrogen core ratio than when they began their lives move closer to the upper edge of the main sequence as they age. Returning now to the sun, we see that this He:H solar core ratio has increased so that half of the initial solar core hydrogen has been transformed into helium. But we know from the 4.6 billion-year age of the sun (based on radioactive dating, i.e., the uranium: lead ratio in uranium ores), that the sun is about halfway through its main sequence phase. In time, of course, all the hydrogen in the solar core will be fused into helium, thus causing very important changes in the structure and overall behavior of the sun.

To follow the overall changes that will occur, we first note that when the hydrogen fuel is exhausted in the core (which we may define as a sphere with a radius of about 30,000 miles), this will not be the end of the proton–proton cycle as a source of nuclear solar energy because the temperature and density in a thin concentric shell (about 1000 miles thick) will then be high enough to sustain the proton—helium chain. But the helium core will then contribute gravitational energy to this hydrogen-shell burning phase of the sun. Because no nuclear energy is generated in the core at this stage, the core will then begin to shrink quite rapidly. The gravitational energy thus liberated will more than make up for the loss of the proton chain energy in the core, so that the luminosity of the sun will increase, and this increase will force the sun to reorganize its structure to accommodate this sudden increase in solar energy.

Without going into the precise details of the dramatic changes that now occur, we note that the sun will have to readjust itself to allow all the energy generated per second in the core and shell to get to the sun's surface. The sun will do this by expanding (increasing in size) so that its surface area will increase while its surface temperature falls. The first of these changes will make the sun more luminous, but the second (the lower temperature) will tend to make it less luminous. In the overall balance, the sun's luminosity will increase while it becomes redder (lower surface temperature). In other words, the sun will have left the main sequence and will be approaching the red end of the giant branch of the H–R diagram. But long before that happens drastic events will change the structure and appearance of the solar system. As the sun's surface moves outward, it will swallow each planet in turn, starting with Mercury. At this point the solar radiation striking the earth will be intensive enough (far exceeding the 2 calories per second per square centimeter we now receive from the sun) to cause the earth's temperatures to approach 1000 degrees fahrenheit. The oceans will boil away leaving a dry barren earth. Even the atmosphere will vanish because the atoms and molecules will be moving fast enough to escape completely from the very hot earth.

But what about the sun? As the solar core temperature continues to rise it will approach 100 million degrees and a nuclear fusion process called the triple helium reaction, first discovered by the Cornell astrophysicist E. Salpeter, will occur. In this reaction three helium nuclei combine to form the carbon 12 nucleus. Because this reaction may set in quite rapidly, it is called the "helium flash," but even though it releases a vast amount of energy in a very short time, it does not alter the overall rate of the steady evolution of stars like the sun; the solarlike star settles down to the transformation of helium to carbon expressed symbolically as 3 He4 → C12.

This is as far as a star like the sun can go in its evolution. It settles down as a red giant, with a small carbon core surrounded by a helium burning shell, which, in turn, is surrounded by a hydrogen-burning shell. The temperature in the core of this evolved solar structure will remain fairly constant because the core cannot collapse enough (its mass is too small) to raise the core temperature sufficiently to produce any nuclear transformations in which the carbon 12 nucleus is built up to such heavier nuclei as O16, Ne20 (neon), and Mg24 (magnesium). The internal structure of the final stages in the life of the sun will consist of a small carbon core and two concentric shells, in the first of which the triple alpha-particle process (3 He → C12) proceeds very slowly and in the second of which the proton–proton chain continues, again at a slower pace.

What will happen to the sun in time? This depends on whether or not its mass remains sufficiently large so that after it has exhausted its nuclear fuel it can contract gravitationally. At the present time, owing to the solar wind, the sun's surface emits a plasma (consisting mostly of protons) which hits the earth's atmosphere at a speed of about 40 million centimeters per second. The total mass of hydrogen carried away from the sun by this wind is about 3.6 million trillion grams per year or 3.6 million billion trillion per billion years. This is so small a fraction of the sun's total mass that we may accept the conclusion that the sun will lose hardly any of its mass as it descends to the final stage of its life—its "white dwarf" stage.

By "white," we mean that it is "white hot" like the filament of an incandescent glowing lightbulb; this can be so only if the surface temperature of the sun reaches about 11,000 degrees kelvin, which is the surface temperature of a typical A-type star such as Sirius A. Interestingly enough, a white dwarf, Sirius B, is circling Sirius A. By "dwarf" we mean a star of low luminosity that is no larger in size than the earth so that the density (mass per cubic centimeter) of an average sample of the white dwarf material may be as large as several tons per cubic centimeter (by comparison, the density of water is 1 gram per cubic centimeter and the density of one of the densest known materials, platinum, is 21 grams per cubic centimeter). We can understand the consternation that the discovery of white dwarfs must have produced among astronomers, for such stars, less luminous than the sun by a factor of 1000 and glowing with a white heat, were difficult to understand and even more difficult to explain in terms of the physical principles that were known at the time.

To see how daunting the white dwarf problem was we note that for a star like the sun to become a white dwarf, it must gravitationally collapse down to the size of the earth. We may express it differently in its most dramatic form by pointing out that since the mass of the sun is some 340,000 times that of the earth, the force of gravity must, in a sense, squeeze 340,000 earths (earth masses) together to produce a single white dwarf. At first this may seem beyond the capacity of the gravitational force, which we know to be quite weak ordinarily. But this is an underestimation of gravity which, under the proper circumstances, can overwhelm all other forces in the universe.

Accepting this idea, we may then wonder why the white dwarfs do not go on contracting forever, ultimately becoming no larger than "points." Here, too, mass and another property of the electrons in atoms play important roles to prevent complete collapse. Although the mass of the sun is large enough to compress it down to a white dwarf, it is not large enough to compress it any further because another important principle (known as the Pauli exclusion principle, already discussed) comes into play by acting like a repulsive force to keep the white dwarf in equilibrium. Under these conditions, the electrons in the white dwarf are said to be in a degenerate state and the white dwarf is then in a state of dynamical equilibrium, because the electron pressure of the degenerate electrons just balances the inward pull of gravity.

We can understand how this degenerate outward electronic pressure arises if we recall that the Pauli exclusion principle states that, owing to the electron's spin, no two electrons can have the same momentum (velocity) if they are within "touching distance" of each other (about a hundredth of a billionth of a centimeter). This means that as more and more electrons are squeezed together by gravity to form the "degenerate" electron state of a white dwarf, more and more of them have to move with increased momentum (increased kinetic energy) to obey the Pauli principle. More and more of them thus exert an outward pressure so that, in time, this pressure prevents any further collapse and the white dwarf is in equilibrium.

We now complete this chapter by describing the evolution of stars that are at least 2 or 3 times as massive as the sun; these stars, particularly those having masses some 15 or 20 times larger than that of the sun, played and still play an important role in the chemistry of stars like the sun and the planets.

It is clear from our discussion of the relationship between stellar masses and their luminosities that these stars (those high up on the main sequence whose energy tracks are governed by the proton–carbon cycle) evolve much more rapidly than solar-type stars; they complete their evolution to the giant and, ultimately, to the supergiant branch in, at most, a few hundred million years, and, during this time, build up the heavier elements in their interiors. Icko Iben has been particularly active in this area of astrophysics, carrying out extensive calculations on the evolutionary tracks in the H–R diagram of stars with masses ranging from 3 solar masses, intermediate mass stars, to very massive stars of 10 to 15 solar masses.

Although Iben has been the driving force behind the rapid growth of the astrophysics of stellar evolution during the last quarter century, it remained for a very imaginative and intellectually daring quartet of astrophysicists, Fred Hoyle, W. Fowler, and Geoffrey and Margaret Burbridge to show that these massive stars build up the heavy elements owing to the rapid rise of their core temperatures to hundreds of millions and even billions of degrees. We saw in our discussion of the final stages in the evolution of stars like the sun that they cannot build up elements heavier than carbon by capturing alpha-particles (helium nuclei). Hoyle was convinced that in the later stages of the evolution of these very massive stars, the very high core temperatures mentioned above do occur and that the carbon barrier to the alpha-particle build up of elements beyond carbon is breached. Some preliminary calculations convinced him of this and his collaborations with Fowler and the Burbridges settled the matter. They showed that in massive stars the core temperatures can rise to a billion degrees or more and that in these cores alpha-particle fusion processes can produce the heavier elements right up to iron 56 (Fe 56).

We now consider this model of the alpha-particle fusion process in somewhat more detail and see what happens to the massive stars in which this kind of evolution occurs. We consider the evolution of the massive stars which were formed billions of years ago (about a million years after the big bang) when only the elements hydrogen and helium (protons and alpha-particles) existed. These earliest massive stars were metal-poor (now called population II stars) and burned hydrogen just as the sun does because very little if any carbon was around to support the carbon cycle. But owing to their massiveness, these stars built up helium and carbon very quickly so that the carbon cycle took over in a relatively short time. As their core temperatures rose, these stars built up the various heavy elements in their cores whose atomic weights are multiples of 4 (O16, Mg24, Si28, S32, Cl35, Ca40, Sc45, Ti48, . . . , Fe56). We stop with iron56 because nuclear physics teaches us that energy is not released when the iron56 nucleus captures an alpha-particle. In fact, this capture causes the iron56 nucleus to break up. This marks the end of this type of stellar evolution because the binding energy per alpha-particle of the iron 56 nucleus (the energy required to tear an alpha-particle away from this nucleus) is larger than that for any other nucleus.

One must then wonder how such elements as cobalt, nickel, copper, etc., were produced. Here again, Hoyle and his team were very imaginative; they suggested that these nuclei could have been produced from iron nuclei by the capture of neutrons. However, since neutrons have a very short lifetime (about 12 minutes), these very heavy elements could have been built up only if a constant source of fresh neutrons had been present in the iron cores of these massive stars. Such neutron sources were pointed out by the Burbridges as having existed in the collapse of these massive stars.

The occurrence of the violent collapse of massive stars has been known since AD 1054 when Chinese astronomers observed the collapse (without knowing it was a collapse) of the star at the center of the Crab Nebula in Taurus. We now call such a stellar collapse, which may be as luminous as a whole galaxy, a supernova, many of which have been observed in galaxies other than our own. The formation of iron cores in the final stages of the evolution of massive stars tells us why such violent stellar collapses occur: the core itself, with nothing to offset the intense inward pull of its own gravity because it can no longer release nuclear energy to this end, collapses violently (free fall) in a matter of seconds, following which, the outer layers of this supergiant containing some 90 percent of its mass collapses violently onto the shrinking core. After the core is compressed to the fullest extent permitted by the physical laws, the collapsing overlying layers "bounce" off the compressed core, moving outward at a speed of thousands of kilometers per second. Such a vast amount of energy is released in a supernova collapse that the supernova itself blazes with the luminosity of a 200-billion-star galaxy. Following this magnificent spectacle, the supernova begins to fade, leaving as a residue, a very hot compressed core.

Because the supernova collapse is the most dramatic event that ever occurs in the universe, presenting to us a cosmic laboratory in which energetic phenomena occur that far exceed anything we can produce in our terrestrial laboratories, it is instructive to study the supernova phenomena in some detail. In the core of the supernova, the degenerate outward electron pressure that keeps white dwarfs in equilibrium is far too small to stop the collapse of a supernova at the "white dwarf" stage; the electron masses are too small for this to occur. The free electrons, having no way out, are forced into the iron nuclei by the vast inward gravitational pressure produced by the collapsing shells. Each electron carries an antineutrino along with it and combines with a proton in the iron nucleus, leaving a neutrino behind, which quickly (at the speed of light) leaves the supernova. The observers on the earth should be able to detect streams of neutrinos coming from any new supernova This did indeed happen when the famous supernova 1987 burst into prominence in February 1987 when different observers on earth detected neutrinos coming from this supernova in the large Magellanic star cloud. Because this star cloud, a small irregular galaxy (a companion of our galaxy containing about 20 billion stars) is 160,000 light years away from us, the 1987 supernova actually occurred some 160,000 years before it was detected on the earth.

Continuing with our analysis of the physical events in the contracting core, we see that as electrons are absorbed by the protons in the iron nuclei in the core, these nuclei become unstable (the balance between the nuclear protons and nuclear neutrons in the iron nuclei is upset) and try to reestablish stability by shedding neutrons. If we were inside the supernova core we would see the nuclei "shedding" neutron "tears" which, as free neutrons, in time, stop the core collapse. Because neutrons have the same spin as electrons they (the neutrons) obey the Pauli exclusion principle. But since a neutron is about 2000 times as massive as an electron, neutrons can be squeezed 2000 times closer to each other than the electrons in a white dwarf. But in time the neutrons form a degenerate neutron gas and begin to exert an outward degenerate pressure to stop the collapse of the supernova. When the diameter of the collapsing core is about 10 miles (the approximate length of Manhattan Island), the neighboring neutrons will be "within their touching distance" and the collapse will cease. The core will then be a "neutron star," such as that which is found at the center of the Crab Nebula.

Can we detect a neutron star directly? Yes, because certain important changes occur in the structural and electromagnetic properties of the neutron star. If a spinning sphere or cylinder is squeezed down so that the mass is brought closer to the axis of rotation, the rate of spin (the rate of its rotation) increases more rapidly than its dimensions decrease as illustrated by a pirouetting skater who increases her rate of spin by bring her arms closer to her body. This manifestation of a very important physical principle or law, known as the principle or law of conservation of angular momentum (or rotation), is also the basis of Kepler's second law of planetary motion (as planets get closer to the sun, they revolve around the sun more rapidly).

This increase in the rate of spin as the nova core contracts, combines with another consequence of the core collapse to produce a very important and remarkable observable phenomenon. As the core collapses, the intensity of any magnetic field at its surface increases so that very strong magnetic fields exist on the surface of the neutron star, and, as the core spins, the magnetic lines of force are dragged around at the rate at which the core is spinning. Now these magnetic lines of force capture electrons which swirl around the lines of force and, therefore, send out radio waves. The rotating neutron star (the supernova core) can be detected by a radio telescope which picks up a rotating radio signal which sweeps past the radio telescope at the same speed as the neutron supernova core is spinning; such signals have been detected.

The story of how these signals were first detected is interesting because their detection had nothing to do with a deliberate attempt to detect the neutron core of a supernova. In fact, the periodic radio waves that were detected quite accidentally were not traced to a spinning neutron supernova core at all. Jocelyn Bell, a British graduate research student at Cambridge University, analyzing radio signals from outer space, discovered that certain signals are arriving at very precisely spaced intervals in time as though emitted by an incredibly accurate pulsating source; the first such regular radio signals, discovered by Ms. Bell in 1967, were coming from the constellation of Vulpecula at 1.33728 second intervals, which means that the time intervals of these signals are as accurately spaced as the signals from an atomic clock. Because these signals are pulsed, their sources were (and still are) called pulsars. When pulsars were first discovered, most astronomers thought they were produced by the radial pulsations of a star, but this idea was discarded when Thomas Gold of Cornell University pointed out that pulsations of the kind proposed could not be produced and that they would die out quickly owing to internal stellar friction. Gold then suggested that the pulsars are rapidly spinning neutron stars, some of which rotate relatively slowly (a few times per second) and some very rapidly (up to thousands of times per second). These very rapidly spinning pulsars are called minipulsars—the term "mini" designating the very short period—a hundredth or a thousandth of a second.

The rotating pulsar produces strong pulsed radio signals, which have the same period as the rotation itself, because the magnetic field of the pulsar is swept around by the rotation as are the resulting radio waves (produced by electrons circling the magnetic field) which then reach us periodically. This is similar to the pulses of light that a ship at sea receives from the spinning search light on top of a lighthouse.

The astrophysics of the evolution of the supergiant stars also explains the presence, in galaxies, of stars that are really pulsating. Such stars, now called "cepheid variables," were named after the typical star in this category, alpha-cephei, first properly labeled as a regular variable in 1784 by J. Goodericke.

In 1912 the Harvard astronomer Henrietta Leavitt, studying the cepheid variables in the Magellanic clouds, discovered the famous period-luminosity law of cepheids which states that the luminosity of a cepheid variable is proportional to its pulsation period; the more slowly it pulsates, the greater its luminosity. This is physically reasonable. The larger a cepheid variable, the more luminous it is and the larger it is, and the more slowly it pulsates (like a swinging pendulum, a long pendulum oscillates more slowly than a short one). The evolutionary paths of giant stars take them into an unstable region in the H–R diagram where they must pulsate as they pass through this region. Thus astrophysicists verified theoretically the existence of cepheid variables.

The discovery of the cepheid period luminosity law opened up a new and very exciting branch of astronomy: the probing of the depths of space—particularly the distances of the most distant galaxies. All one has to do now to find the distance of a galaxy is to pick out a cepheid variable in the galaxy and measure its pulsation period, which can be done very accurately. From that measurement the luminosity of the cepheid can be found from the period-luminosity law. From this luminosity and the cepheid's observed brightness, the distance of this cepheid and, hence, of its galaxy can be found.

In this chapter we have taken the reader on one of the most exciting intellectual trips in the history of thought, revealing a panorama of events that ultimately led to the evolution of human intelligence. Life could not have originated in the universe without the formation of heavy elements, but only in the cores of massive stars could such nuclear syntheses have occurred. When these stars collapsed and subsequently exploded, they spewed into space the raw material (hydrogen and helium introduced with traces of heavy elements, particularly carbon and oxygen) necessary for stars like the sun and their planets to be formed.

CHAPTER 17

# Beyond the Stars: The Galaxies

*Silently one by one, in the infinite meadows of heaven*
*Blossomed the lovely stars, the forget-me-nots of the angels.*

—HENRY WADSWORTH LONGFELLOW

With Galileo's discovery that the Milky Way (our galaxy) consists of countless bright tiny dots (which he correctly interpreted as stars at vast distances) astronomy took its first step beyond the naked-eye stars that seem to fill the sky every moonless night. But until large, optically accurate telescopes were constructed nothing beyond Galileo's speculations about the dynamics and structure of the Milky Way could be deduced nor could Galileo's speculations be confirmed. But this did not deter people who gloried in observing the stars with their unaided eyes from drawing conclusions about what they saw. Thus, as previously noted, Thomas Wright, a sailor, whose own travels had acquainted him thoroughly with the visible stars and the Milky Way, proposed in 1740 that the Milky Way was, indeed, a collection of stars. Wright concluded that the stars around us in the sky are not distributed equally in all directions and throughout all of space but are concentrated in a relatively thin band, and that our sun is a member of this remarkable collection of stars.

He correctly concluded that the way the stellar distribution appears to us is due to the fact that we are within that distribution and see the nearby naked-eye stars as all around us and the very distant stars as forming a band that we call the "Milky Way." He went on to conclude that we do not see any distant stars (points of light in the form of a band) in a direction 90 degrees away from the direction to the Milky Way because there are no such stars. In short, our stellar system is not spherical in shape but more like a pancake.

This revolutionary concept of a flat distribution of stars was accepted by the transcendental German philosopher Immanuael Kant—who, in his 1759 book *General Natural History and the Theory of the Heavens*, argued that the Milky Way was formed from a spinning nebula and that the stars in the Milky Way are revolving around some hidden center deep within the stellar distribution. Kant went on further to argue that the Milky Way is but one of innumerable such nebulae throughout the universe. This marked the beginning of the study of cosmology and of the extragalactic galaxies—originally called "island universes." This concept stemmed from the general idea that our galaxy (the Milky Way) is the "main" or "principal universe" and all other such galaxies are minor island universes somewhere "offshore" of our universe in the infinite realm of space.

As ever larger telescopes were constructed and turned to the heavens, an ever increasing amount of data was amassed, which verified Wright's and Kant's hypotheses. Astronomers began to accept the concept that extragalactic nebulae exist in the universe. In this astronomical research, Sir William Herschel was the first to show conclusively that the distribution of the stars in space is not spherical but unidirectional. He did this by using a procedure he called "star gauging"—that is, actually counting the distribution of stars in all possible directions in space. In 1785 he published his "gauge observations" which encompassed star counts in 683 selected regions scattered over the sky. His star counts ranged from one or two in some regions to a thousand or more in others. This work clearly established the asymmetry of the distribution of stars in the Milky Way: Herschel's star counts showed, very early in this astronomical game, that the counts thin out (fall off) very quickly as one counts in directions at an angle to the plane of the Milky Way (galaxy) and increase very quickly as one counts in the direction of the Milky Way. But as one counts stars in the plane of the Milky Way, the star counts fall off dramatically. Herschel's conclusion was that the stars that we can see with our naked eyes and with telescopes form a thin, "grindstone-like" system with the sun somewhere near the center. The grindstone-like picture is correct, but placing the sun near the bulging center is incorrect, as we shall soon see.

To explain why the distribution of the naked-eye stars in our sky is spherical, whereas the star counts show that most stars are distributed in a "thin" stellar disk, we note that the naked-eye stars are nearby stars and our sun is in the midst of these stars. We compare this idea to our perception of how people would appear to be all around us if we were all standing close together along a stretch of a north–south avenue in a city. Our nearby neighbors in this stretch would appear to be (would, indeed, be) uniformly distributed in all directions. But, if we looked beyond our neighbors, we would immediately see that the people thin out in the east–west direction and we would then deduce that the people were standing in a north–south band. This is exactly the conclusion to which star counts lead us concerning the shape of the Milky Way: it is a thin disk of stars which, if seen edgewise, looks like a "grindstone" or a lens seen edgewise which bulges at the center.

The precise distribution of the stars and the sun's position in this distribution was not fully and correctly described until the Harvard astronomer Harlow Shapley (1885–1972) completed his investigation of the distribution of globular stellar clusters with respect to the Milky Way. Globular clusters are large spherical distributions containing tens of thousands (up to a few hundred thousand) of stars forming a spherical halo or shell around the Milky Way. That they do not lie in our galaxy (ie., in the plane of the Milky Way) is indicated by our ability to see them at great distances at an angle to the plane of the Milky Way. In this way, they differ from the invisibility of the center of the Milky Way (the center of the galaxy) which is obscured from us by the dust (the fog) in our galaxy which cuts off our view at a distance of about 6000 light years.

That dust grains, in addition to atoms and molecules, are present in our galactic interstellar medium was already known in the early part of the twentieth century through the observations of the reddening of distant stars. As we have seen, astronomers have divided stars into color groups according to the spectral features which are related to their colors in the H–R diagrams. Astronomers then found that the colors of the most distant stars in the galaxy appear redder than they should be. This means that the light from these distant stars is altered by the interstellar medium in its trip to the earth. The dust in the interstellar medium scatters out the blue rays making the light redder. From this reddening and the distances of the distant stars we can determine the amount of dust present in the galactic interstellar medium.

Although this dust accounts for only 1 percent of the interstellar medium it plays an important role in the structure of the galaxy, accounting for its spiral arms and for the presence of dark nebulae in our galaxy. These regions may be the birthplaces of new stars. In fact, astronomers in the last quarter of the twentieth century have detected concentrated globules in these dust clouds which can be interpreted as the forerunners of stars.

Returning now to globular clusters, we note that Shapley discovered that the globular clusters appear to be concentrated in one direction (about 150 of them); none can be seen in the opposite direction. From this asymmetry of the globular distribution relative to our solar system, Shapley correctly concluded that our solar system is not at the center of the Milky Way.

Shapley went further in this analysis of the globular cluster distribution to determine the distance of the center of the galaxy from our solar system. Again, using the correct assumption that the center of the galaxy coincides with the center of the halo in which the globular clusters lie and which surrounds the galaxy, Shapley calculated the distance of the center of the galaxy. From his observations of the distances of the closest globular clusters (about 35,000 light years when viewed obliquely with respect to the plane of the Milky Way) and most distant ones (more than 70,000 light years), he concluded that the center of the globular cluster halo and hence of the Milky Way is about 30,000 light years away in the direction of the constellation of Sagittarius. These deductions were later confirmed completely by direct observations of radiation (radio waves) from the center of the galaxy but these observations required radio telescopes which we discuss later.

Star count technology also led to the discovery of the spiral structure of the galaxy. Although we cannot pursue individual star counts beyond 6000 light years owing to the dust in the galaxy which becomes increasingly thicker as we look toward Sagittarius (toward the center), observations of clusters of stars enable us to go beyond 6000 light years and almost out to a distance of 15,000 light years. Careful star counts have shown that two breaks or gaps occur in these counts as we look toward the center, the first at about 1000 light years away and the second at about 6000 light years. As we look in a direction opposite to that of the center of the Milky Way, we find another gap. These gaps were correctly interpreted by the star count astronomers as having been produced by spiral arms of stars. Our Milky Way was thus discovered to be a "spiral nebula" (really a spiral galaxy and not, correctly speaking, a nebula, which is a gaseous structure).

The star count analysis led astronomers to the conclusion that our galaxy consists of three distinct spiral arms: the "Orion" arm, in which the solar system lies at about 1000 light years from the inner edge, the "Perseus" arm further out from the galactic center than the Orion arm, and an inner arm which passes through the region of Sagittarius and Scorpius. The discovery of the spiral structure of the Milky Way (the galaxy) is one of the great achievements of twentieth century astronomy and a tribute to mankind's pursuit of science not for the sake of material gain but for the sake of pure knowledge. But more was to come: the discovery of the size of the galaxy, the number of stars within it, and, finally, its speed of rotation.

With the discovery that the globular clusters are in the nature of a "boundary" around the galaxy, astronomers were quick to see that the dimensions of this boundary can be used to measure the size of the galaxy. It was found to be about 100,000 light years in diameter as measured in the plane of the Milky Way. But the thickness of the galaxy varies (increases) from 1000 light years near its edge, where the solar system lies, to about 15,000 light years through its central bulge (7500 light years on each side of the plane of the Milky Way). Determining the number of stars in the Milky Way is a little bit more tricky using star counts, but it can be done by assuming that the distribution of stars around the solar system is representative of this distribution with respect to any star at the same distance from the galactic center. We can then multiply the star count in a small region of the sky in our part of the galaxy by the number of regions all around the galaxy and we then discover that our galaxy contains about 200 billion stars, ranging from very faint ones to the superluminous ones such as the stars of Orion. Star counts also teach us that the very luminous stars are rare compared to stars like the sun and those less luminous that the sun. Thus among the 50 nearest stars (lying within a distance of about 18 light years) only 7 are more luminous than the sun. To pick up a star as luminous as Betelgeuse we must go out to about 600 light years; this means that to find one supergiant we must survey a volume of space 27,000 times as large to find some 50 stars about as luminous as or less luminous than the sun. Thus the supergiants are rare, indeed.

Starting with Sir Edmund Halley who, in 1718, observed that Arcturus and Sirius had changed their positions in the sky by a full degree and half a degree, respectively, compared to their positions given by Ptolemy, astronomers began to study stellar motions in great detail and with great precision. Since Halley's time hundreds of observations of stellar motions show that stars are moving about, relative to the sun (which, we assume to be at rest) with speeds of about 30 to 40 km/sec. Here we must be more specific about the meaning of stellar motion as introduced here because motion involves direction as well as magnitude. Ptolemy, Halley, and other astronomers measured, or detected only what we now call the "transverse" motion, that is, the apparent motion of a star transverse to (perpendicular to) the line of sight. That is all the apparent change in the position of a star can give us. This motion is now called "proper motion." But the proper motion tells us nothing about the star's radial motion (motion toward or away from us) which we discuss later.

Proper motion measurements were very tedious and extremely time consuming before photography was introduced into astronomy, because it involved comparing positions of the star being studied with positions of nearby stars or with positions of other stars in the sky. With images of the same star on photographic plates taken weeks, months, or even years apart, the proper motion of the star can be measured just by comparing the time-sequence images on the photographic plates. Today catalogues of stellar positions prepared over many years reveal the proper motions of many stars.

The radial motions (motions away from or toward the sun) of stars (relative to the sun) presented an entirely different problem to the astronomers who were studying stellar motions. Changes in the positions of stars reveal nothing about their radial motions, but it was originally thought that changes in the brightnesses of stars would reveal their radial velocities. Unfortunately, this idea had to be discarded when the vast distances of the stars became apparent. At these vast distances, the radial motions of the stars can change their brightnesses only after centuries have passed. However, the day was saved with the discovery of the Doppler effect. In 1842, as previously mentioned, Christian Doppler discovered that if a source of light is moving away from an observer, the observer finds that all the colors in the spectrum of the light are redder (longer waves). If the source is moving toward the observer, however, the observer finds the light bluer; the wavelengths of the various colors are decreased. Here only the relative motion of the source to the observer matters: if the distance between source and observer is increasing, the light is reddened (wavelength is increased). If this distance is decreasing, the light is bluer. The amount by which the wavelength is increased depends on the speed of recession in the first case and the amount by which the wavelength is decreased depends on the speed of approach in the second case.

In any case the radial velocities of stars (recession or approach) can be measured using this Doppler effect. Combining the transverse velocity with the radial velocity of a star we can obtain the true velocity of the star with respect to the sun. We then find that these velocities are very nearly the same. But we must conclude that the sun itself is also moving with the same average speed as all the stars around it. With this understood the question arose as to whether or not this motion can be interpreted in terms of a general motion of the stars in the galaxy. This question was answered in the affirmative by one of the great twentieth century astronomers, Jan Oort.

Oort had become interested in constructing a model of the galaxy. He had analyzed star counts to show that the galaxy may be pictured as consisting of layers of stars lying in planes parallel to the plane of the Milky Way. These planes thin out with increasing distance from the plane of the Milky Way. This led him to the study of the motions of the stars in the galactic solar neighborhood. Oort reasoned that all the stars in the galaxy are revolving around the center of the galaxy according to Kepler's laws of motions for the motions of the planets around the sun. This means that the stars closer to the center of the galaxy are moving faster than the sun and those more distant stars are moving slower than the sun. Oort verified this relationship by measuring the velocities of stars relative to the sun both between the center of the galaxy and the sun and between the sun and the outer edge of the galaxy. These velocities supported his hypothesis which led him to the conclusion that the galaxy is rotating. Oort found that the sun is revolving around the center of the galaxy at a speed of about 215 km/sec. We may interpret this finding as meaning the galaxy rotates once every 250 million years.

The detailed study of the galaxy led to the discovery of extragalactic galaxies or the distant "spiral nebulae," as they were originally called. Their discovery led to a considerable controversy as to their nature and as to whether or not they are within our galaxy or far outside it. This controversy was not resolved until their correct distances could be measured. From 1786 to 1802 William Herschel had presented to the Royal Society three catalogues containing 2500 nebulae. In 1888, J.L.E. Dryer published the first *General Catalogue of Nebulae* listing about 8000 nebulae. This number grew to 15,000 in a subsequent 1908 catalogue, in which some of these nebulae were identified with stars, such as the Orion nebula.

The difficulty in correctly identifying these objects and determining whether they are within or beyond the galaxy arose because gaseous nebulae (large clouds of gas and dust) do lie in our galaxy. The final answer to the question of the distances of these nebulae was finally given with the construction of the 100-inch Mount Wilson telescope in Pasadena, California, which showed that the distant nebulae are huge agglomerations of stars like the Milky Way, some with spiral structures and spiral arms, containing arrays of stars, and gaseous nebulae, like those in our galaxy.

That the spiral nebulae are not members of the Milky Way was already indicated by what we call the "zone of avoidance"; we can see no spiral nebula close to the plane of the Milky Way. The Milky Way itself blocks our view so that we cannot see the distant galaxies in that direction; but they show up immediately if we look obliquely across the core of the Milky Way. The most beautiful of these spiral nebulae is the Great Nebula in Andromeda (in the direction of the constellation Andromeda) some 2 million light years away. This remarkable stellar structure is a huge disk with spiral arms and a large central core. It is about twice the size of our galaxy and contains some 400 billion stars.

If our study of the extragalactic galaxies had depended entirely on optical telescopes we would not be as far advanced as we are in our understanding of the nature, structure, and distribution of these galaxies. Nor would we have as great a knowledge about the much broader subject of the structure, the origin, and, ultimately, the evolution of the universe. But with the development of electromagnetic technology, other windows into space were discovered and used extensively. Radio astronomy was the most important of these new technologies because the optical part of the electromagnetic spectrum (to which the human eye responds) is only a small portion (one octave—from red to violet) of what we might use for peering into space. Imagine the wonders that could be discovered if we had some techniques for using longer wavelengths of light than those of the visible red part of the spectrum. This goal was realized with the introduction of radio telescopes, which can pick up a much wider range of electromagnetic wavelengths, particularly the range from a few centimeters to 21 centimeters and beyond.

Radio astronomy did not begin as the brainchild of astronomers because they themselves never suspected that radio waves contain vast treasures of astronomical information. Indeed, astronomers at the beginning of the fourth decade of the twentieth century when cosmic electromagnetic waves (radio waves) were first detected, did not even think of astronomical bodies such as stars and galaxies as generators of radio waves. Communication technicians concerned with improving radio reception knew that terrestrial radio reception was strongly affected during intense solar activity, but they did not, in general, carry this knowledge over to astronomy. Fortunately, one such technician, Karl Jansky of the Bell Telephone

Laboratories, who was concerned with improving the radio antennas used for long-range radio communication, discovered that he could not eliminate a persistent background interference (static) or locate the source of the static. He then discovered that the strongest signals from this source arrived 4 minutes earlier each successive day, and correctly concluded that this periodicity is related to the rotation of the earth. He went on further to assert that the unknown, interfering radio source behaves like one of the "fixed stars" which also rise 4 minutes earlier each day. This was the beginning of radio astronomy.

In 1936, an amateur astronomer and radio ham operator named Grote Reber built the first radio telescope, which was essentially a spherical, concave metal dish with a reflector and an electronic detector, to record the reflected radio waves. Reber built several such radio telescopes to pursue his study of cosmic radio waves, discovering that most of the sources of such waves are discrete, like the stars. After 1942, when radio signals were detected from the sun, astronomers began to study Reber's radio sources in earnest. This was the beginning of radio astronomy as an important branch of astronomy, and many nations began to build radio telescopes and radio astronomical observatories. It became clear to astronomers that radio astronomy would be much less expensive to pursue than optical astronomy. Radio dishes do not require very accurately polished surfaces and can be made as large as may be desired. Whereas the mirror of a large optical telescope such as the Mount Wilson 100 inch and the Palomar 200 inch mirror must be exquisitely ground and polished, costing millions of dollars and requiring constant care, a radio reflecting telescope consists of a parabolic dish that can be made of solid metal, strands of metal, or a fine wire mesh such as chicken wire. These radio dishes are mounted so that they can be pointed in any direction in space or they can be mounted to point in a fixed direction.

Radio telescopes have an important advantage over optical telescopes in that they can be mounted in long arrays which aggregate the sensitivity of the system so that all the telescopes in the array augment the receiving power of the array. The signals received by radio telescopes often contain patterns which have very important information about the source of such radio signals, particularly the size of the source. Radio telescopes can also be made more sensitive than optical telescopes. This increased sensitivity may be further enhanced by increasing the size of the radio telescope dish, which can be as large as 100 feet across. Such a dish can pick up radio signals which may be thousands of times fainter than those signals which can be detected by optical telescopes. Radio telescopes can record radio signals that are so weak they would have to impinge on one square centimeter of surface for about 70 thousand trillion years to transport to that square centimeter just 1 calorie of heat.

Radio astronomy received a great boost after World War II owing to the rapid development of radar, which is essentially a radio technology in which short wave radio signals are directed to and then reflected from the surfaces of nearby celestial objects such as the moon, the planets, comets, and meteors. One of the most remarkable astronomical discoveries made with radar waves reflected from the planets was that the surface of Venus is very hot, with a temperature of about 700 degrees kelvin (about 900 degrees fahrenheit). No surface water can be present on Venus's surface at such high temperatures and the atmospheric pressure there is about 100 times that on the earth. This very high surface temperature is accounted for by a high atmospheric concentration of carbon dioxide (about 95%). This produced the "greenhouse effect" in the atmosphere. The carbon dioxide molecules in the atmosphere do not block the passage of the visible rays of the sun so that these rays, absorbed by the surface of Venus, heat this surface, keeping it at a high temperature. But the surface cannot cool itself by radiating heat (infrared) rays because these rays are blocked and radiated back to the surface by the carbon dioxide molecules.

The astrophysicist I. I. Shapiro of the Harvard-Smithsonian Institute used radar to test the general theory of relativity, according to which a beam is slowed down as it traverses a gravitational field (the amount of slowing down depends, of course, on the strength of the gravitational field and, hence, on the mass of the source of the field). Shapiro tested this theory by bouncing radar beams (from an earth source) off the various planets and measuring the times of the traversal of these beams from earth to planet and back. The amounts they were delayed corresponded exactly to those reductions in the speed of light that were predicted by the general theory of relativity.

The discovery of the hydrogen 21-centimeter (wavelength) radio line was another remarkable and extremely important event owing to its great usefulness in plotting the hydrogen atoms in our galaxy and in other galaxies. The existence of such a spectral line was first predicted by the Dutch astronomer H.C. van de Hulst in 1944. When we find a line in the spectrum of an atom we associate the emission of radiation having the wavelength of this line with a definite jump in an electron in the atom from one state of energy to a lower state of energy. The difference in the atom's energy then comes out of the atom as a photon having the frequency of the observed line.

The situation with the hydrogen 21-centimeter line was quite intriguing to begin with because this line is not visible. It is in the radio part of the electromagnetic spectrum. Moreover, van de Hulst's task was not to explain a "known" or observed "line"; the discovery or observation of the 21-centimeter line came later. But van de Hulst showed that such a line must exist if we accept prevailing atomic theory. In testing the electrodynamics of the hydrogen atom we must take into account the magnetic properties of the proton and electron. Each of these particles is a tiny spinning magnet and, like all magnets, they pull on each other magnetically if they are lined up so that their magnetic poles are opposite (north next to south and vice versa). If their neighboring poles are the same they tend to push each other away. Van de Hulst pointed out that if the electron and proton in the hydrogen atom are in this repulsive state they can flip over to the attractive magnetic state by emitting a single photon with a wavelength of 21 centimeters.

The 21-centimeter signal from a single hydrogen atoms is, of course, unobservable, but this did not daunt van de Hulst, who argued that the number of hydrogen atoms is so vast that with the appropriate radio detectors the hydrogen 21-centimeter line should be detected. This sort of radio detector was not available in 1944, but some 7 years later, Ewen and Purcell of Harvard University detected the hydrogen 21-centimeter line precisely as predicted by van de Hulst.

The study of the 21-centimeter line, emitted from different parts of our galaxy, became a new and very important phase of galactic astronomy because it revealed the center of the galaxy, which is not accessible to optical telescopes owing to the opaque dust clouds that cut off our view of the galactic center at a distance of 6000 light years. Radio waves are not blocked from detectors so that we can survey the inner regions of the galactic core right down to its center using the 21-centimeter line. Because hydrogen is the most abundant element in the universe, it must dominate the constitution of the galactic spiral arms. The 21-centimeter line of the galaxy, and the Doppler shifts produced in the 21-centimeter line by the motions of the spiral arms confirm that the galaxy is rotating with a period of 250 million years. Moreover, a detailed analysis of the spiral arms using the 21-centimeter line shows that the innermost spiral arm is no more than 9000 light years from the center.

Using the optical Doppler shift alone applied to the radiation from stars in the solar galactic neighborhood, we discover that most stars in this neighborhood are moving with speeds of the order of 100 km/sec or less relative to the sun. From this observation we reason that they are moving around the core of the galaxy with speeds about equal to that of the sun around the galactic core. This is in line with the gravitational dynamics and stability of the galaxy for if such high velocity stars had been present in our galaxy in the past, they would have escaped from the galaxy millions of years ago. In fact, from the knowledge of the mass of the galaxy and the distance of the sun from the galactic center, astronomers have calculated that the speed of escape from the galaxy for stars in the neighborhood of the sun is about 180 km/sec, well above the measured speeds of these stars. Astronomers discovered a very remarkable property of these stellar velocities: they do not decrease for stars more distant than the sun from the galactic center. This discovery was quite disturbing and puzzling because it seemed to conflict with the law of gravity from which one deduces that the revolving speed of the galactic stars around the core of the galaxy should decrease with the increasing distance of these stars from the core (essentially Kepler's third law of planetary motion). That this is contradicted by the observational evidence can be explained only by the assumption that our galaxy contains a great amount of invisible matter that extends far beyond the edges of the galaxy. This was the first indication of a much broader problem which became known as the problem of the "missing mass" or the "dark matter" problem.

The designation of this matter as the "missing mass" is an oxymoron because if mass is missing one cannot refer to it at all. What is meant here is that the observed speeds of revolution of the stars require a larger amount of gravity than that provided by the mass of the visible stars in the galaxy. Because this mass cannot be observed, it is described as missing. A more meaningful definition of it is "invisible" or "dark" matter. We return to the dark matter problem later on, noting here that its nature and origin are still two of the most perplexing problems in modern astronomy.

The detailed study of the interstellar medium in our galaxy using optical and radio technology has revealed that this interstellar medium contains clouds of dust and gas (now called bright and dark nebulae) and atoms and molecules distributed fairly uniformly throughout the galactic interstellar space. The interstellar atoms and molecules are calcium, neutral and ionized hydrogen, and such molecules as OH, CO, and titanium oxide. The dust is in the form of elongated grains, probably carbon grains. About 1 percent of the total mass of the galaxy is in the form of grains of dust and molecular clouds. Because the grains of dust are lined up perpendicular to the plane of the galaxy we deduce that a weak galactic magnetic field exists. But this dust is not completely opaque: certain of the galactic nebulae such as the Orion nebula, the most dramatic nebula in our galaxy, can easily be seen even with a small telescope. But the presence of dark nodules in such nebulae may herald the birth of new stars.

The detailed study of these diffuse nebulae showed that they are quite different from the spiral nebulae that we observe beyond our galaxy; the study of these nebulae with large telescopes revealed not only that they are extragalactic, as we have already noted, but also that they are stellar conglomerates with a variety of structures, ranging from ellipsoidal, amorphous armless structures through spiral structures with well-defined spiral arms and containing easily observed stars, as best illustrated by the Andromeda spiral nebula, to irregular structures such as the Magellanic star clouds, some 170,000 light years away and containing some 20 billion stars each. These star clouds are close enough to our galaxy to be considered as companions of or, alternatively, gravitationally attached to our galaxy.

The development of large optical telescopes set in motion what is probably the most important phase of twentieth century astronomy, the study of the distribution, dynamics, and recession of the distant galaxies. This study finally blossomed into the most exciting and challenging branch of astronomy called cosmology.

The study of the distant galaxies was dominated by Edwin P. Hubble who gave up a boxing and a law career to study astronomy. In 1919 he joined the staff of the Mount Wilson Observatory and, a few years later, became its research director. Hubble's studies convinced him, and in time, all astronomers that the extragalactic galaxies are, indeed, aggregates of stars like our own. His galactic studies also proved to him that the distant galaxies are distributed throughout space the way they are distributed around our galaxy. To him this meant that space and its galactic building blocks would look the same no matter where we were in it. This is direct evidence of Einstein's famous "cosmological principle" which Einstein stated as follows: "aside from random fluctuations which may occur locally, the universe must appear the same for all observers no matter where in the universe they may be."

The study of the distant galaxies led very early in the game to two very important discoveries: the galaxies are not universally distributed as individuals, equally spaced from each other as one might have, at first, expected, but they tend to cluster. Hubble expressed this important discovery as follows:

> While the large-scale distribution appears to be essentially uniform, the small scale distribution is very appreciably influenced by the well-known tendency to cluster. The phenomena might be rightly represented by an originally uniform distri-

Edwin P. Hubble (1889–1953) (Courtesy AIP Emilio Segrè Visual Archives; Hale Observatories)

> bution from which nebulae have tended to gather about various points until now they are found in all stages from random scattering to groups of various sizes, up to occasional great clusters.

A careful study of the galaxies within 1 and 2 million light years (distances measured using the period luminosity law of the Cepheid variables in these galaxies) shows that about 10 galaxies lie within 1 million light years of our galaxy (the closest are the Magellanic star clouds) and that about 10 more lie within 1 to 2 million light years. These 20-odd galaxies constitute what is known as the "local group" of galaxies. Such clusters, some containing hundreds and even thousands of galaxies, are found in all directions in space. Each such cluster is given the name of the constellation in which the cluster appears to lie in the sky. Thus, to name but a few, we have the Virgo cluster, at 50 million light years, containing 2500 galaxies, the Pegasus cluster, at a distance of 130 million light years with 100 galaxies, the Perseus cluster with 500 galaxies at 160 million light years, and the Centaurus cluster with 300 galaxies at 500 million light years.

The study of such clusters has led to a deeper insight into the importance of the "missing mass" (the dark matter) in the universe. That a great deal more matter is present in each cluster of galaxies is clear if one determines the mass: luminosity ratio in each cluster. The luminosity is given by the total number of galaxies in the cluster if we simply assign a single luminosity (a luminosity of 1) to each galaxy. If we now assign a unit mass to each galaxy, the total mass of the cluster should also equal the number of galaxies, but the measured mass: luminosity relationship is considerably larger than one in every cluster. In other words, each cluster has much more mass than that amount deduced from the cluster's luminosity. The German-American astronomer, Fritz Zwicky, was the first to apply a famous theorem in gravitational dynamics to determine the masses of the clusters. This theorem states that a cluster of objects gravitationally bound to each other will remain intact only if the cluster has enough matter in it (enough galaxies) to bind gravitationally any galaxy strongly enough to prevent that galaxy from leaving the cluster.

This theorem was best illustrated by the dynamical analysis of the Virgo cluster by Oort. As we have already noted, this cluster contains 2500 visible galaxies, each moving with a speed of about 750 km/sec within the cluster. Oort showed that the 2500 visible galaxies together do not exert enough of a gravitational pull on the member galaxies to keep them from flying apart. About 100 times as many galaxies as the observed 2500 are needed to keep the Virgo cluster intact. Astronomers now accept the concept of "dark matter," as being appropriate for both individual clusters of galaxies as well as the universe. Indeed, most astronomers now believe that the universe consists mostly of dark (invisible) matter.

The second great discovery made by astronomers studying galaxies was the recession of the distant galaxies, which led to the much larger and much more important cosmological study of the beginning and end of the universe. This great discovery began with the cosmic "shaking" discovered by Vesto M. Slipher in 1912 that the spectral lines of the light from the stars in the galaxy M31 reveal an enormous Doppler shift toward the blue end of the spectrum. This large blue shift means that M31 is approaching the sun at a speed of about 300 km/sec. This is so large compared to the speeds of stars in the solar neighborhood in our galaxy that Slipher decided to check the spectra of other galaxies. Over the next two decades Slipher studied the lines in the spectra of some 40 nebulae, finding that most of the spectral lines are shifted toward the red, meaning that these nebulae are receding from the sun.

To explain the relationship between the Doppler effect (the shifting of the spectral lines) and the relative motion (motion with respect to the observer) of the source of light, we note that if the source is approaching the observer (we may take the observer as fixed and the source as moving or vice versa), the waves of light are crowded together. The wavelengths (as seen by the observer) therefore appear shorter and the spectral lines thus appear bluer than normal. If, however, the source and observer are receding (separately) from each other the waves are stretched out so that the wavelengths are longer, thus shifting the spectral lines toward the red end of the spectrum. The amount by which the wavelengths are shifted from their normal spectral positions are a measure of the speed of approach or of recession of the respective nebulae (galaxies). Slipher discovered that the fainter (hence, more distant) the galaxies are, the larger the Doppler shift toward the red end of the spectrum, meaning that all but two of the galaxies studied are receding from us. Most of the extragalactic galaxies are receding from our galaxy at speeds that increase with their distances (the more distant galaxies are receding more rapidly than the nearby ones).

The unexpected and unexplained sizes of the radial velocities of the distant galaxies and their one sidedness (departure from random motions) led the remarkable team of Hubble and Milton Humason (a mule train driver, who evolved into an expert observer at Mount Wilson) to undertake a systematic investigation of the very distant galaxies. Because the meaningful analysis of the relationship between the distance of a galaxy and the Doppler red shift of spectral lines involved not only a careful study of the spectral lines but also an accurate determination of its distance, if a meaningful relationship between distance and speed of recession was to be found, Hubble set himself the difficult task of finding the distances of the galaxies while Humason measured the spectral line Doppler shifts.

In 1929, Hubble published his findings as a paper in the Proceedings of the National Academy of Sciences in which he demonstrated conclusively that the radial velocities (motions away from the sun) of galaxies studied by Humason are proportional to their distances: a galaxy B twice the distance from the sun as a galaxy A is receding from the sun twice as fast as galaxy A. All that remained to be done then was to find the proportionality constant, now known as the "Hubble constant." Even though this constant was still to be determined, the relationship between distance and velocity of recession became known as the Hubble law and is now a fully accepted cosmological principal. The content of the Hubble law was published jointly by Hubble and Humason in 1931 in the *Astrophysical Journal*, and became (and still remains) the basis of the theory of the expanding universe and the concept of the big bang. Humason extended his spectral analysis to the most distant galaxies that could be observed in 1935, confirming Hubble's law out to galaxies receding at speeds of about 40,000 km/sec. To measure the distances of these distant galaxies, Hubble had to use a reliable "standard candle"; a source of light within each galaxy whose luminosity was known. From the known luminosity of this source (standard candle) and its measured brightness, one can find the distance of the galaxy by dividing the known luminosity $L$ by the observed brightness $B$ (written as $L/B$) and taking the square root of this ratio.

The standard candle that Hubble used in this procedure was and still is the Cepheid period luminosity law, which permits us to determine the luminosity of any observable Cepheid variable in the galaxy from its observed period of pulsation. The Hubble law is written as $Hr = cz$, where $H$ is Hubble's constant, $r$ is the distance of the galaxy, $c$ is the speed of light, and $z$ is the Doppler shift.

To apply this formula, one must, of course, know the numerical value of the Hubble constant, which is expressed as "speed per distance" and gives us the rate of expansion of the universe. From Hubble's constant, we can also determine the "age of the universe," that is, the time it took the galaxies to separate from each other to the extent they are now observed to be. Hubble's initial determination of the value of the constant $H$ was much too high (about 30 km/sec per million light years); this constant has been reduced, from more accurate measurements of the distances of the extragalactic nebulae, to about 17 km/sec per million light years. Before leaving this study of the structure of the galaxies, we point out an interesting feature of stars that this study reveals and which now plays an important part in our classification of stars. This feature was discovered by Walter Baade, a contemporary and friend of Hubble's and a member of the Mount Wilson staff when Hubble was doing his work. Baade, at the time, was studying the great Andromeda galaxy. He was puzzled by his inability to see (photograph) individual stars in the center of this galaxy although he could clearly see the blue-white stars in the spiral arms of the galaxy. Finally realizing that the core (center) of Andromeda is reddish, he decided to photograph the galaxy's center with red-sensitive film and quickly discovered the stars there are, indeed, reddish stars.

Using these data about the color differences of stars in the core and stars in the arms of the Andromeda galaxy, Baade proposed the concept of stellar populations in galaxies: population I stars (the stars in the arms) and population II stars (the stars in the core). This concept of stellar populations has played an important role in our understanding of stellar evolution. We know now that the population I stars, like the sun, are relatively "rich" in metals and iron, whereas population II stars are relatively "metal poor." By relatively "metal rich," we mean that the metal content of these stars is about 3 percent, and by "metal poor" we mean that the metal content is no more than 1 percent. Finally, we note that the population I stars are young, like the sun (about 5 billion years old) and the population II stars are old, ranging from 10 billion to 15 billion years old.

CHAPTER 18

# Cosmology

*And God said: "Let there be light," and there was light.*

—GENESIS

We begin this chapter by recalling the basic features of physics that are important in astronomy. The first three decades of the twentieth century were the most remarkable and revolutionary in the history of civilization, and, most certainly, in the history of science. By the year 1930 all the great revolutions in the physical sciences had been completed. Max Planck had propounded his law of blackbody radiation; Einstein had presented his special theory of relativity, his theory of the photon, his famous formula $E = mc^2$, his explanation of the photoelectric effect, his introduction of the concept of the corpuscle–wave dualism; and, in 1915, he had presented his general theory of relativity, which was to usher in cosmology in the 1930s. In parallel developments, Heinrich Hertz had verified experimentally Clerk Maxwell's electromagnetic theory of light, Sir J.J. Thomson had discovered the electron and proton, Ernest Rutherford had brought order into the theory of radioactivity, and had proved experimentally that the atom has a nuclear structure, with a positively charged, compact nucleus at its center surrounded by a revolving cloud of negatively charged electrons, with one electron for each positive charge on the nucleus. This nuclear model of the atom and Planck's quantum theory led Niels Bohr in 1913 to his theoretical model of discrete electronic orbits. Each of these orbits differs from the neighboring orbit below it and that above it by a unit of action, in accordance with Planck's quantum concept that action comes in discrete units which Planck called $h$, the famous Planck constant.

In 1917, in his last great paper on radiation, Einstein deduced the Planck radiation formula from the Bohr model of the atom, and in doing so, had to introduce the revolutionary concept of "stimulated emission" of radiation which is the theoretical basis of the modern laser. Einstein introduced another revolutionary idea in this remarkable paper, that of the momentum of the photon which emphasized the particle aspect of the photon, even though the photon is massless.

This stimulated the French physicist Louis de Broglie in 1920 to suggest that particles, such as electrons and protons, have wave characteristics: associated with each such particle is a wave—now known as its de Broglie wave—which has no physical reality, but must be included in the description of the physical behavior of the electron. This closed the theoretical circle that Planck and Einstein had begun to trace: just as the electromagnetic waves have their particle aspects (the photons), so the physical particles in nature, the electrons, protons, etc., have their de Broglie waves, so that the wave–particle dualism is universal.

Because waves tend to spread out, physicists such as Werner Heisenberg discarded the classical idea that electrons can be described as being in a particular position in space. Instead, this concept of localization of action of a particle must be replaced by what has since been called Heisenberg's "uncertainty" or "indeterminacy" principle: if one tries to measure the precise position of a particle one loses all knowledge of its motion and vice versa. This led Heisenberg to a model of the atom which departs from the Bohr model. The Heisenberg model pictures an electron in an atom as being simultaneously in all possible Bohr orbits, but not in each with equal probability, in line with Heisenberg's picture of an electron as being spread out. This theory of an atom, which incorporates the uncertainty principle, is called "quantum mechanics"; it was developed by Heisenberg, Max Born, Jordan, and Wolfgang Pauli to its greatest extent. The mathematical skills this required placed this technology beyond the use of most physicists.

There is a very simple way of viewing the Heisenberg uncertainty principle, which shows that it stems from the quantum concept, on which Planck's great discovery of the quantum theory rests: the quantum concept means that a certain basic physical entity called "action" comes in units (such as the units in a coinage system: in our American coinage system, the unit is the "penny" because no coin smaller in value than a penny exists). No action smaller than this unit, called $h$, exists. Because action in physics is defined as the product of the position of a particle and its motion (position × motion), we see that if we try to determine the position of a particle as being exactly at a given point, let us say at a distance zero from our position, the product of position × motion $v$ would become exceedingly small and less than the unit of action $h$, unless the motion owing to our measurement became infinitely large and hence beyond our knowledge.

The mathematics required by the Born, Heisenberg, and Pauli version of quantum mechanics (called "matrix mechanics" because the physical quantities such as the position and motion of a particle are represented by matrices), is well beyond the capabilities of most physicists and even beyond those of good mathematicians. It appeared, therefore, in those early years of the history of quantum mechanics, that this remarkable theory would rarely be used in studying atomic structure. But then the great Austrian physicist Erwin Schrödinger (1887–1961) simplified the mathematics of quantum mechanics by introducing the "wave mechanics" which quickly replaced the "matrix mechanics."

Schrödinger approached the quantum mechanics from the wave aspect (the de Broglie point of view) of the electron (or of any subatomic particle). He reasoned that if the electron is dual (particle and wave together), then all the particle features of an electron should be deducible from the wave representation. But this required the formulation of a wave equation for the electron; Schrödinger discovered this wave equation, now called the Schrödinger wave equation. Because wave equations (called "partial differential equations" by mathematicians) were well known at the time, Schrödinger had little difficulty in altering the standard wave equation used in acoustics, optics, and hydrodynamics to represent the de Broglie wave of an electron.

To physicists, the Schrödinger equation is in the nature of a miracle or a magic wand: all one has to do is wave this "wand" to solve any problem in atomic physics (at least, in principle). Of course, the wave equation one has to set up for any such problem may be very complicated, but once the equation is solved (by computers, if necessary) all the properties of the atomic configuration (of the electron, or electrons that are involved in the problem) can be deduced. For this reason physicists say that the wave function that the solution of the wave equation gives us describes the state of the electron or electrons in the atomic configuration.

The solutions of the Schrödinger equations for all kinds of atomic configurations show that each such configuration (or state) can be represented (or described) by a set of three numbers (integers) for each electron in a given state. These numbers, previously discussed, are called the quantum numbers of the electrons. Each electron is identified by its quantum numbers. Here as previously stated a most remarkable discovery was made by Pauli which is the basis of all chemistry and of the Mendeleev periodic table of the chemical elements; this discovery is now called the Pauli exclusion principle. It states that no two electrons in the same atom can (or may) have the same set of quantum numbers. This means that all the electrons inside an atom, like carbon (6 electrons) or oxygen (8 electrons) cannot crowd themselves into the same lowest state (the orbit closest to the nucleus) but must arrange themselves in successively more distant orbits to obey the Pauli principle. The electrons in the most distant orbit, called the "valence electrons," determine (or produce) the chemical properties of the atom.

A final remarkable contribution to this rapid development of the quantum theory of the atom was made by the British theoretical physicist Paul Dirac. This discovery is now called the Dirac relativistic equation of the electron. Dirac had readily accepted the wave–particle dualism of the electron but he was very unhappy with the Schrödinger wave equation because, as he pointed out, this equation does not meet the requirement that it be consistent with Einstein's theory of relativity. Dirac then proceeded to change the Schrödinger wave equation to make it relativistically invariant.

As was pointed out by Dirac himself, shortly after promulgating his relativistic wave equation, this equation predicts or reveals two new properties of the electron (and, for that matter, of the proton also): (1) the electron is spinning like a top and this spin must appear as a fourth quantum number of the electron; (2) an antiparticle, called the positron, exists with a positive electric charge (the electron has a negative electric charge).

During this exciting epoch two other important discoveries were made which were to give us a deeper insight into the structure and dynamics of stellar interiors than we would have without them: (1) the discovery of a massless chargeless particle called the neutrino; and (2) the discovery of a neutral (chargeless) massive particle called the neutron.

The neutrino was introduced into physics by Pauli in 1930 to explain a certain difficulty physicists encountered in trying to explain the radioactivity of certain heavy nuclei such as cobalt 60 (with a mass of 60 atomic units), which decay by emitting electrons called beta rays when so emitted. When these particles (beta rays) are emitted, they come from the given nuclei not always with the same speeds but with a range of speeds from zero to a maximum value. Now the principle of the conservation of energy dictates that if only one particle is emitted from the radioactive nucleus, its speed should always be the same and equal the maximum speed permitted by the conservation of energy. Pauli therefore suggested that the visible electron (the beta ray) is always accompanied by an invisible companion—now called the neutrino—which carries away part of the energy that the radioactive nucleus gives up when it decays. In 1956 the experimentalists Cowan and Reines announced that they had discovered neutrinos emitted from the Savannah River nuclear reactor.

The neutron, discovered in 1932 by the British experimental physicist Sir James Chadwick, is the "missing link" that physicists were seeking to explain the structures of nuclei and, in particular, nuclear isotopes. The neutron and the proton are considered by physicists as non-identical twins in the realm of particle physics, and, so, both are called nucleons with the neutron the uncharged (electrically-neutral) member of this duo. Because the neutron is slightly more massive than the proton, it is not stable unless it is within a nucleus. If it is outside a nucleus it lives for about 12 minutes and then decays into a proton, emitting an electron and a neutrino in the process.

Without neutrons, nature could not have built up heavy nuclei just by using protons. Thus the carbon nucleus has six protons and six neutrons to account for its atomic mass of 12, and the oxygen nucleus has eight protons and eight neutrons. The protons, which repel each other, give the nucleus its positive electric charge so that it can hold on to negatively charged electrons to produce a neutral atom; a nucleus consisting of protons alone would explode. The neutrons in the nucleus prevent this explosion by pulling on the protons and on themselves to keep the nucleus intact. The nonelectrical force with which protons and neutrons pull each other together is more than 100 times as strong as the electrical repulsion between the protons. Hence this nucleon–nucleon force is called the strong nuclear force.

Returning now to the role of the neutron in constructing heavy nuclei in conjunction with protons, we note that a proton and neutron can combine to form heavy hydrogen, called the deuteron, an isotope of hydrogen; two protons and two neutrons form helium 4 (the nucleus of ordinary helium); helium 6, an isotope of helium, consisting of two protons and four neutrons also exists. Thus isotopes of an atom are variations of the given atom with the same chemical properties of the given atom but with a heavier nucleus, owing to the presence in this nucleus of one or two neutrons more than in the normal nucleus. The number of neutrons that can exist in a nucleus is strictly limited by the tendency of neutrons to emit electrons if they are not completely held captive in the nucleus by the strong force.

The reader may wonder how neutrons could have played any role at all in the build-up of the heavy nuclei. The answer to this question, as we have already seen in our discussion of the generation of energy by the nuclear fusion of protons to form helium, is that if the protons are packed very closely together, as they are inside the sun, and are moving fast enough to interpenetrate in spite of their natural mutual repulsion, one of them can change into a neutron, and then be captured by the other to form a heavy hydrogen nucleus called the deuteron; these deuterons can then go on to form helium nuclei with the release of the solar energy we receive on the earth. As we have seen, neutrinos play an important role in this process.

All of these revolutionary developments in physics did not have a direct impact on astronomy immediately, although by the end of the 1920s astronomers had incorporated into the study of stellar radiation most of what the quantum mechanics of the atom had revealed about spectroscopy. Indeed, the stars became the spectroscopic laboratories of the physicists themselves, for no laboratory on earth could produce the extremes of pressures, densities, and temperatures found in the stars. And, as nuclear physics grew in our earth-based laboratories, it became clear that the ultimate verification of nuclear theories would be found in the dynamics of stellar interiors.

That astronomy can contribute significantly to the test of our physical theories had already been demonstrated by the astronomical verification of the three crucial tests of the general theory of relativity: (1) that massive bodies alter the path of a ray of light grazing these bodies; (2) that the light leaving the surface of a highly compressed massive body (e.g., a white dwarf or a pulsar) is reddened (the Einstein red shift); (3) that the orbit of the planet Mercury is not fixed, but rotates with the planet (the advance of the perihelion of Mercury where the perihelion is the point where Mercury is closest to the sun).

All three of these predictions of the general theory of relativity were confirmed in time by astronomical observations. The most famous of these predictions, the bending of the path of a ray of light from a distant star grazing the sun, was confirmed in 1919 by the total solar eclipse expedition led by the British astronomer Sir Arthur Eddington. By comparing the photographs of groups of stars on both sides of the sun during the total eclipse of the sun (when the stars are suddenly visible for a short period of time) with a photograph of the same groups of stars taken 6 months later when they were visible in the nighttime sky, Eddington found that the two groups (one on the right and the other on the left) appeared during the eclipse further apart than they appeared 6 months later. This means that the paths of the light from these stars (on both sides) were bent by the sun's gravitational field, as predicted by Einstein.

This discovery was announced to the British public the following day by huge headlines in the London dailies stating that Newton had been "dethroned." This was but the first instance of the strong intimacy that had developed between physics and astronomy and that went far beyond that which had been generated by stellar spectroscopy. Moreover, it convinced astronomers that the understanding of cosmic events requires a mastery of the most advanced physical theories, particularly the general theory of relativity.

During this period the parochial astronomy of the nearby stars and that of our galaxy was giving way to the cosmic astronomy of the distant galaxies and to cosmology. By this time, owing to the brilliant work of Harlow Shapley and Edwin Hubble, nobody doubted that the distant nebulae are galaxies. Their study became one of the most active branches of astronomy; this was also the period during which new windows into space were opened, stimulated by the discoveries of new kinds of "rays" coming from all parts of space. In particular, "cosmic rays," first correctly described by the Austrian physicist Victor Hess in 1911 as emanating from outer space and not from the earth, were among the most intriguing of these new kinds of rays. After a lengthy debate about the nature of these particles everyone agreed that they are very energetic charged particles, primarily protons (intermixed with heavy nuclei) and not electromagnetic rays (e.g., X rays and gamma rays). The speeds with which these cosmic rays strike the earth indicate that they are produced, in some way or other, by the most energetic processes in the universe, exceeding the energies we can produce in our laboratories by trillions of times. The origin of cosmic rays is still not understood or even known.

The study of cosmic rays led astronomers and physicists to the first venture into the construction of a self-consistent cosmology and to the careful study of phenomena not immediately ascribable to the stars or even to the galaxies. The primary question the cosmic rays presented dealt with their vast energies (speeds) which far exceed the energies which stars can generate. Of course, vast energies are released when a giant star collapses to become a supernova but supernovas are rare and are formed in galaxies, whereas cosmic rays come from all regions of space, where no galaxies are found.

Because stars like the sun emit constant streams of charged particle ions called stellar winds, usually protons, one might think that cosmic rays can be accounted for as components of such winds. But this is not so because the particles that constitute the solar wind have energies far below cosmic ray energies.

The search for other "rays" from outer space that were blocked from reaching the surface of the earth by the earth's atmosphere led astronomers to develop and launch space probes that can rise far above the earth's atmosphere and make observations that are not distorted by it. The first of these were short-lived rockets that reached a height of a few hundred miles and fell back to earth. Rockets of this sort were greatly improved during the Second World War, and they became a fairly common astronomical tool, reaching their greatest achievements with the launching of the Soviet artificial satellite, Sputnik, which opened up a new and exciting branch of observational astronomy.

Following the launching of the first man in space, artificial satellites took on another dimension with the launching of men to the moon, of a satellite to the martian surface, of a satellite to probe Saturn and its rings, of a Jupiter probe, and so on. This finally culminated in the launching of the Hubble telescope and the construction of a permanent Soviet space station thousands of miles above the earth's surface. This was designed to house astronauts for months at a time who could then carry on their observations, unhindered, above the earth's atmosphere. This space station is still operating in the mid-1990s.

Among the more important cosmological probes was the Cosmic Background Explorer (COBE) to determine the temperature of the universe, that is, the absolute (kelvin) temperature of intergalactic space, which we discuss later in connection with cosmology. In this discussion we shall have to define these terms as precisely as possible. Cosmology, as we define it now, is far different from what the ancient Greeks, the Romans, and the pre-Renaissance scholastics (e.g., Thomas Aquinas) called cosmology, and is even different from the cosmologies of Copernicus, Tycho Brahe, Kepler, Galileo, and Newton. Among these, Newton was the only one who pointed to a serious difficulty associated with his law of gravity and, what appeared to him to be, the static, uniform distribution of the "fixed stars" around the solar system. Kepler, and Galileo also, appeared to be content with their model of the solar system, although Galileo was probably puzzled by his discovery that the Milky Way consists of myriads of faint stars at vast distances. How the solar system, with its sun and planets, fits into such a one-sided distribution of stars was a mystery to him and quite inexplicable to Newton who viewed the problem as being of a more technical nature: if all the stars pull upon each other gravitationally, why have they all not fallen together (collapsed) to form a single massive sphere? Because nothing was then known about the stellar motions, this was a reasonable suggestion.

But other questions arose which pointed to what appeared to be irreconcilable contradictions or paradoxes, the most famous of which is the Olbers' paradox. Heinrich Oblers, a German physician and astronomer of the late nineteenth century, puzzled by the faintness of starlight and the darkness of the night sky, argued that this faintness and darkness could not occur if the stars and the distant galaxies were distributed uniformly in all directions. If the stars are distributed uniformly in all directions we may picture the earth as being at the center of such a distribution and the stars lying uniformly on the surfaces of larger and larger concentric shells surrounding the earth. Therefore the line of sight of any sky observer is bound to hit a star however far away it may be, or, to put it differently, a ray from that star is bound to enter the observer's eye. Hence the observer should see each point of the sky as being as bright as the surface of the sun. Because this is not true, we conclude that the cosmic galactic distribution is not infinite. The universe is therefore finite.

With these various difficulties facing cosmologists at the beginning of the twentieth century, it is no wonder that some 30 years were still to elapse before cosmology became a serious branch of astronomy and before cosmologists began to construct acceptable models of the universe. But in spite of these stumbling blocks and dead ends that faced cosmologists, Einstein, who never allowed such things to deter him from the pursuit of the truth, saw in the solution of the cosmological problem a cosmic application of his general theory of relativity. He approached the cosmological problem from the most elementary but, physically, most unassailable way. He pictured all the matter in the universe as smeared out into a thin fog, occupying and filling every cubic centimeter of space uniformly.

This implies that Einstein had assumed that the universe does not extend to infinity but is confined within finite boundaries (like the surface of a sphere). Indeed, he went on to introduce the "radius" of the universe as though the universe were a sphere. He was led to this model of the universe by his general theory, which is really a theory of gravity that supersedes Newton's law of gravity. In this theory (now universally accepted), the force of gravity is replaced by a curvature of space-time produced by matter. He proposed, pursuing his picture of the universe filled with a thin "fog of matter," that at each point in such a universe the matter cause space to curve inward so that a closed surface (a three-dimensional or hypersurface) is formed. This, of course, is very difficult for us to perceive because in our three-dimensional world we are aware of two-dimensional surfaces in three-dimensional space and no higher dimensional surfaces.

This did not deter Einstein because he had available to him a fourth dimension, time, which can easily be incorporated with space into a four-dimensional space–time manifold. From a mathematical point of view, the concept of a three-dimensional curved surface in a four-dimensional space–time manifold is easy to formulate and comprehend. In 1917 Einstein published his historic paper on cosmology, in which he sought to construct a closed spherical model of the universe. But he ran into considerable difficulties because his equations did not lead to a closed universe with radius $R$. He therefore altered his equations somewhat by adding a term (the famous cosmological constant) to obtain a static universe, for he had no reason to suppose that the universe is either expanding or contracting. This turned out to be a misjudgment, as we shall see.

In spite of this flaw, Einstein's first venture into cosmology was revolutionary, for it opened the floodgates of cosmological model building. Those cosmologists who followed Einstein used the Einstein cosmological equations as their guide. Some, such as the Dutch cosmologist Wilhelm de Sitter and the British astrophysicist Arthur Eddington, accepted Einstein's small addition, the cosmological constant. Others rejected it, as Einstein later did, calling it the "biggest blunder" of his life. The de Sitter model of the universe was readily accepted by Eddington and others because the cosmological constant acts like a repulsive force which not only prevents the universe from collapsing, but actually gives it an expansion. In fact, de Sitter showed that the cosmological constant allows an expanding space to exist even if it contains no matter.

All of this, of course, was quite ad hoc and arbitrary and, hence, contrary to the spirit of the scientific enterprise. However, the concept of the expanding universe became accepted as fact with Hubble's observations of the recession of the distant galaxies and his verification (by careful observations) in 1929 of his "law of recession." But some important questions remained to be answered before such a strange concept as an "expanding universe," could be accepted. These questions were, and some still are, most difficult to answer. The first of these questions dealt with the theoretical nature of the expansion. As the only basic theory one can use to describe or analyze the universe is Einstein's general theory of relativity and the cosmological equations that stem from it, one had to endow these equations with the necessary mathematical properties to be applicable to cosmology. This was done in 1921 by the Russian mathematician Alexander Friedmann, who showed that Einstein's equations can be used to describe an expanding universe if the parameters in these equations that describe the universe at any moment in its history (its size, its mass distribution, etc.) are permitted to change with time. Expressed mathematically, they are functions of time. This gave the theory of the expanding universe and the cosmologists themselves an enormous boost.

But even with the basic theory of an expanding universe in good shape, the concept itself still had several difficulties. To see the nature of the difficulty that most people (including some science teachers) encounter in discussing the expanding universe, we try to go back to some initial moment when we may say that the expansion began. Clearly, at that moment with the entire universe (every bit of matter and energy) concentrated in a very tiny volume, the temperature, pressure, and density must have been exceedingly high (e.g., the temperature at about a trillion trillion trillion degrees). It follows, therefore, that the expansion must have begun as a vast, very hot explosion which Fred Hoyle nicknamed the "big bang." This name has stuck and is commonly used in all literature (scientific and general) referring to what is also called the "birth of the universe." Indeed, the "age of the universe" is reckoned as the time from that "initial moment" to the present, which is just the arithmetic reciprocal of the Hubble constant of expansion. Unfortunately, this picture of the origin of the universe leads people, in general, to believe that the universe originated as a vast explosion that occurred billions of years ago at some point in space. But this is wrong for no space existed before the "big bang"; if it did exist at all, it was confined in the tiny nodule in which all the matter and energy were confined. Indeed, the expansion of the universe means that space itself is expanding and that it is meaningless to speak of the expansion of the universe as though it were expanding into something. Indeed, the concept of "something" outside the universe into which the universe is expanding is a contradiction in itself because the universe is defined as everything, so that no "something" can be beyond it.

To make the expansion of the universe more understandable we use the earth and our perception of it as a metaphor for the universe. Here we imagine ourselves as two-dimensional creatures so that our perception is entirely two-dimensional (we have only horizontal, not vertical, perception) so that we cannot perceive anything within the earth or outside it; we live on a two-dimensional surface. To draw the proper analogy with the universe, we picture the three-dimensional space of the universe as being a three-dimensional surface, a so-called "hypersurface" in which time is the fourth-dimension. We can move in all directions in space (along our hypersurface) but we cannot move along the time direction.

To pursue this analogy further we picture the size of the earth as doubling overnight (in about 10 hours) so that at the end of that interval the radius of the earth has increased to 8000 miles and the diameter to 16,000 miles. This means that all distances on the earth have doubled and the surface area has become fourfold larger. Suppose, now, that while this was happening we in New York were receiving radio signals from a Chicago station (1000 miles away) and a San Francisco station (3000 miles away). During the 10 hour interval of the earth's hypothetical expansion, the Chicago station would be receding from us at 100 miles per hour whereas the San Francisco station would be receding from us three times faster, that is, at 300 miles per hour. These speeds of recession would show up as Doppler increases in the wavelengths of the respective radio signals. We could not receive these radio signals during the expansion at the exact same radio dial positions as before the expansion. This essentially would be the Hubble law for the expanding earth and the Hubble constant would be 1/10 mile per hour per mile for this fantasy. Fantasy though it is, it has in it all the basic physical elements that we find in the expansion of the universe. Accepting the big bang theory and the resulting expansion of the universe as facts, we are immediately confronted with very difficult and probably unanswerable questions: What produced the big bang? How will the universe finally end? The first question is extremely difficult to answer because it requires that we introduce the concept of the pre-big bang state of the universe, about which we can say very little in any nonarbitrary fashion.

Considering the second question, we have two possible answers: (1) the universe will go on expanding forever or (2) it will stop expanding and collapse to some previous initial state. Here, of course, we must define the term "initial state," which we discuss later. We can answer this second question if we know something very important about the mass of the universe and its distribution in the universe. Clearly, if the universe is to stop expanding and begin to collapse it can do so only if the gravitational pulls of all the stars and galaxies on each other in the universe are large enough to halt the expansion and cause all the matter in the universe to begin to collapse.

Here we again draw an analogy with the earth. We know that if we throw an object upward, it rises to a certain height and then falls back to the earth again. We also know that the faster it leaves our hand, the higher it rises before it returns to the ground. Can we then throw it so fast that it never comes back? The answer is yes; if it leaves our hand at a speed of 7 miles per second, it will move off and never return. This speed, called the "speed of escape" from the earth, depends on only two terrestrial parameters: the mass of the earth and the radius of the earth (the distance of the earth's surface from its center). The larger the mass, the larger is the speed of escape, and the larger the earth's radius, the smaller is the speed of escape. We could apply these same criteria to the expanding universe if we knew the total mass of the universe and its radius (if we assume it to have been a sphere at the big bang) at the moment of the big bang, but we know neither of these parameters.

Concerning the total mass of the universe we immediately face a difficulty owing to the presence of "dark matter" which, we estimate, accounts for about 96 percent of the total mass of the universe, with the visible matter (stars, interstellar dust, and certain very luminous intergalactic objects called quasars) accounting for no more than 3 percent. Because we cannot determine the total mass of the universe directly, astronomers have introduced another criterion for determining whether the universe is expanding faster than the speed of escape (an open universe that will expand forever) or slower than the speed of escape (a closed universe that will, in time, stop expanding and begin to collapse). This criterion is called the "critical density" of the universe. Density is defined as the amount of matter or mass concentrated in one cubic centimeter of space. This critical density is 100 thousandths of a trillionth of a trillionth of a gram. The visible matter in the universe gives a density that falls below this critical value by almost a factor of 100, so that if only the visible matter were present in the universe, the universe would go on expanding forever (an open universe), but the dark matter, if counted, gives a density well above the critical value so that we must conclude that the universe will stop expanding and begin to collapse after some 60 billion years.

Accepting the universe as a closed universe we may first consider how long it has been expanding (the age of the universe) since the big bang. This is just the arithmetic reciprocal of the Hubble constant, as we have already stated, and the radius of this universe is of the order of 10 billion light years. This means that if we could see out to a distance of 10 billion light years we would see the universe as it was just after the big bang. Our earth-based telescopes and the orbiting Hubble telescope have not yet revealed this distant horizon to us, but other observations, not of the matter in the universe but of the cosmic background radiation, give us an almost complete picture of the conditions in the universe shortly after the big bang (about 500,000 years). These observations were made and recorded by a special satellite, the Cosmic Background Radiation Experiment (COBE) designed to detect very weak, long wavelength microwave radiation from all directions. This experiment was successful, proving, conclusively, that a weak background radiation exists, reaching the earth from all directions. This radiation is homogeneous but not quite isotropic (the same from all directions) because of the motion of the solar system through space. Indeed, this background radiation is found to be more intense coming along a direction opposite to that along which the solar system is moving and somewhat weaker from the opposite direction. This line of the solar system's motion determined by the cosmic background radiation agrees with that found from the apparent motions of the stars in our neighborhood in the galaxy.

The conclusions drawn from the COBE data are that the cosmic background radiation corresponds exactly to what we would expect if the radiation were coming from a furnace at a temperature about 3 degrees above absolute zero, which is 273 degrees below zero centigrade. With this discovery of the cosmic radiation the big bang became a fact because the cold cosmic background radiation is the cosmic remnant radiation of what, originally, was extremely hot radiation. Here we have introduced the concept of the temperature of the radiation, which implies that the radiation (also called "thermal radiation") is "blackbody radiation," which was described by Max Planck in his black body radiation formula. This formula tells us that the nature of this type of radiation does not change, whether it is emitted from a very hot furnace or a cool furnace. All that changes is the relative intensities of the various components (colors) in the radiation. The temperature of this radiation is thus given by the most intense color (wavelength) in the radiation.

With this in mind we trace the story of the cosmic background radiation from the time of the big bang to the present. Initially this radiation was extremely hot (trillions of trillions of degrees) and was mixed together with protons, electrons, and alpha particles (helium nuclei) with the protons constituting about 77 percent of the total mass. Because the protons, electrons, and alpha particles are electrically charged particles, the big bang radiation interacted with these particles, exerting a pressure on them, causing them to move along with the radiation so that matter and radiation were expanding more or less together. Cosmologists refer to this stage of the expansion as the "radiation dominated" phase, since the particles of matter were very much like bits of straw in a strong wind. This radiation dominated phase lasted for about a million years, until the temperature of the whole universe had dropped to about 3000 degrees kelvin, at which point the protons and alpha particles began to capture electrons to form neutral hydrogen and helium atoms, so that the radiation could no longer exert a pressure on these atoms nor tear the electrons out of them. Cosmologists refer to this as the "decoupling stage" of the expansion which was then followed by the "matter dominated" stage, which brings us to where we left off earlier in this chapter in our discussion of cosmology.

From this discussion we can understand why the initial radiation had to cool off to a temperature of the order of a few thousand degrees before neutral atoms could be formed and before these neutral atoms, primarily hydrogen and helium, would then settle down, gravitationally, to form stars. These stars combined gravitationally to form the galaxies, which in turn, again, under the action of gravity, arranged themselves into clusters of galaxies, which now form what cosmologists call the "large-scale structure" of the universe.

Here we pause for a moment to consider the role that electric charge played in the formation of stars. One might suppose that stars could have been formed gravitationally from clouds of particles, with no internal electromagnetic structures, such as we know exists in neutral atoms (protons and electrons). But this is not so because of the constraints imposed on all dynamic phenomena by the principle of the conservation of energy. To see what is involved here we consider as previously described the material of a star, such as the sun, spread out through all of space like a thin cloud of molecules so that its total energy (kinetic plus potential energy) is zero. Suppose now that all the particles in this thin cloud begin to move toward each other under the action of their mutual gravitational attractions, thus acquiring kinetic energy. But the conservation of energy demands that their total energy (kinetic plus potential) still remain zero. This can only be so if their potential energy becomes less than zero (negative). It is clear from this that these particles can never form a stable gravitationally bound system (e.g., a star) unless they have some mechanism for losing energy so that their total energy can drop below zero, and remain that way. In other words, a gravitationally bound structure can remain so bound only if its total energy (kinetic plus potential) somehow or other becomes negative. The electric charge provides that mechanism in a very beautiful and simple way.

We again picture the particles (now neutral atoms) in the cloud, moving with ever increasing speeds, toward each other until they begin to collide violently. These collisions cause the electrons in the atoms to jump to higher orbits in the atoms (higher energy states) but they jump back to their lowest orbits (states) again, releasing the energies they had acquired but now in the form of discrete photons which then leave the star in streams of radiation. This radiation cools the star, reducing its total energy below zero so that it remains stable and continues to collapse.

It may seem to the reader that we have tied up all loose cosmological ends with this discussion, giving us a beautiful self-consistent cosmological model of the universe, and yet not all cosmologists are happy with this picture. This picture was and still is distasteful to astronomers such as Fred Hoyle, Herman Bondi, and Thomas Gold, who proposed what they called the "continuous creation" or "steady state" cosmology. This cosmology, which argues that the universe has always been the same, drops the big bang, replacing it with a mechanism which "creates protons in each cubic centimeter of space" at a steady rate to replace the distant galaxies that continually disappear owing to their recession from each other. This cosmology is unacceptable to most astronomers for two reasons: it requires the introduction into Einstein's field equations of a "continuous creation operator" (an additional tensor) thus spoiling the simplicity and beauty of

Einstein's equations, and it cannot account for the observed background cosmic radiation.

Another cosmology that was vigorously supported initially by a number of leading cosmologists but is no longer as popular as it was initially, was introduced in the 1970s by the MIT physicist Alan Guth, who was concerned with certain esoteric features of the observable universe that seemed to contradict the big bang cosmology. Guth called his cosmology the "inflationary" universe cosmology. Using the Einstein–de Sitter universe as a starting point, with the cosmological constant which leads to an expanding empty space, Guth argued that the empty universe expanded initially by many orders of magnitude (a hundred thousand trillion trillion times) thus "supercooling" itself and then releasing vast amounts of energy which became the big bang. This theory, like the continuous creation theory, has lost favor because it is just too complex a scenario for nature to have followed. Why could not nature have produced the big bang directly instead of going to the trouble of a vast inflation first and then a big bang? One important feature that we have learned about nature is that it performs its "miracles" minimally, using only as much technology as it needs.

Having presented the various cosmologies that have been invented to account for the origin and evolution of our universe, we return to the universe, as we see it, and describe some of its mysteries that still remain to be explained, and we return to "cosmic rays," a rain of very energetic particles that strike the earth from all directions. No one suspected the existence of such rays until the Viennese physicist Victor Hess in 1911, in a famous paper on the emission of gamma rays from radioactive nuclei (e.g., uranium, radium, etc.) in the earth's surface, pointed out that not all the radiation detected on the earth's surface can be accounted for by the gamma rays from terrestrial radioactive nuclei. Hess used the data gathered by balloon flights which showed that a component of the radiation detected on the earth does not fall off with distance from the earth as the gamma rays from the terrestrial radioactive nuclei do.

For some time the nature of the cosmic rays was unknown; all that is known about them even now is that they are extremely penetrating and can be detected in the deepest mines, but we also know that cosmic rays are primarily very energetic protons intermixed with other heavier nuclei. These particles have energies which exceed those of any other known phenomena in the universe. The origin of these rays is still not known.

Another mystery was presented to cosmologists in 1963 by the discovery of certain radio sources which appeared like ordinary faint stars on photographic plates. But from the careful observations of a group of California astronomers including J. Greenstein, T. Mathews, M. Schmidt, and A. Sandage, it was discovered that these objects are at vast distances (billions of light years away) and, therefore, cannot be stars. These objects, now called "quasars," are small, compact extremely luminous objects, emitting as much energy per second as ten trillion suns or a few hundred galaxies.

Quasars first caught the attention of astronomers because they are sources of intense radio waves emitting enormous quantities of radio energy even though they appear like ordinary faint Milky Way stars on photographic plates. Because ordinary stars, like the sun, emit only weak radio waves, the discovery of a "star" that is a source of strong radio waves was an exciting event and a source of wonder. These objects are now labeled 3C48 or 3C273 where 3C refers to the third Cambridge catalog and the numbers 48 and 273 to the listing of these objects in that catalog. Some 600 quasars are now known. The resemblance of quasars to stars is startling: they are starlike; they are strong radio sources; they are variable (their visible luminosities change periodically—often weekly or monthly—showing that they are very small compared to galaxies, less than one thousandth the size of a galaxy); they emit vast amounts of ultraviolet and infrared radiation; their visible radiation contains broad spectral emission lines which show that they are receding from us at very high speeds. Hubble's law of recession then tells us that the quasars are as distant as billions of light years, from which we deduce that they are extraordinarily luminous, as previously stated. At this point in our story we can say very little, if anything, about how these objects arose or how they generate their vast flux of radiation.

In the 1960s the American Vela satellites were orbited to detect gamma ray flashes as signatures of nuclear bomb explosions because gamma rays are emitted by nuclei. These satellites did, indeed, record such flashes but it soon became clear that these gamma ray bursts were not produced by exploding bombs but were natural phenomena. This was announced in 1973 in the prestigious *Astrophysical Journal*. These gamma ray bursts have been studied extensively since then and they have been found to occur throughout the universe. These bursts may last for several seconds, and if the gamma ray energy released were emitted in the visible part of the spectrum, each such burst would appear brighter than any other object in the sky outside the solar system. If these events are phenomena beyond the Milky Way, each flash consists of an amount of energy equivalent to that emitted by the sun over billions of years.

We finally describe the problem associated with the observed distribution of matter in the visible universe, which is called the large scale structure of the universe. From general symmetry considerations and from the symmetry of the gravitational force that governs this large scale structure, we expect the galaxies (and clusters of galaxies) to be distributed uniformly throughout space. But the observations of the galaxies out to great distances do not validate this "reasonable" assumption. Indeed, the observations show that the galaxies are distributed linearly, arranging themselves in filaments that appear to lie on the surfaces of hollow spheres so that the large scale structure of the universe looks like bubbles in a soapy mixture. In other words, the universe looks something like a slab of swiss cheese and not like a chocolate chip cup cake with the chocolate chips representing the galaxies and the dough representing the empty space.

Here we have presented the basic cosmological problems that face the current cosmologists who, it seems to us, have tried to solve these problems separately with separate unrelated solutions for each problem. We believe that this approach is futile; the problems are interrelated and should therefore have a single solution. From general physical principles developed in a series of published papers, one of the authors (Motz) has shown that the basic particle in nature is what has become known as the Planck mass particle with a mass equal to that of a grain of sand (about 10 million trillion times the mass of the proton) and that the universe began as a small, cold, dense sphere, tightly packed with these particles (called unitons by Motz). These unitons coalesced gravitationally to form protons (triplet rotators, like molecules) with an explosive release of energy (the big bang). This energy was just the transformation of the masses of the unitons into energy in accordance with Einstein's famous formula $E = mc^2$.

Because not all unitons coalesced in this way, a few were left over—one for every hundred thousand trillion protons formed. These extra unitons account for the dark matter, and close the universe (produce a finite spherical universe), will stop the expansion of the universe, and will cause it, ultimately, to collapse, gravitationally, back to its original cold, pure uniton stage. This will be followed by another big bang. One can show that all the energetic phenomena described in the previous paragraphs and the large scale structure of the universe can be accounted for by the free unitons that are still around. And, so, we end our discussion of cosmology on the hopeful note that we will in time obtain a much deeper understanding of the universe than we now have.

# Epilogue

The story of the birth of the universe and its evolution from chaos to order, told here, spans some 15 billion years. As one contemplates this vast panorama, in which a world of order, symmetry, and intelligent life arose from disorder, as the culmination of a series of natural events, one is struck by the incredible contrast, bordering on what appears to be incompatibility and antagonism, between one realm of nature, life, and the antagonistic realm of the inanimate universe surrounding us. Every step in the formation of the complex organic molecules that constitute a living organism can be explained in terms of natural laws and the basic electromagnetic forces that govern all chemistry, but an understanding of the essence of life eludes us; therein lies the apparent contradiction between the living and the nonliving. Life is not simply the sum of all the molecules and atoms in a living creature at any one moment and the forces that keep these particles together, as is a stone, but rather a complex, unchanging dynamical pattern—a kind of molecular dance with very precise and well-laid-out steps—that maintains itself from generation to generation. The sameness of the species in each generation and of every single living being from day to day is preserved, even though the molecules in every living organism are constantly being replaced by new ones of the same kind, which take their proper places in this molecular dance of life. Now the overwhelming mystery and apparent contradiction presented by these everlasting and precise dynamical patterns, without which life would be impossible, is that they coexist with, indeed, that they give rise to, patterns in living creatures that appear to be imprecise, unpredictable, capricious, and without any fixed structure. The reference here is to what is usually called free will, the freedom of choice that every human has to perform or not to perform a variety of apparently unrelated acts. All the molecules in one's body are in accordance with the laws of the quantum theory in well-defined ways that, although not strictly deterministic, owing to the quantum principle of indeterminacy, are statistically precise. Yet one is not governed by this precision, for one can move his or her muscles as he or she pleases, think what he or she pleases, and respond to sensations in a variety of ways. How can such macroscopic unpredictability stem from microscopic predictability? Because the fluctuations arising from quantum indeterminacy can be shown to be negligible when large masses, such as the human brain, are concerned. One cannot ascribe the acts of free will to quantum indeterminacy; they must stem from a qualitative change in the behavior of molecules when they form large aggregates.

But these various random and unpredictable acts that all human beings perform from moment to moment are actually less capricious than they appear, for although thoughts of all sorts flit through people's minds constantly and all kinds of minor things are done by people in a random way, they all add up to overall patterns that are not only predictable but common to all people. Moreover, these overall predictable behavioral patterns distinguish man from all other living species, each of which is marked by its own set of behavioral patterns, which also emerge from a sea of random acts. However randomly one idea follows upon another in the thoughts of each person, and however arbitrary and unpredictable the momentary actions of each may be, there is a quality common to all that people do and think that permits each to relate his or her individual consciousness to that of everyone else. Without this commonality or sense of identity, communication between individuals would be impossible and life would cease.

In recognizing the existence of these overall patterns, one becomes aware of two distinct groups of such patterns; one of these groups is essential not only to the survival of the individual but to life itself, but the other, having no apparent survival value, presents the greatest mystery of all. The first group includes all those responses to the environment that are necessary for the protection and sustenance of life, such as the avoidance of danger, the escape from pain, and the satisfaction of hunger and sexual desires. It is easy to understand and to classify such drives in terms of survival value, and one can see how they fit into nature's plan for the propagation of life, but these survival responses and the incredible clinging to life that are common to all living beings do not explain life or justify its existence. But an explanation of and reason for life may be found in the second group of patterns, which are most pronounced and highly developed in man, but are probably present on a very much smaller scale in lower animals: all those activities and human urges that are associated with truth, love, beauty, justice, charity, mercy, a compassion for living things in general, and other virtues that are often pursued and practiced at a loss to one's own well-being. Here, also, in these purely cerebral activities, one perceives patterns common to all human beings; that great musicians, artists, scientists, poets, and philosophers can communicate their ideas to others and generate in others similar thoughts establishes that universality of all minds which coexists with individual identities.

Of all of these virtues, the search for truth is the most remarkable and is, perhaps, the reason for intelligent life itself, as though nature were seeking to satisfy its own intellectual curiosity by examining and comprehending the universe through the eyes of a thinking being. Truth here does not mean the truth thought of in one's daily activities, where it is essential to know the "truth" about the time, the weather, one's health, or one's financial state. The virtue of such pragmatic "truths" is obvious, for without them the constant uncertainty about routine events would make life intolerable. Rather, it means that truth expressed in the laws that govern the universe and reveal to man the universal properties of space, time, and matter. These properties are often so obscure and hidden from direct view that they cannot be discerned in isolated facts or deduced from unrelated events but must be derived from fundamental laws. Now the curious thing about this aspect of truth is the remarkable appeal that it has for the human mind; and its pursuit, which motivates humanity more powerfully than anything else, clearly distinguishes mankind from the rest of the animal kingdom. This pursuit is the most glorious and exalting experience of which man is capable and such truth possesses elements of beauty. Emily Dickinson speaks of "one who died for beauty" as being a brother to "one who died for truth," since the "two are one." And Edna St. Vincent Millay opens her sonnet on Euclid with the line "Euclid alone has looked on beauty bare," thus equating the truth of euclidean geometry with the purest form of beauty. Again, Shakespeare, in one of his sonnets, notes that truth enhances beauty:

> O, how much more doth beauty beauteous seem
> By that sweet ornament which truth doth give.

Man also assigns a quality of truth to the beautiful and tends to equate any intellectual or emotional experience that exalts him to a revelation of some great truth. Men thus speak of a "valid" experience or a "moment of truth" or the "validity of a religious revelation." But this exaltation of emotion and sensation to the level of great truth obscures the nature of scientific truth and opens the door to mysticism and metaphysics, which have no place in science.

That such drives toward truth and beauty exist and stem from the same molecular patterns that drive people to satisfy their hunger is the greatest of all mysteries, unless one supposes that these very mysteries are not to be explained by all the events that preceded the emergence of life but rather, as the most sublime manifestations of life, explain and give reason for everything else that went before. One would then have to say that the big bang, the expansion of the primordial matter stemming from it, the formation of the galaxies and the oldest stars, the nucleosynthesis of heavy elements from hydrogen and helium in the hot interiors of these stars, the nova outbursts on the death of these stars, and the consequent formation of planetary systems like the earth's, were all in preparation for the greatest event of all—the appearance of intelligent life.

But one may ask why, if intelligent life is the reason for the universe, was it necessary for such vast and complex cosmic preparations before life emerged? Scientists often say that nature "solves all its equations at once and exactly" and that everything moves and behaves in accordance with these equations. Why, then, having the solutions of all its equations at hand, and presumably knowing the end results, could not nature have set the stage for a cosmic genesis, and created a complete intelligent being in a single act? The reason lies in the quality of the natural laws themselves and in the mathematical properties of the equations; the laws are quantum laws and the equations are differential equations. Atoms and molecules can exist only because action is quantized; but this very important feature of the universe, which is essential for the existence of the great variety of matter, introduces an ineluctable indeterminacy (the famous Heisenberg principle, which states that the position and motion of a particle—for example, of an electron—cannot both be known simultaneously with infinite accuracy) that prevents precise knowledge of future events, even if the solutions of all the equations that govern all events are known. The reason for this is that solutions of differential equations can predict the future only if the present is precisely known; but quantum indeterminacy, which is the essential part of nature, prevents this from occurring. Now this indeterminacy is not only a restriction on man's ability to make precise measurements; it is a restriction built into the laws of nature themselves, so that there can be no infinite intelligence that can know all things and predict all future events. This means that nature, with all its wisdom, cannot construct a perfect being in a single act because quantum indeterminacy makes the knowledge of what is perfect unattainable even to an infinite intelligence, if there were such an entity in the universe.

The development of beings of ever-increasing intelligence must therefore occur in that piecemeal manner called evolution, with the quantum fluctuations at each stage of development determining the new forms that can arise from the old ones. It is customary at this point to invoke such concepts from the theory of evolution as random mutations, natural selection, and survival values to explain the emergence of complex living forms from simple forms, but such concepts cannot possibly explain many highly developed and extremely complex attributes and characteristics of advanced living organisms that neither have survival value now for the organisms nor had any such value at any stage of their development. Even where some particular constituent of an organism, such as the venom of a snake, has obvious survival value, random mutations and natural selection cannot explain its final emergence from earlier stages in its development when it had no survival value, when it was a nonpoisonous protein. How could random mutations and natural selection have forged the simple inoffensive proteins of a nonpoisonous presnake organism into the virulent venom of a rattlesnake or a cobra without some kind of prescience about the end product? If one accepts such prescience, one is thrown into the realm of mysticism, which is contrary to the whole concept of evolution; but if one does not accept this idea, one cannot understand, using the present concepts of evolution, how there can be any direction in the evolution of a constituent of an organism during periods when the constituent has no survival value.

There is a way out of this difficulty that not only explains the evolution of biological components, such as venomous proteins from nonpoisonous proteins, but also the development of the intellectual attributes of human beings. Random mutations and natural selection play an important role in evolution but by no means the dominant role, which must be assigned to what might be called the potential for change or evolution in a certain direction. It may be that programmed into each gene or perhaps each DNA molecule in the nucleus of a germ cell is a potential for change in a certain preferred direction (which is programmed into the molecule itself as the result of the changes up to that point). As evolutionary changes in some particular gene occur, the symmetry, and hence the potential for additional change, in that gene may be enhanced or diminished; but as time goes on, a point of saturation is approached, and changes become less and less frequent until all significant changes stop. Thus one may compare the vast evolutionary proliferation of living organisms to a huge maze with many bypaths that lead to dead ends but with one main, open-ended track, along which mankind is progressing.

The evolutionary process suggested here is similar to the process that leads from simple atoms to complex molecules in a given mixture of various kinds of atoms and molecules under appropriate conditions. As noted, a hierarchy of complex molecules evolves quite naturally out of interacting atoms if they have enough energy and are allowed to come into contact with each other. An examination of such a mixture reveals a whole range of molecules, some of which undergo no changes at all while others change into more complex molecules; each molecule that is formed either carries with it the potential for combining with other molecules and becoming more complex or has exhausted that potential and proceeds no further. Applying this idea on a much vaster scale, to the evolution of living organisms, one sees that the direction in which an organism will evolve is dictated by its previous history and development, and among all such organisms there will always be one species—the main track in the maze—in which the probability for evolution to higher and higher forms will be a maximum. Human beings are this species on the earth, and their development toward perfection, a state that cannot be defined at any stage in the development of human beings, are inevitable. This is entirely a consequence of the symmetry and three-dimensional structure of human DNA molecules, which in turn stem from the basic forces in nature, and which at each stage are a product of the changes that were dictated by the symmetry of the molecules at a previous stage of their development. Thus the very chance and randomness that prevent man from predicting the future are the tools that nature uses to insure not only the emergence of life in each expansion cycle of the universe but also the approach to perfection and complete harmony as more and more complex forms of life evolve.

We present one more attribute of the universe as a whole, but one which is most pronounced in living organisms, particularly in the human brain: that of self-organization. This self-organization does not require any act of creation; it simply stems from the laws of nature and the basic forces governed by these laws. Though the self-organization of such inanimate structures as galaxies, stars, and planets are qualitatively different from those of living structures (plants and animals), we can point to no natural law or force that produces this difference; we must conclude then that these living structures as parts of the universe are governed by the same laws that govern the universe and that societal structures (e.g., families, tribes, cities, states, etc.) as parts of the universe, are governed by the same laws that govern the universe. Although we do not know the precise nature of the force that cements societal structures, we may surmise that it is some variation of the electromagnetic force flavored in some way by the quantum mechanics. This surmise is prompted by our conjecture that thought (mind) is a manifestation of the quantum mechanical waves (called de Broglie waves, after the French physicist Louis de Broglie, who proposed them in 1923) produced by the motions of electrons in our brains.

The human body provides an interesting metaphor to understand better how mutually beneficial cooperation, prompted in some unknown way, by the fundamental forces of the universe, occurs in nature. The trillions of cells and complex organs in our bodies operate together to produce healthy bodies. Each organ absorbs from the blood exactly what it needs (no more, no less) to operate most efficiently. But the self-organization of our bodies also provides us with powers far greater than those given to other animals on earth. In our perception of disease, for example, we differ drastically from lower animals. Whereas such animals perceive pain, they are unable to understand that the pain is associated with some type of disease. Our awareness of the concept of disease may ultimately lead to the evolution of the brain to a point where it can construct molecular healing devices, specifically designed to move from certain healing cells where they are produced to the diseased cells.

We complete this story with Lucretius' paean of praise to science:

> Science lift aloud thy voice
> That through the conscience thrills.
> Thrills through the conscience the news of peace.
> How beautiful thy voice sounds from the hills.

# Bibliography


Abetti, Giovanni, *Exploration of the Universe*. London: Faber & Faber, 1968.

Aller, L. H., *Atoms, Stars and Nebulae*. Cambridge, Mass.: Harvard University Press, 1971.

Baade, Walter, "Stellar Populations." In *Evolution of Stars and Galaxies*, ed. Cecilia Payne-Gaposchkin. Cambridge, Mass.: Harvard University Press, 1963.

Bergamini, David, *The Universe*. New York: Time-Life Books, 1962.

Bok, B. J., and Bok. P. F., *The Milky Way*. Cambridge, Mass.: Harvard University Press, 1957.

Bondi, Herman, *Cosmology*. Cambridge: Cambridge University Press, 1960.

Bondi, Herman, *Rival Theories of Cosmology*. Oxford: Oxford University Press, 1960.

Burbidge, Geoffrey, and Burbidge, Margaret, *Quasi-Stellar Objects*. San Francisco: Freeman, 1967.

Chandrasekhar, S., *Stellar Structure*. Dover, 1957.

Dicke, R. H., *Gravitation and the Universe*. Philadelphia: American Philosophical Society, 1970.

Eddington, Sir A. S., *The Internal Constitution of the Stars*. Dover, 1959

Fowler, W. A., *Nuclear Astrophysics*. Philadelphia: American Philosophical Society, 1967.

Gamow, George, *The Creation of the Universe*. New York: Viking, 1952.

Glasby, J. S., *Boundaries of the Universe*. Cambridge, Mass.: Harvard University Press, 1971.

Hawkins, Gerald S., *Splendor in the Sky*. New York: Harper & Row, 1969.

Hoyle, Fred, *Galaxies, Nuclei and Quasars*. New York: Harper & Row, 1965.

Hoyle, Fred, *The Nature of the Universe*. Oxford: Oxford University Press, 1960.

Hoyle, Fred, *Frontiers of Astronomy*. New York: Harper, 1961.

Hubble, Edwin P., *The Realm of the Nebulae*. New York: Dover, 1958.

Jeans, Sir James, *Astronomy and Cosmogony*. Dover, 1961.

Kopal, Zdenek, *Man and His Universe*. New York: Morrow, 1972.

Lovell, A. C. B., *Exploration of Outer Space*. New York: Harper, 1962.

Lovell, A. C. B., *Our Present Knowledge of the Universe*. Cambridge, Mass.: Harvard University Press, 1967.

Lovell, A. C. B., *et al., The New Universe.* New York: Giniger-Rand McNally, 1968. See especially articles by R. H. Dicke, J. L. Greenstein, Frederick Reines, and Maarten Schmidt.

Menzel, Donald, Whipple, F. L., and Vaucouleurs, Gerard de, *Survey of the Universe.* New York: Prentice-Hall, 1970.

Misner, C. W., Thorne, K. S., and Wheeler, J. A., *Gravitation.* San Francisco: Freeman, 1973.

Motz, Lloyd, *This Is Outer Space.* New York: New American Library, 1962.

Motz, Lloyd, and Duveen, Anneta, *Essentials of Astronomy.* New York: Columbia University Press, 1971.

Motz, L. and Nathanson, C., *The Constellations.* Doubleday, 1988.

Motz, L. and Weaver, J. H., *The Unfolding Universe.* Plenum Press, 1989.

Motz, L. and Weaver, J. H., *This Is Astronomy.* New York: Columbia University Press, 1963.

Motz, L. and Weaver, J. H., *Astrophysics and Stellar Structure.* Waltham, Mass.: Ginn-Xerox, 1970.

Motz, L. and Weaver, J. H. (ed.), *Astronomy A to Z.* New York: Grosset and Dunlap, 1964.

Russell, H. N., *The Solar System and Its Origin.* New York: Macmillan, 1935.

Shapley, Harlow, *Galaxies.* Cambridge, Mass.: Harvard University Press, 1961.

Singh, Yagit, *Great Ideas and Theories in Modern Cosmology.* New York: Dover, 1961.

Sitter, Willem de, *Kosmos.* Cambridge, Mass.: Harvard University Press, 1951.

Urey, H. C., *The Planets, Their Origin and Development.* New Haven, Conn.: Yale University Press, 1952.

Weinberg, Steven, *The First Three Minutes.* Basic Books, 1977.

Whipple, F. L., *Earth, Moon and Planets.* Cambridge, Mass.: Harvard University Press, 1963.

Woltjer, Lodewijk (ed.), *Galaxies and the Universe.* New York: Columbia University Press, 1968.

Young, Louise B. (ed.), *Exploring the Universe.* New York: McGraw-Hill, 1963.

Young, Louise B. (ed.), *The Mystery of Matter.* New York: Oxford University Press, 1965.

# Index

Absolute motion, 251, 253
Absolute space, 243, 244–245
Absolute time, 243–244
Absorption lines
  Balmer lines, 265
  Lyman series, 264
  scattering process of, 261–262
  *See also* Fraunhofer lines
Academia dei Lincei, 132
Academia del Cimento, 132
Academy of Science, Munich, 173
Academy of Sciences, French, 108, 109–110, 112, 132, 148, 149, 188
Acceleration
  of gravity, 120
  Newton's concept of, 94, 119, 120–122, 188
  principle of equivalence of, 253, 254
  relationship to force, 121–122
Achromat, 173
Action, 217, 218–219
  as physical entity, 332–333
  Planck's constant of, 209, 224–225, 331
  as universal constant, 241
*Advancement of Learning, The* (Bacon), 113
Advance of the perihelion, 256
Agriculture, 3
Air, as basic element, 24–25
Airy, George Biddell, 139–140, 258–259
Airy disk, 139–140, 258–259
Albertus Magnus, 53, 57
Alcor, 18
Aldebaran, 133
Alexander the Great, 5, 32
Alexandria, Egypt, 32–33, 34–35
Alexandrian Library, 33, 52
Alfonsine Tables, 70, 71
Alfonso X, King of Spain, 70
*Almagest, The* (Ptolemy), 37, 46, 49, 67, 71
Alpha Centauri, 183
Alpha particle(s), 224, 346–347
Alpha-particle fusion process, 302–303
Alpha rays, 284, 285
Altitude, 112
  of sun, 8–9
Amber, electricity generated by, 186, 187
Amberger, Paulus, 91
American Astronomical Society, 203
*American Ephemeris and Nautical Almanac, The*, 203
Amiri, Giovanni Battista, 58–59
*Analytical Mechanics* (Euler), 144
Anaxagoras, 22
Anaximander, 21–22, 35
Anaximenes of Miletus, 23
Anderson, Carl, 248
Andromeda, 322–323, 328–329
  Great Nebula of, 317
Angle, as measure of rotation, 70
Angstrom, Anders Jonas, 175, 260
Angstrom (A) unit, 175
Angular momentum, law of conservation of, 86, 305–306
Animals, as zodiac figures, 18
*Annalen der physik*, 233
Antares, luminosity, 269–270
"Antiearth," 25–26
Apollonius of Perge, 36–37, 39–40

Aquinas, Thomas, 53, 57
Arabs
  astronomy of, 54–55
  star names used by, 18
Arc, minutes of, 70–71
Archimedes, 15, 17, 32, 35
Arcturus, 133, 266, 314
Argelander, Friedrich, 259
Aristarchus of Samos, 17, 27, 30, 31–32, 62, 133
  estimation of size of sun and moon by, 48
  influence on Hipparchus, 39–40
  *On the Dimensions and Distances of the Sun and Moon*, 39
Aristophanes, 33
Aristotelianism, 76, 77
  Christianity's acceptance of, 53, 54
  Galileo's disagreement with, 93–94, 100, 103, 105–106
Aristotle, 22, 28, 29–30, 31, 51
  influence of Pythagoreans on, 26
  *Meteorologica*, 29
  *On the Heavens*, 29
  planetary motion theory of, 29, 117
  as Plato's disciple, 27
Asteroids, 176–178
  of Jupiter, 147
Astrology, 1, 9–11, 14, 69–70, 75
*Astronomea Nova* (Kepler), 97–98
Astronomers
  first female, 160
  priests as, 7–8
  role of, as defined by Copernicus, 65
  royal, 130, 132, 132–136
    Airy, 139
    Bradley, 134
    Brahe, 75–76
    Flamsteed, 130, 132–133
    Halley, 133
    Hamilton, 202
*Astronomical Journal*, 203
*Astronomical Message* (Galileo), 99
Astronomical triangle, 131
Astronomical units, definition, 163
*Astronomische Gesellschaft*, 259
Astronomy
  dynamical, 132–133
  internationalization of, 259
  during Middle Ages, 51–56
  naked-eye, 89, 259
  observational: *See* Observational astronomy
  optical, comparison with radio astronomy, 318
  origins of, 1–12
  positional: *See* Observational astronomy
  radio, 317–322
  unity in, 23
*Astronomy and Cosmogony* (Jeans), 276–277
*Astrophysical Journal*, 203, 328, 351
Astrophysics, 185, 257–278
  beginning of, 176
  complete law of gases of, 140–141
  definition, 257
  electromagnetic theory and, 192
  mass-luminosity law of, 278
  quantum theory and, 257
  solar, 273–276
  stellar evolution, 279–308
  stellar physical analysis, 267–276
    Hertz–Russell (H–R) diagram, 268–270, 278
  stellar spectral classification, 261–267
  telescope use in, 260
  theory of relativity and, 257
  thermodynamic theory and, 192–200
Athens, 23
  decline of, 32–33
  Golden Age of, 32
  Plato's Academy in, 27, 32, 52
Atom(s)
  absorption lines, 261–262
  effect of magnetic fields on, 227–228
Atomic models, 207
  Bohr's, 221–222, 223, 224–227, 233, 263, 331, 332
  Heisenberg's, 332
  implications for astrophysics, 208
  nuclear, 331
  Rutherford's, 223–224

Atomic theory, 198
ancient Greek, 22, 23
Atomic weights, 199, 208
Autumnal equinox, 42, 44, 45
Avogadro, Amedeo, 199
Avogadro's number, 199

Baade, Walter, 328–329
Babylonians, 1–2
astrology of, 9
astronomy of, 36, 45
map making by, 8
Bacon, Francis, 113, 132
Bacon, Roger, 53, 113, 114
Beauty, relationship to truth, 356
Bell, Jocelyn, 306
Bell Telephone Laboratories, 317–318
Benatky, Bohemia, 76, 82, 108
Benedetti, Giovanni Battista, 74
Benedictbeuern, Germany, optical institute at, 173
Bentley, Richard, 125, 154
Berlin Academy, 151
Berossus, 4
Bessel, Friedrich Wilhelm, 135, 181–183
Beta rays, 284, 285, 335
Betelgeuse
color, 265
luminosity, 269–270, 314
Bethe, Hans, 289, 290
Biela, Wilhelm von, 178
Big bang theory, 327, 342–343, 344, 345, 346–347, 348, 349, 352, 356
Big Dipper: *See* Ursa Major
Binomial theorem, 115
Blackbody radiation, 209, 211–216, 219–220, 225, 331, 346
Blackbody radiation curve formula, 213, 216–217, 225, 346
Black holes, 249
Blake, William, 156
Blood, circulation of, 114
Bode, J.E., 177
Bode–Titius law, 177, 178
Bohr, Niels Henrik
atomic model of, 221–222, 223, 224–227, 233, 263, 331, 332
portrait, 226
Bologna, University of, 59
Boltzmann, Ludwig, 200, 201
Bondi, Herman, 348
Bonn Observatory, 259
*Book of Genesis*, 2
Born, Max, 233, 332, 333
Boyle, Robert, 140
Boyle's law of gases, 140
Boyle Lecturer, 125
Bradley, James, 130, 134–136, 163
Brahe, Tycho, 10, 31, 67, 70–78, 89
Copernicus and, 64, 70–75
Kepler and, 81–83, 98
observatories of, 75–76, 107, 108
Britons, ancient, astronomy of, 11–12
Broglie, Louis de, 332
Brudzewski, Albert, 59
Bruno, Giordano, 74
Burbridge, Geoffrey, 302, 303, 304
Burbridge, Margaret, 302, 303, 304

Calcagnini, Celio, 58–59
Calculus, 237
differential, 115, 118–119, 129, 132
integral, 115
of variations, 145
Calendar
of astrological forecasts, 10
Babylonian, 1, 7
development of, 3, 5, 6–8
Egyptian, 7
lunar periodicity as basis of, 6
precursors of, 3
Pythagorean, 26
Callinachus, 33
Calorie, 271
Cambridge University, 115, 306
Cannon, Annie J., 267
Canopus, spectra, 267
Capra, Baldassar, 93–94
Carbon, electron arrangements, 230–231

Carbon cycle, of stellar energy generation, 289–290, 303
Carnegie Observatories, 292
Carnot, Nicholas Sadi, 192, 193, 195–196
Cartography, 22
Cassini, Giovanni, 108, 109–110
Cathedral of Frauenburg, 59
Cathode rays, 206
Catholic Chruch, influence on astronomy, 21, 64
  opposition to Galileo, 103, 105–106
Celestial bodies, spherical shape of, 27, 30
Celestial equator, 41, 44
Celestial navigation, 9
Celestial neighborhood, 137
Central fire concept, of earth and lunar rotation, 25, 26
Centrifugal force, 125, 148–149
Cepheid period luminosity law, 307, 325, 328
Cepheid variables, 307
Ceres, 177–178
Chadwick, James, 283, 335
Chaldeans, astrology of, 9
Change, evolutionary, 358
Charles, Alexander Cesar, 140–141
Charles II, King of England, 129–130
Chemical elements, periodic table of, 231, 334
Chinese
  astrology of, 9
  astronomy of, 10, 11, 54, 55
  knowledge of magnetism 189–190
  star names of, 17, 18
  supernova collapse observed by, 304
  zodiac figures of, 18
Christianity
  as obstacle to development of astronomy, 51–52
  *See also* Catholic Church
Christian IV, King of Denmark, 75–76
Chromatic aberration, 109
Chronometer, 130
Cicero, 62
Circle
  points on circumference of, 237
  table of chords of, 48
Clairaut, Alexis Claude, 143, 148
Claudius Ptolemaus: *See* Ptolemy
Clausius, Rudolf Julius, 192, 195, 196, 197
Cleopatra VII, Queen of Egypt, 33
Clock, invention of, 4
Color
  Newton's theory of, 115
  wavelengths, 174–175
Color defect: *See* Chromatic aberration
Comets
  Biela's, 178–179
  discovered by first female astronomer, 160
  Encke's, 178
  Halley's, 130, 133–134, 149
  of 1577, 77
  orbits, 77, 133–134, 178
  Pons', 178
  tails, 179
*Commentariolus* (Copernicus), 64
Compass, magnetic, 57, 190
Concentric model, of the universe, 22
Conic sections, 36–37
Conservation of energy, 146, 285
Conservation of spin, 286
Conservation principles, 86, 217–218
Constant of action, 209, 220, 224–225
Constellations
  circumpolar, 9
  images and names of, 17–18
  as "signs of the zodiac," 5, 10, 18
  *See also* names of specific constellations
Convection, 275–276
"Conversation with the Star Messenger" (Kepler), 99–100
Coordinate systems. 9, 236–237, 245
Copernicus, Nicolaus, 58, 59–67, 69–70
  *Commentariolus*, 64
  *De Revolutionibus*, 64–67, 71–72, 75

Copernicus, Nicolaus (*cont.*)
errors of, 69, 72
heliocentric model of, 32, 61–67, 69, 73
Brahe's modification of, 71–75
Catholic Church's acceptance of, 64
Galileo's acceptance of, 91–92, 93, 95, 100, 105–106
Kepler's acceptance of, 78
religious opposition to, 73–75
Martin Luther on, 74–75
portrait, 60
Cornell University, 306
Corpuscle-wave concept, 233, 331
Corpuscular theory, of light, 138, 139, 180
Cosmic Background Radiation Experiment (COBE), 339, 346
Cosmic harmony, 15, 24, 25–26, 27
*Cosmic Mystery* (Kepler), 81, 91
Cosmic rays, 338–339, 349–350
Cosmogony, 14, 16, 151
Cosmological constant, 323, 341–342, 349
Cosmology, 331–352
ancient, 1–2, 13–19, 339
Babylonian, 13–14
Egyptian, 14–15
Greek, 15–17, 23–26
Sumerian, 13
astronomical photography and, 260
beginning of, 125, 138, 164, 183, 310
continuous creation (steady-state), 348–349
definition of, 1, 151
development of, 323
Einstein's contributions to, 340–342
paradoxes of, 340
physics and, 331–338
problems of, 349–352
quantum theory and, 257
theological, 1–2
theory of relativity and, 257
Cosmos, 24
Coulomb, Charles Augustine de, 187, 189
Coulomb's law of force, 187, 189
"Counterearth," 25–26
Cowan, 335
Crab Nebulae, 55, 304, 305
Cracow, University of, 59
Creation
Egyptian beliefs about, 14–15
of universe, 2
*See also* Big bang theory; Cosmology
Creationists, 196
Crookes, William, 206
Crookes tube, 206, 207, 222–223
Cube, 80

Dalton, John, 208
Dante Alighieri, , 53, 54
Dark lines: *See* Fraunhofer lines
Dark matter, 322, 325–326, 345
unitons as, 352
Day length, 4
increase of, 134
sidereal, 5
solar, 4, 5
De Broglie, Louis, 332
De Brogli wave, 332, 333
*Defense Against the Calumnies and Impostures of Baldassar Capra, A* (Galileo), 93–94
Degree
length of, 110, 149
as unit of rotation, 70
*De Magnete* (Gilbert), 72
De Maupertius, Pierre Louis Morean, 143, 149–150, 151, 218–219
D'Alembert, Claude Jean, 143
D'Alembert principle, 148
Democritus, 22, 23, 198
*De Revolutionibus* (Copernicus), 64–67, 71–72, 75
Descartes, Rene, 236–237
*De Stellar Nova* (Brahe), 76
Deuterons, 288, 336
*Dialogue on the Two Chief Systems of the World, The* (Galileo), 106
Dickinson, Emily, 355–356
Differential equations, 143–145, 357

"Differential Geometry of Space Curves" (Clairaut), 148
Digges, Thomas, 72
*Dimensions and Distances of the Sun and Moon*, (Aristarchus of Samos), 31
Dinar I, Caliph, 33
Dirac, Paul, 248–249, 334–335
Dirac relativistic equation, 334–335
Distance
  as coordinate system invariant, 236–239
  between two points, 238–239
*Divine Comedy* (Dante), 53
DNA (deoxyribonucleic acid), 358, 359
Dodecahedron, 80
Doppler, Christian, 111–112, 183–184
Doppler effect, 111–112, 183–184, 315, 326–327
Dreyer, J.L.E., 62, 316
Duns Scotus, 57
Dust, interstellar, 311–312, 322
Dynamics, 94, 122, 125
Dyne, 188–189

Earth
  atmosphere, effect on astronomical observational precision, 134–135
  axis, nutation of, 136
  circumference, 33–35, 110
  diameter, 30
  ellipsoid shape, 125, 148–149
  future destruction by sun, 299
  oblateness, 125, 149
  orbit, 135–136
  primitive concepts of, 1–2
  rotation
    axis of, 41
    ancient Greek theories of, 25, 30
    Pythagorean concept of, 25
    sidereal period of, 131
  size, 35
  speed of escape from, 344–345
  spherical shape, 9, 25, 30, 53
  surface gravity, 270
Earth (soil), as basic element, 24–25
Eclipse
  ancient observations of, 10, 12, 22
  appearance of stars during, 256, 337
  prediction of, 10, 12
Ecliptic
  obliquity of, 21–22, 33, 41, 43
  plane of, 41
  polar axis of, 43
Eddington, Arthur, 256, 276, 277–278, 337, 341–342
Egyptians, ancient
  astrology of, 9
  astronomy of, 3, 11, 45
  calendar of, 7
  zodiac figures of, 18
Einstein, Albert, 170, 233, 331
  contributions to cosmology, 340–342
  contributions to quantum theory, 221
  cosmological principle of, 323, 341–342, 349
  field equations of, 348–349
  mass-energy ($E = mc^2$) formula of, 246–249, 272
  portrait, 252
  "Propagation of Electromagnetic Phenomena Through Rapidly Moving Media," 233
  *See also* Theory of relativity
Einstein–Lorentz law of the contraction of moving bodies, 255
Einstein red shift, 255–256, 337
Ekphantes, 62
Electric, 186
Electric battery, 189
Electric charges, 186–187, 205–207, 223
  Coulomb's law of force of, 189, 190
Electric current, 189
Electric forces, 114
Electricity, 186–187
  first experiments in, 113–114
  watt as measure of, 271

Electromagnetic fields, equations of, 191–192
Electromagnetic induction, 190–191
Electromagnetic radiation, 281–282
Electromagnetic spectrum, 317
Electromagnetic theory, of light, 185–186, 191–192, 205, 207, 219, 223, 331
Electromagnetism, 185–192
Electrons
in atomic models, 223, 224–225
discovery, 206, 331
distribution in stellar atmospheres, 232
law of motion of, 217
negative energy states, 248–249
Pauli exclusion principle of, 229–231, 334
valence, 229, 230, 231, 263
wave–particle dualism, 332, 333
Dirac relativistic equation, 334–335
Schrodinger wave equation, 333–334
Electron-volt, 271–272
Electrostatics, 187
Elevator thought experiment, of Einstein, 254
Empedocles of Sicily, 23
Encke, Johann Franz, 178
Energy
conservation of, 146, 217–218
Lagrange's theory of, 145–146
Energy-frequency equation, 220
England
astronomy development in, 113–114
name of Ursa Major (Big Dipper) in, 17
observational astronomy in, 129–130
Entropy, 195–198, 283
Epicycle theory
of Anaximander, 22
of Copernicus, 66
of Herakleides, 30–31, 36
of Hipparchus, 47
of Ptolemy, 22, 31, 36–37, 47, 49–51
Equator, celestial, 41, 44
Equinoxes
definition, 41
precession of, 41–45, 46, 47, 48, 59, 136
Equivalence, principle of, 253–254
Eratosthenes, 17, 33–35, 41, 110
Erg, 271, 272, 287
Ether, 219, 241–242, 243
Etruscans, astrology of, 9
Euclid, 15, 27
sonnet to, 356
Euclidean geometry, five solids of, 80–81
Eudemus of Smyrna, 40
Eudoxus, 27–29, 30
Euler, Leonhard, 143, 144, 151, 154, 155
Euler equation, 144–145
Evening Star (Venus), 25, 50
Evolution
creationism and, 196
stellar, 279–308
electrical forces in, 281–282
of giant stars, 302–307
gravity in, 280–281
of main sequence stars, 290–308
nuclear energy in, 282–290
Ewen, Harold Irving, 320–321
Experimentation, 140
*Exposition du System du Monde* (Laplace), 151

Falling bodies
acceleration: *See* Acceleration
law of, 91, 93
Faraday, Michael, 190–191, 207
Ferdinand the Great, 151
Ferdinand II, Grand Duke, 132
Fermat, Pierre de, 150
Fermat's last theorem, 150
Field, John, 72
Figures, Pythagorean theory of, 24–25
Fire, as basic element, 24–25
"First principles," 23
Fitzgerald, George, 242
Fizeau, A.H., 180

Flamsteed, John, 130, 132–133
Fluxions, Newton's theory of, 118–119, 120–121
Force, 120, 121
  Coulomb's law of, 187, 189
  relationship to acceleration, 121–122
  units of, 188
Foules, W., 302, 303
Four elements theory, of matter, 24–25
Fourier, Joseph, 51
Fourier series, 51
Frames of reference, observational, 241, 253–254
Franklin, Benjamin, 186–187
Fraunhofer, Josef von, 173–177, 181, 258, 262–263
  spectroscope of, 181
Fraunhofer lines, 174–176, 192, 208, 221–222, 232, 260, 261
Frederick II, King of Denmark, 75, 108
French Academy of Sciences, 108, 109–110, 112, 132, 148, 149, 188
Fresnel, Augustine, 180, 181
Friedmann, Alexander, 342

Galaxies, 309–329
  clusters, 347
  distance, 307
  distribution, 232, 309–311, 325, 351
  formation of, 347
  galactic: *See* Milky Way galaxy
  local groups, 325
  mass, 325–326
  recession of, 326–328, 342
  as spiral nebulae, 168, 316–317
Galileo Galilei, 18, 67, 89–106, 240–241
  Academia dei Lincei membership of, 132
  *A Defense Against the Calumnies and Impostures of Baldassar Capra*, 93–94
  contributions to natural philosophy, 93–94
  death of, 107
  *The Dialogue on the Two Chief Systems of the World*, 106
Galileo Galilei (*cont.*)
inertia concept of, 29, 40
  inertial/gravitational mass equivalence theory of, 253
  influence on observational astronomy, 107–108
  Kepler and, 81–82, 86
  mathematical career, 91
  portrait, 90
  *The Starry Messenger*, 97–103
Galvani, Luigi, 189
Gamma rays, 249, 284–285, 349, 351
Gases
  kinetic theory of, 198–200
  laws of, 193–195
    Boyle's, 140
    complete law of, 140–141
    of stellar interiors, 291
  solar, 274–275
Gauss, Karl F., 177, 181
Gay-Lussac, Joseph-Louis, 140–141
*General Catalogue of Nebulae* (Dreyer), 316
*General Natural History and the Theory of the Heavens* (Kant), 138, 310
Geocentric model, of the solar system, 5–6
Geographers, 52
Geography, ancient Greeks' knowledge of, 15–16
*Geometrie* (Descartes), 236
Geometry
  absolutes of, 239
  analytic, 237
  ancient Greek, 15–16, 24–25, 356
  euclidean, 356
  general theory of relativity and, 251
  Newtonian compared with Einsteinian, 244–248
  Pythagorean, 24–25
  of space-time curvature, 254–255
Georgium Sidus (Uranus), 162
Gerbert, 52
Germany, 17, 173
Gilbert, William, 72, 85, 113–114, 186

God(s)
  Babylonian, 14
  as creator, 2
  Egyptian, 15
  Greek, 15, 16
  sky as dwelling place of, 1
Gold, Thomas, 306–307, 348
Goodericke, J., 307
Goudsmit, Samuel, 228
Gould, B.A., 203
Gracastoro, Girolamo, 58–59
Gram, 188–189
Gram-atomic weight, 199
Gram-molecular weight, 199
Gravitational dynamics, 185
Gravitational fields, 154–155, 253, 255–256
Gravitational forces, 122–124, 253, 254
Gravitational interaction, energetics of, 280–281
Gravitational mass, 253
Gravitational potential, 154–155
Gravity
  acceleration of, 120
  action at a distance of, 154–155
  Einstein's geometric theory of: *See* Theory of relativity, general theory
  Herschel's theory of, 164–166
  Hooke's claim to discovery of, 136
  Kepler's theory of, 81, 84–85
  Newtonian law of, 81, 89, 115, 122–124, 125, 151–152, 205, 237, 240–241
  universal constant of, 209, 220, 241
Great Nebula, 317
*Great Syntax, The*: *See Almagest, The*
Greek astronomy, 11
  obstacles to development of, 40
  relationship to Greek philosophy, 21
Greek philosopher-astronomers, 11, 21–37
  Anaxagoras, 22
  Anaximander, 21–22, 35
  Anaximenes of Miletus, 23
  Apollonius of Perge, 36–37, 39–40
Greek philosopher-astronomers (*cont.*)
  Archimedes, 15, 17, 32, 35
  Aristarchus of Samos, 15, 17, 27, 30, 31–32, 39–40, 48, 62, 133
  Aristotle, 22, 26, 27, 28, 29–30, 31, 51, 117
  Democritus, 22, 23, 198
  Empedocles of Sicily, 23
  Eratosthenes, 17, 33–35, 41, 110
  Eudoxus, 27–29, 30
  Heracleitus of Epheus, 23
  Herakleides, 27, 30–31, 50, 62
  knowledge of electricity, 186
  knowledge of magnetism, 190
  Leucippus, 22
  Metrodeus of Chios, 23
  myth use by, 16–17
  Parmenides of Elea, 23
  Philolaus, 25–26
  Plato, 26, 27–28, 29, 32, 52
  Pythagoras and Pythagoreans, 23–26
  Renaissance interest in, 57
  Thales of Miletus, 21–22, 23, 35, 40, 186
  Xenophanes of Kalophon, 23
Greek philosophy
  Christian condemnation of, 54
  relationship to Greek astronomy, 21
Greenhouse effect, on Venus, 319
Greenstein, J., 350
Greenwich, England, geographical importance of, 235
Greenwich Observatory, 129–130, 134, 137, 165, 258
Gresham College, London, 137
Guth, Alan, 349

Hadrian, 49
Hale, George E., 203
Halley, Edmund, 130, 133–134, 314
Halley's comet, 130, 133–134, 149
Hamilton, William Rowan, 202, 218–219
*Harmonice Mundi* (Kepler), 86

Harmonic law, of planetary motion, 24, 82, 86–87
Harmony, cosmic, 24, 25–26, 27
Harmony of the spheres, 15, 24
Harun Al Rashid, 54
Harvard College Observatory, 267
Harvey, William, 113, 114
Heat
as energy, 193
transformation into work, 195–196, 197
Heavy elements, stellar creation of, 303–304, 308
Heavy nuclei, 335–336
Heisenberg, Werner, 154, 221, 332, 333
Heisenberg uncertainty principle, 154, 283, 357
Heliocentric model: *See* Solar system, heliocentric model
Heliopolis, 27
Helios, 16
Helium
absorption lines, 263
atomic structure, 336
nuclear mass, 283
as primordial material, 280
as solar component, 273
as stellar component, 347
stellar absorption spectrum, 263, 265, 266
Helium 6, atomic structure, 336
Helium flash, 299–300
Helium nuclei, solar production of, 287–289
*Henry Draper Catalogue*, 267
Heracleitus of Ephesus, 23
Herakleides, 27, 30–31, 50, 62
Hercules (constellation), 165
Herschel, Alexander, 160
Herschel, Caroline, 160
Herschel, John, 167–168
Herschel, William, 156–157, 159–168, 171, 174, 310, 316
discovery of Uranus by, 162
observational technique, 162
portrait, 161
Herschel, William (*cont.*)
star distance scale of, 167
stellar research of, 162–166, 167, 310, 316
telescope design and construction by, 159–160, 167, 173
Hertz, Heinrich, 191–192, 219, 331
Hertzsprung–Russell (H–R) diagrams, 268–270, 278, 279, 290–298
of galactic clusters, 292–297
Hesperus, 50
Hess, Victor, 338, 349
Hevelius, Johannes, 108, 109
Higgins, William, 184, 262–263
Hindus
astrology of, 9
astronomy of, 10, 54
name for Ursa Major (Big Dipper) in, 17
Hipparchus, 40–49, 136, 169
Copernicus and, 69–70
epicycle concept of, 22, 49
equinoxal precession discovery by, 41–45, 46, 47, 48
influences on, 39–40
observational precision of, 40–41, 45–46, 48
observation of supernova by, 76
star catalogue of, 46–47
stellar magnitude scale of, 257
Holy Roman Empire, 51, 54
Homeric poems, 15–16
Homocentric sphere model, of planetary motion, 28–29
Hooke, Robert, 136–137
Hooke's law of the elasticity of solids, 137, 140
Hour, 4
Hourglass, 4
House of Solomon, 132
Hoyle, Fred, 302–304, 343, 348
H–R diagrams: *See* Hertzsprung–Russell diagrams
Hubble, Edwin P., 323–325, 338
Hubble constant, 327, 328, 345
Hubble law, 327–328, 342, 350

Hubble telescope, 339, 345
Human nature, 353–356
Humason, Milton, 328
Huygens, Christian, 108, 112–113, 138–139, 180
Hyades cluster, 259–260
Hydrogen
  absorption lines, 263
  atomic weight, 208
  electron arrangement, 229, 231
  heavy, 336
  planetary model, 223, 224–225
  as primordial material, 280
  as solar component, 273
  stellar absorption spectrum, 263–265, 266–267
  as stellar component, 347
Hydrogen 21 centimeter radio line, 320–321
Hydrostatic balance, 91, 94
Hypotheses
  Copernicus on, 65
  Newton on, 125, 166

Iben, Icko, 298, 302
Icosahedron, 80–81
Illinois, University of, 298
Indeterminacy principle, 332, 353–354, 357
Industrial Revolution, 194
Inertia, 29, 40, 91, 95, 119–120
Inertial forces, 253, 254
Inertial mass, 253
Infinite series, 115
Ingolstadt, University of, 104
Inquisition, 74, 95, 106
Intelligent life, evolution of, 356–360
Interferometer, Michelson, 242
*Internal Constitution of Stars, The* (Eddington), 276
International Astronomical Union, 203–204
Interstellar medium
  absolute (Kelvin) temperature of, 339
  composition of, 322
Ionian School, 22
Ishak ben Said, 70
Island of St. Helena, 133
Island universes, 164, 310; *see also* Galaxies
Isotopes, 336
Israelites, zodiac figures of, 18

Jansky, Karl, 317–318
Jeans, James, 276–278
Jehuda ben Mose Cohen, 70
Jesuits, 104
Jordan, Max, 332
Joule, 271
Joule, James, 271
Journals, astronomical, 203
Julius Caesar, 33
Juno, 178
Jupiter
  apparent brightness, 172
  asteroids, 147
  in Chinese astrology, 11
  distance from sun, 66
  epicycles, 30–31, 36
  equatorial dark band, 110
  influence on orbits of comets, 178
  moons, 110–111
  rotation period, 110
  satellites, Galileo's observations of, 99, 100, 102, 105

Kant, Immanuel, 138, 152, 310
Keeler, James E.K., 203
Kelvin, Baron, 277, 192, 195, 197–198
Kepler, Johannes, 10, 18, 67, 77–87, 89
  *Astronomea Nova*, 97–98
  Brahe and, 74, 77–78, 81–83
  "Conversation with the Star Messenger," 99–100
  *Cosmic Mystery*, 81, 91
  Galileo and, 89, 91–93, 99–100, 102–103, 104
  *Harmonice Mundi*, 86
  influence on observational astronomy, 107–108

Kepler, Johannes (*cont.*)
  *The New Astronomy Based on Causation or Physics of the Sky Derived from the Investigations of the Motions of Mars Founded on the Observations of the Noble Tycho Brahe*, 86
  planetary motion theory of: *See* Planetary motion, Kepler's laws of
  portrait, 79
Kepler problem, 82–83
Kinetic energy, 145–146, 217, 271, 280, 347–348
Kinetic theory, of gases, 198–200
Kirchhoff, Gustav, 209, 211, 260, 261
Kirchhoff's law of radiation, 211
Konigsberg observatory, 181
Kosmos, 27

Lagrange, Joseph Louis, 143, 145–148
Lagrangian of a system of bodies, 145–146
Landgrave Wilhelm IV of Hesse-Cassel, 73–74
Langley, Samuel P., 260
Laplace, Pierre Simon, 143, 151–155, 202
Lapland, de Maupertuis' expedition to, 149, 151
Lasers, 225, 331–332
Latitude, 8–9, 66, 130, 235, 238
Law of action and reaction: *See* Motion, Newton's laws of, third law
Law of recession, 342, 350
Laws of nature, 355
  indeterminacy of, 357
  invariant, 240
  universal, 240–241
Leaning tower of Pisa, 93
Least action, principle of, 149–150, 151, 218–219
Least time, principle of, 150
Leavitt, Henrietta, 307
Lenses, 138
Leopold, 132
Leucippus, 22
Light
  aberration, 135
  "bending," 256, 337
  diffraction, 139, 180, 181
  Doppler effect, 111–112, 183–184, 315, 326–327
  Einstein red shift, 255–256, 337
  electromagnetic theory of, 185–186, 191–192, 200, 205, 207, 219, 223, 331
  Huygen's wave theory of, 139–140, 180
  Newtonian corpuscular theory of, 138, 139, 180
  refraction, 134–135, 138, 150
  speed, 139, 175, 180
    Einstein's theory of, 209
    Roemer's measurement of, 110–111
    as universal constant, 241
  stellar, effect of sun's gravtitational field on, 337
  wavelengths of, 174–175
Lightning, 186–187
Light years, 167
Lippershay, Johann, 95–96
Lithium, 230
Lodestone, 190
Logarithms, tables of, 83
Longitude, 8–9, 66, 112, 130, 235, 238
Lorentz, H.A., 191, 242
Louis XIV, King of France, 108, 110
Lucius Domitus Aeoreliones, 33
Lucretius, 198, 360
Lunar nodes, regression of, 28
Luther, Martin, 74–75

Maestlin, Michael, 78, 83, 91
Magellanic star clouds, 305, 307, 322–323, 325
*Magnete, De* (Gilbert), 72
Magnetic compass, 57, 190
Magnetic fields, 114, 322
Magnetic forces, 190
Magnetic poles, 133, 190, 191
Magnetism, 85, 189–190
  first experiments in, 113–114

*Magnets, Magnetic Bodies, and the Great Magnet of the Earth* (Gilbert), 114
Mamun, Caliph, 54
Map making, 8
Marconi, Guglielmo, 192
Marco Polo, 55
Mariotte, Edme, 140
Mars
  Brahe's observations of, 107
  brightness, 63
  canals, 179–180
  Cassini's observations of, 110
  distance from sun, 66, 110
  as evening star, 63
  as morning star, 63
  orbit, 36–37, 82–83
  polar ice caps, 110, 180
  rotation period, 110
Mass
  distinguished from weight, 188–189
  effect on space-time curvature, 254–255
  units of, 188
Mass-energy ($E = mc^2$) formula, 246–249, 272
Mass-luminosity law, 278, 292
Mathematics, astronomical applications of
  by ancient Greeks, 15, 27, 35–36, 48
  by Arabs, 54
  in observational astronomy, 141–142
  in post-Newtonian astronomy, 143–157
  in theoretical astronomy, 129, 132
Mathews, T., 350
Matrix mechanics, 333
Matter
  Anaxagoras' theory of, 22
  electrical structure of, 205–208
  four elements theory of, 24–25
Maupertuis, Pierre de, 149–150, 151, 218–219,
Maxwell, James Clerk, 185–186, 191, 200–201, 241–242, 331
Mayans, 12
Mayer, Julius, 192–193
*Mecanique Analytique* (Lagrange), 145
Mechanical materialism, 153–154
Medicean Stars, 98, 99, 102
Medici(s), 132
Medici, Cosimo de, 98–99, 100, 104
Medici, Julian de, 99, 103
Mediterranean civilizations, astronomy development by, 12
Megaliths, astronomical use of, 11–12
Melanchthon, Philipp, 75
Mendeleev, Dmitri Ivanovich, 231
Mendeleev periodic table, 231, 334
Mercury
  distance from sun, 66, 177
  epicycles, 49
  future destruction by sun, 299
  Hevelius' observations of, 109
  motion, 63
  orbit, 31, 66, 337
*Meteorologica* (Aristotle), 29
Meteor showers, 179
Meter, 188
Metric system, 188
Metrodeus of Chios, 23
Metronomes, 91
M51 galaxy, 78
Michelson, Albert, 242
Michelson interferometer, 242
Michelson–Morley observations, on speed of light, 242–243
Middle Ages, astronomy during, 51–56
Milky Way galaxy
  in ancient cosmologies, 16
  center of, 312, 321
    stellar velocities in, 321–322
  diagram, 165
  distribution of stars in, 137–138, 309–310, 311
  Galileo's observations of, 101–102, 309
  globular stellar clusters in, 311, 312
  Herschel's observations of, 165
  Kant's theory of, 310
  location of solar system in, 137–138

Milky Way galaxy (*cont.*)
rotation, 316
spiral arms, 321
as spiral nebula, 313
star counts in, 312–314
Wright's observations of, 137–138, 309, 310
Millay, Edna St. Vincent, 356
Millikan, Robert, 207
Mind, 360
Minimal principle, 217
Minipulsar, 306–307
Minkowski, Hermann, 245–246
Minkowskian equation, 245–246, 254–255
Minutes, 4–5
of arcs, 70–71
"Missing mass" problem: *See* Dark matter
Mizar, 18
Molecular weight, 199
Molecules
formation of, 358–359
organic, 353
Momentum, 217
of a particle, 150
conservation of, 218
Monte Regio, Johannes de, 58–59
Month, lunar, 6, 7, 26
Moon
acceleration, 124
craters, 167
distance from earth, 31, 35, 47–48, 134
effect on tides, 125
Eudoxus' model of, 28–29
full, apparent brightness, 172
Galileo's description of, 101
librations of, 109
mountains, 167
phases, 26, 31–32
sidereal period, 7
size, 35, 47–48, 234
speed of orbit, 134
velocity, 120, 121
Morley, Edward, 242
Morning Star (Venus), 25, 50
Mose Cohen, Jehuda ben, 70
Motion, 40
absolute, 251, 253
Aristotle's description of, 117
differential equations of, 143–144
Newton's laws of, 89, 115, 117–118, 237, 240–241
first, 95, 119–120
implications for nebular hypothesis, 151–152
relationship to principle of least action, 151
second, 120–121, 124, 144–145, 188, 216–217
third, 123
*See also* Planetary motion
Motz, Lloyd, 351–352
Mount Wilson Observatory, 316, 318, 323, 327, 328
Mutations, random, 357–358
*Mysterious Cosmographium* (Kepler), 81
Mysticism, 356, 358
Mythology, Greek, 15, 16–17

Napier, John, 83
*Natural History* (Pliny), 46, 76
Natural selection, 357–358
Nature
as cosmological concept, 1
influence on development of astronomy, 2–3
laws of, 240–241, 355, 357
Nautical almanacs, 131
Navigation, 8–9, 130–131
celestial, 9
measurement units in, 188
Navigational triangle, 131
*n*-body gravitational problem, 142, 146, 147
Nebulae
as birthplace of stars, 312, 322
bright, 322
catalogues of, 316
dark, 322
Galileo's observations of, 102
gaseous, 102, 316

Nebulae (*cont.*)
  Herschel's observations of, 168, 172
  Kant's theory of, 310
  magnitude number of, 171–172
  Rosse's observations of, 168
  spiral, 168, 316
Nebular hypothesis, 151–154, 202
Neon, electron arrangements in, 231
Neutrinos, 284–286, 288–289, 335, 336
Neutrons, 283, 284, 335–336
  decay of, 285–286
  supernova emission of, 304–305
Neutron stars, 305–307
*New Astronomy Based on Causation or Physics of the Sky Derived from the Investigations of the Motions of Mars Founded on the Observations of the Noble Tycho Brahe* (Keper), 86
*New Atlantis, The* (Bacon), 132
Newton, Isaac, 18, 107–108, 114–128, 174, 198
  birth, 107
  contributions to astronomy, 129
  contributions to optics, 125–127
  corpuscular theory of light, 138, 139, 180
  fluxion theory, 118–119, 120–121
  on hypotheses formation, 141, 166
  laws of gravity, 81, 89, 185, 240–241
  laws of motion, 89, 115, 122–124, 125, 151–152, 205, 237, 240–241
    first, 95, 119–120
    Galileo's contributions to, 91
    second, 120–121, 124, 144–145, 188, 216–217
    third, 123
  personal characteristics, 115
  portrait, 116
  *Principia*, 117, 134, 141
  telescope design of, 115, 125–127
  two-body gravitational problem, 142, 144
  universal constant of gravity, 123–124, 209, 220, 241
  velocity theory, 118, 120–121
Newtonian dynamics, 205
  astronomical basis of, 185
  three-dimensional, 244–248
  units of force and mass in, 188–189
Nicetus, 62
Nicholas of Cusa, 58
Nile River, 3, 7–8, 15, 16
Nobel Prize winners, 206, 223, 229, 242
Noon, solar, 4
North celestial pole, 41
North Star, 41, 172
Novara, Domeniro Maria de, 59
*Novum Organum* (Bacon), 113
Nuclear energy, in stellar evolution, 277, 282–290
Nuclear physics, 223–224
Nucleons, 335, 336
Numbers, Pythagorean theory of, 23–24
Numerology, 10, 23–24, 26
Nutation, of earth's axis, 136

Oblers, Heinrich, 340
Oblers' paradox, 340
Observational astronomy
  ancient Greek, 39–56
  application to navigation, 8–9, 130–131
  beginning of, 23
  Bradley's contributions to, 134–136
  Copernicus and, 69–70
  culmination of, 23
  development of, 129–133
  effect of Newtonian laws on, 143
  Flamsteed's contributions to, 132–133
  Herschel's contributions to, 167, 168
  Newton's contributions to, 134, 137, 138, 140, 141–142
  optical and photographic procedures in, 156, 258–260; *see also* Photography, astronomical

Observational astronomy (*cont.*)
precision in, 39, 134, 136
Bessel's contributions to, 181–182
Galileo's contributions to, 94
Hipparchus' contributions to, 40–41, 45, 46, 48
lack of, 69–71
use of space probes in, 339
Observations, frame of reference of, 235, 236, 240
Observatories
Benathky, Bohemia, 76, 82, 108
Bonn, Gemany, 259
Carnegie, 292
clocks used by, 1301–31
development of, 129
Dunsink, Ireland, 202
of French Academy, 109–110
Greenwich, England, 165, 203
Harvard College, 267
Hven, Denmark, 75–76, 107, 108
Konigsberg, Prussia, 181
medieval Arab, 54
Mount Wilson, 316, 318, 323, 327, 328
national, 203
Palomar, 318
Peking, China, 75
post-Galilean, 107
Radcliffe, 206
radio, 318
of Tycho Brahe, 75–76, 107, 108
university-affiliated, 178, 203
of University of Göttingen, 178
Oceans, in ancient cosmologies, 1–2
Octahedron, 80
Odysseus, 16
Oersted, Hans Christian, 189–190, 191
Okeanos, 16
Olbers, Heinrich, 178, 179, 181
Old Testament, 2
Omayyad Caliphs, 54
*On the Dimensions and and Distances of the Sun and Moon* (Aristarchus), 39
*On the Face in the Disk of the Moon* (Plutarch), 32
*On the Heavens* (Aristotle), 29
"On the Nature of Things" (Lucretius), 198
*On Velocities* (Eudoxus), 28
Oort, Jan, 315–316, 326
Optical spectrum, 126
Optics
Newton's contributions to, 125–127
physical, 139
physics of, 185
*Opus Magnus* (Bacon), 113
Orbits
advance of the perihelion of, 256
of asteroids, 177, 178
circular, 47, 66, 84
of comets, 77, 133–134, 178
discrete electronic, 331
elliptical, 39, 83–84, 93, 144, 146
hyperbolic, 146
Keplerian theory of, 80–81, 82, 132
mathematical determination of, 89, 146, 177
Newtonian deduction of, 89
parabolic, 146
Order
relationship to entropy, 196–197
of universe, 353
Orion
as arm of Milky Way, 313–314
Galileo's description of, 102
in Greek mythology, 17
nebula in, 316, 322
Osiander, Andreas, 64–65
Oxford University, Savilian professorship of astronomy, 137
Oxygen, electron arrangements, 231

Padua, University of, 61, 91, 97
Pallas, 178
Palomar Observatory, optical telescope of, 318
Parallax, of stars: *See* Stars, parallaxes
Parmenides of Elea, 23
Parsec, 173
Parsons, William, Earl of Rosse, 168

Particles
average velocity, 200–201
wave characteristics, 332
Pauli, Wolfgang, 229, 284, 285, 332, 333, 335
Pauli exclusion principle, 229–231, 301–302, 305, 334
Pegasus cluster, 325
Peking, China, Jesuit observatory in, 75
Pendula, 91
Pendulum clock, 94
Perception, 234
Pericles, 32
Perseid meteor shower, 179
Perseus, 179, 313
Perseus cluster, 323
Persia, 54
Perturbations theory, 144, 146–147, 148, 149, 151
Phaeton, 16
Philolaus, 25–26, 62
Philosophy
modern, founder of, 74
as obstacle to development of astronomy, 51
*See also* Aristotelianism; Greek philosopher-astronomers
Phosphorus, 50
Photoelectric effect, of photons, 221, 331, 332
Photography, astronomical, 96, 156, 168–173
application to star magnitude measurements, 169–173
application to stellar spectral analysis, 261
image formation process of, 170
time-sequence, 259–260, 314
Photometry, application to star brightness measurement, 171
Photons
absorption and emission by electrons, 227
Einstein's theory of, 331, 332
momentum, 332
photoelectric effect, 221, 331, 332
Photons (*cont.*)
in photographic image formation, 170
Photosphere, 175
Physics, relationship to astronomy, 185–204
application to observational astronomy, 141–142
general theory of relativity and, 251
"new," 205–232
solid state, 140
theoretical, 114
Piazzi, Guiseppe, 176–177
Pickering, E. C., 267
Pisa
leaning tower of, 93
university of, 91
Planck, Max
blackbody radiation curve formula of, 213, 216–217, 225, 346
photon concept of, 170
portrait, 210
quantum theory of, 219–221, 332
Planck's constant of action, 209, 224–225, 241, 331
Planck mass, 220
Planck mass particle, 351–352
Planetary motion
as "apparent motion," 49
Archimedes' theory of, 35
Aristotle's theory of, 29
cosmic harmony theory of, 15, 24, 25–26, 27
early research on, 132
epicycle concept of, 22, 30–31, 36, 47, 49–51
homocentric sphere model of, 28–29
Kepler's laws of, 24, 71, 77, 89, 107, 142, 156
first law, 83–84, 86
Newton's deduction of, 237
second law, 84–86, 305–306
third law, 86–87, 124–125, 256, 321–322
in latitude, 66
in longitude, 66

Planetary motion (*cont.*)
  Newton's theory of, 124–125, 237
  Plato's theory of, 27
  retrograde, 30–31
Planetary tables
  Alfonsine, 70, 71
  Kepler's, 73
  Ptolemic, 71
Planets
  astrologers' observations of, 69–70
  distance between, 27
  distance from sun, Bode–Titius law of, 177
  future destruction by sun, 299
  number of, Kepler's theory of, 80–81
  *See also* names of individual planets
Plasma, stellar, 276
Plato
  Aristotle as student of, 29
  Athenian Academy of, 27, 32, 52
  Eudoxus and, 27–28
  influence of Pythagoreans on, 26–27
Pleiades, 11, 164
  color-luminosity diagram of, 297
  Galileo's description of, 102
Pliny, 52, 82
Plutarch, 25, 27–28, 39, 62
  *On the Face in the Disk of the Moon*, 32
Pogson, Norman, 257
Pogson magnitude scale, 46–47, 170, 171, 172, 173
Poisson, Simeon Denis, 143
Polaris (North Star), magnitude, 41, 172
Poles, 41, 136
Polygon, 24–25
Pompey the Great, 33
Pons, Jean Louis, 178
Pope Paul III, 62, 63, 64
Pope Sylvester III, 52
Poseidonius, 48
Positrons, 2482–49
Potential energy, 280–281
Praesepe nebula, 102
Precession, of the equinoxes, 41–45, 46, 47, 48, 59, 136
Prejudices, 113
Priests, as early astronomers, 7–8
*Principia* (Newton), 117, 134, 141
Principle of least action, 149–150, 151, 218–219
Principle of least time, 150
Printing press, 57
Prism, 126
  objective, 181, 267–268
Probability theory, 151
Procyon, spectra, 267
Projectiles
  laws of, 91
  parabolas of, 94
"Propagation of Electromagnetic Phenomena Through Rapidly Moving Media" (Einstein), 233
Proton–proton chain reaction, 288–289, 298–299
Protons
  in atomic models, 223, 225
  discovery, 331
  formation, 348, 352
  law of motion of, 217
  mass, 207
  relationship to electrons, 206–207
  relationship to neutrons, 335–336
  as stellar wind component, 338
Ptolemies, 33
Ptolemy, 6, 40, 49–51
  Copernicus and, 69–70
  epicycle concept of, 22, 31, 36–37, 47, 49–51
  position of Arcturus and Sirius determined by, 314
  stellar catalogue of, 49
  stellar magnitude scale of, 46–47, 169, 171, 172, 257
Ptolemy I Soter, 33
Ptolemy II Philadelphus, 33
Ptolemy III, 7
Pulsars, 306–307
Purcell, Edward Mills, 320–321

Pythagoras, 23–26, 50
  harmony of the spheres theory, 15, 24
  theorem of, 237–238, 239
Pythagoreans, 23–26

Quanta, 219–220, 221
Quantum mechanics, 332–335
  application to stellar radiation research, 336–337
  definition, 332
  matrix mechanics of, 333
  wave mechanics of, 333–334
Quantum numbers, 225, 228–229, 334
  magnetic, 227–228
  spin, 228
Quantum theory, 224–225, 233, 331
  of blackbody radiation, 219–221
  indeterminacy principle of, 332, 353–354, 357
  relationship to astrophysics, 257
Quasars, 350

Ra (Egyptian god), 15
Radar, astronomical applications of, 319–320
Radiation
  blackbody, 209, 211–216, 219–220, 331
    radiation curve formula, 213, 216–217, 225, 346
  cosmic background, 212, 346–347, 348–349
  solar, 167, 215–216, 273, 275–276
  spectral analysis of, 209, 211–216, 219–221
  stellar, 276, 277
    quantum mechanics of, 336–337
  stimulated emission of, 331–332
Radioactivity, theory of, 331
Radio telescopes, 318–319
Radiowaves, quasar-produced, 350
Rainbows, 174
Ramus, Petrus, 72–73
Reber, Grote, 318
Recorde, Robert, 72
Record-keeping, 6–8
Red giants, 290–291, 300
Red shift, 255–256, 337
Reines, 335
Reinhold, Erasmus, 71
Relativism, 233–234
Religion
  as obstacle to development of astronomy, 51–52, 105–106
  as precursor of astronomy, 1–2
  *See also* Catholic Church; Christianity
Renaissance, 57–58
Reuerbach, George, 58–59
*Revolutionibus, De* (Copernicus), 64–67, 71–72, 75
Rheticus, Georg Joachim, 64–65
Rhodes, Greece, 32–33, 40
Rigel
  color, 266, 267
  evolution, 292
  luminosity, 269–270, 289, 290, 292
  mass, 292
Right triangle, square of the hypotenuse of, 25
Rigid bodies, motions of, 145, 148
Roche, E., 201–202
Roche limit, 201–202
Roemer, Olaus, 108, 110–112
Roman Empire, destruction of, 51
Romans, knowledge of magnetism, 190
Rome, University of, 61
Rothmann, 73–74
Rowland, Henry Augustus, 191, 260
Royal Astronomical Society, 130, 203
Royal Greenwich Observatory, 165, 203
Royal Society, England, 132
Royal Society of Science, Uppsala, 175
Rudolph II, Holy Roman Emperor, 76, 108
Russell, Henry Norris, 290–291
Rutherford, Ernest, 223–224, 284–285, 331

Sagittarius, 312, 313
Said, Ishak ben, 70
Salpeter, E., 299–300
Sandage, Alan, 292–295, 350
Satellites, artificial, 339
Saturn
  distance from sun, 66
  epicycles, 30–31, 36
  Galileo's observation of, 103, 105
  influence on orbits of comets, 178
  rings, 103, 138–139, 201–202
    Cassini divisions of, 110
  satellite of, 138–139
Scalar, 246–247
Scandinavia, name for Big Dipper in, 17
Scheiner, Christopher, 104
Schiaparelli, Giovanni, 179–180
Schmidt, M., 350
Scholasticism, 53, 57–58
Schrödinger, Erwin, 333
Schrödinger wave equation, 333–334
Schwarzschild, Karl, 255, 256, 297–298
Schwarzschild, Martin, 297–298
Science
  Lucretius' poem about, 360
  unified theory of, 153–154
Scorpius, 313
Scriptures, as obstacle to development of astronomy, 74–75, 76, 77
Secchi, Angelo, 263
Second, 4–5, 188
Self-organization, of the universe, 359–360
Sextant, 8–9
Shakespeare, William, 358
Shapiro, I.I., 310–320
Shapley, Harlow, 311, 312, 338
Sidereal day, 5
Sidereal time, 5, 130–131
Sidereal year, 6–7, 44, 45–46
Sirius, 3, 7–8
  apparent brightness, 172
  color, 265
  distance, 183
Sirius (*cont.*)
  luminosity, 269–270
  parallax, 183
  positional change, 133, 184, 314
  spectra, 267
  twin paradox applied to, 250–251
Sirius A, 300
Sirius B, 300
Sitter, Wilhelm de, 341–342
Sky, religious significance of, 1–2
Slipher, Vesto M., 326
Societies, astronomical, 130, 203–204
Socrates, 57
Sodium, electron arrangements, 231
Sodium spectral lines, 176, 222, 228
Solar day, 4, 5
Solar energy, 336
Solar hour, 4
Solar radiation, infrared, 167
Solar spectrum
  definition, 174
  Fraunhofer lines of, 174–176
  wavelengths of, 174–175
Solar system
  Copernican model, 32, 61–67, 69, 73
    Brahe's rejection of, 73–74
    Catholic Church's acceptance of, 64
    Galileo's acceptance of, 91–92, 93, 95, 100, 105–106
    Kepler's acceptance of 78
    Scriptural conflict with, 73–74
  distribution of mass in, 153
  Galileo's model, 339–340
  geocentric model, 5–6
  heliocentric model, 31, 32, 58, 62; *see also* Solar system, Copernican model
  Kepler's model, 339–340
  location in Milky Way, 312, 1371–38
  nebular hypothesis of, 151–154, 202
  Philolaus' theory of, 25–26
  planetary distances in, 27
  Plato's model, 27

Solar system (*cont.*)
Ptolemaic model, 69
Bacon's knowledge of, 53
Christianity's acceptance of, 54
Copernicus' acceptance of, 59
Copernicus' critique of, 62, 63
as obstacle to development of astronomy, 55–56
Pythagorean model, 25–26, 27
stability of, 147–148, 151
Tychonic model, 31, 74, 77, 78, 81
Solar time, 4–5, 130–131
Solar wind, 179, 338–339
Solstices, 42
prediction of, 12
Space
absolute, Einstein's critique of, 243, 244–245
expansion of, 343
Space probes, 339
Space quantizations, 228
Space station, Soviet, 339
Space-time curvature, 254–255
Space-time manifold, 244–246, 341
Spectral analysis
atomic, 225, 227–228
of galactic recession, 326–328
of radiation, 209, 211–216, 219–221, 232
stellar, 176, 232, 261–270
photosphere surface temperature in, 263–267
by star type, 263, 266–267, 269
Spectral theory, 221–223
Spectroscope, 125–126, 174, 176, 180, 184, 262
Speed, Newton's equation of, 118–119
Spheres, Pythagorean theory of, 24–25
Spherical aberration, 109
Spinoza, Benedict, 76
Sputnik, 339
Stadia, as measurement unit, 34–35
Standard star fields, 165–166
Star catalogues, 46–47, 49, 259
Star counts, 165–166, 312–314
Star gauging, 310
Stargazers, of ancient cultures, 2, 5
Star magnitude, definition, 171
Star magnitude scales, 169
of Hipparchus, 46, 169, 171–172
photographic, 169–173
of Pogson, 46–47, 170, 171, 172, 173
of Ptolemy, 46–47, 169, 171, 172, 257
*Starry Messenger, The* (Galileo), 97–103
Stars
age, 283
biblical description of, 2
binary (double), 18, 162–164
brightness
absolute, 171, 173
apparent, 170–171, 172, 173
effect of radial motion on, 315
photographic measurement of, 169
*See also* Star magnitude scales
of "celestial neighborhood," 137
classification
by brightness, 257
by luminosity, 257–258
spectral, 258; *see also* Spectral analysis, stellar
by stellar parallaxes, 257
color
effect of interstellar dust on, 311–312
index of, 172–173
distance scale, 167
distribution in space, 309–311, 340; *see also* Stellar clusters
diurnal rising, 131
energy generation by, 272, 277
with carbon cycle, 289–290, 303
with helium cycle, 286–289
with proton–proton chain reaction, 288–289, 298–299
evolution of: *see* Evolution, stellar
"fixed," 133
formation of, 347–348
galactic (local) clusters
age, 294–295, 297
H–R diagram of, 292–293
Galileo's observations of, 102
internal temperature, 198

Stars (*cont.*)
  luminosity
    classification of, 257–258
    definition, 170
    H–R diagram of, 290
    photographic measurement, 169
    radiation and, 195
    rarity, 314
    relationship to absolute brightness, 171
    relationship to absolute magnitude, 173
    Stefan–Boltzmann law of, 201
  mass luminosity law of, 278
  motion of: *See* Stellar motion
  names of, 17–18
  nuclear reactions in, 272
  parallaxes, 73
    of binary stars, 162–163
    definition, 163
    Kepler's theory of, 92–93
    measurement of, 92, 169, 182, 257
  population I, 292
  population II, 30, 303
  positions
    angles of, 70
    effect of light refraction on, 134–135
    navigational applications of, 130, 131
    *See also* Observational astronomy
  radiation emission, 194–195
  radii, 256
  spectral analysis of: *See* Spectral analysis, stellar
  stability of, 283
  structural models of: *See* Astrophysics
  surface temperature, 201, 215–216, 263–267
  "twinkling" of, 29
Star 61 Cygni, 135, 182–183
Statistical mechanics, 200, 201
Stefan–Boltzmann law, 201, 215–216
Stellar clusters, 259–260
  globular, 164, 311, 312
  Herschel's study of, 164
Stellar motion, 260
  Bessel's observations of, 182–183
  catalogue of, 314
  of local/galactic clusters, 259–260
  photographic measurement, 169, 314
  proper, 314
  radial, 183–184, 315
  relationship to solar motion, 315–316
  relationship to solar system motion, 183–184
  transverse, 314, 315
*Stellar Nova, De* (Brahe), 76
Stellar population, 328–329
Stellar proper motions, 165
Stellar wind, 338–339
Stonehenge, 11–12
Struve, F.G.W., 183
Successive approximation theory, 144, 146–147
Sumerians, 11
Summer solstice, 42
Sun
  age, 280, 286, 298
  altitude, 8–9
  apparent brightness, 172
  apparent motion, 5–6
  color, 265
  Copernicus' panegyric to, 62
  density, 270
  distance from earth, 31, 35, 47–48, 110–111
  distance of planets from, 177
  diurnal rising, 131
  eclipses, 234
  as energy source, 272–273, 277, 286–290
  evolution, 280–282, 298–301
  expansion, 299
  gravitational force, 124–125
  heat of, 29–30
  helium/hydrogen core ratio, 298–300
  internal temperature, 275–276, 278
  luminosity, 269–270, 273, 278, 287–289, 299

Sun (*cont.*)
as main sequence star, 298
mass, 270, 301
noontide altitude, 43
radiation emmitted by, 167, 215–216, 273, 275–276
radio signals from, 318
radius, 270
size, 35, 47–48
relationship to size of moon, 32
speed of escape from, 270–271
surface gravity, 270
surface temperature, 270
theoretical models, 273–276, 277–278
tidal action, 178–179
as white dwarf, 300–301
Sundial, 4, 22
Sunspots, 95, 109, 167, 228
Galileo's observations of, 103, 104–105
Supernovas
collapse, 303–304, 305, 306
ejecta of, as basic material of solar system, 280, 281–282
energy generated by, 338
neutrino emissions of, 304–305
Tycho nova, 76
white dwarfs as, 304, 305
Survival responses, 354–355
Swedenborg, Emanuel, 152
Sytria, 78, 82

Taurus, 17
Hyades cluster of, 259–260
Telegraphy, 192
Telescopes, 57
achromat of, 173
Airy disk for, 139–140, 258–259
in astrophysics, 260
Bessel's design, 182
development of, 129, 130
directional positioning of, 135
Dorpot's design, 173–174
equatorial mount of, 112
eye pieces of, 138–139
Galileo's design, 89, 95–97, 98, 99, 101, 102, 103, 107
Telescopes (*cont.*)
Herschel's design, 159–160, 167, 173
Hubble, 339, 345
image formation by, 139–140
of Konigsberg, 173–174
lens, 108–109
chromatic aberration of, 109
spherical aberration of, 109
meridian circle, 112
Newton's design, 115, 125–127
objective prism of, 181, 267–268
optical, comparison with radio telescopes, 318–319
optical definition of, 138–139
optical principle of, 96–97
Parson's design, 168
radio, 317–319
relationship to cosmology, 323
reflecting, 125–127
as "twice built" instruments, 182
von Fraunhofer's design, 173–174
Tetrahedron, 80
Thales of Miletus, 21–22, 23, 35, 40, 186
Theodosius I, 33
Theology, relationship to astronomy, 1–2
*See also* Catholic Church; Christianity; Religion
Theoretical astronomy
Apollonius as father of, 37
beginning of, 124
Newton's contributions to, 129
*Theorie Analytique des Probabilites* (Laplace), 151
Theory of relativity, 233–256, 331
distinguished from relativism, 233–234
general theory, 233, 238–239, 251, 253–256, 331
application to cosmology, 340–342
astronomical confirmation of, 136, 337–338
radar test of, 319–320
relationship to law of falling bodies, 93

Theory of relativity (*cont.*)
invariance principle of, 239–240, 251, 253–254
mass-energy formula ($E = mc^2$), 246–249, 272
relationship to astrophysics, 257
special theory, 233, 240, 243–249, 331
comparison with Newtonian three-dimensional dynamics, 244–248
transformation of coordinate concept of, 245
twin paradox of, 250–251
*Theory of the Shape of the Earth* (Clairaut), 148–149
Thermodynamics, 192–200
first law of, 193–195
second law of, 195–198, 282
Thermometer, 94
Thirty Years War, 87
Thomson, Joseph John, 206–207, 331
Thomson, William (Baron Kelvin), 192, 277
Three-body gravitational problem, 144, 147, 148, 149
restricted, 147
Tidal action, gravitational, 201–202
Tides, 3, 125
"Tilt of the earth's axis," 41, 43
Time
absolute, Einstein's critique of, 243–244
solar, 4–5, 130–131
Timekeeping, 4–8
Time-telling devices, development of, 4
Titius, Daniel, 177
Totemism, 18
Townes, Charles, 225
*Traite de Mécanique Veleste* (Laplace), 151
*Treatise on Dynamics* (d'Alembert), 148
Trigonometry
Hipparchus' contributions to, 48
Newton's contributions to, 115
Triple helium reaction, 299–300
Truth, search for, 355–356
Twin paradox, 250–251
Two-body gravitational problem, 142, 144
Tycho nova, 76

Uhlenbeck, George, 228
Uncertainty principle, 154, 221, 332–333
Unified theory, of science, 153–154
U.S. Naval Observatory, 203
Unitons, 351–352
Units, astronomical, 187–189
Universal constant of gravity, 123–124, 209, 220, 241
Universal laws of nature, 240–241
Universe
age, 328, 345–346
collapse, 344, 352
concentric cylinder model, 22
dark matter of, 322, 325–326, 345, 352
density, 345
de Sitter's model, 341–342, 349
Einstein's model, 340–341, 349
expanding, 327, 342–346, 349
finiteness, 340–341
initial state, 343, 344
intelligent life in, 356–359
large-scale structure, 347, 351
origin of
ancient Greek theories of, 22
big bang theory, 327, 342–343, 344, 345, 346–347, 348, 349, 352, 356
radius, 341, 345
self-organization, 359–360
static, 341
temperature, 339, 346–347
total mass, 345
visible matter, 345
Universities
astronomy departments of, 203
Bologna, 59
Cambridge, 115, 306
Cracow, 59
Göttingen, 178

Universities (*cont.*)
  Illinois, 298
  Ingolstadt, 104
  Oxford, 137
  Padua, 61, 91, 97
  Pisa, 91
  Rome, 61
Urania observatory, Hven, Denmark, 75–76, 107, 108
Uranus
  discovery, 156–157
  distance from sun, 177
  as Georgium Sidus, 162
  influence on orbits of comets, 178
Ursa Major (Big Dipper), 9, 17, 41, 260

Van de Hulst, H.C., 320
Vega, 183
Vela satellites, 351
Velocity
  constant, 239–240
  Maxwell distribution of, 201
  Newton's concept of, 118, 120–121
  of observers, 240, 241, 243–244, 249–251
  stellar, 321–322
Venus
  apparent brightness, 172
  in Babylonian astrology, 11, 11
  distance from sun, 66, 177
  epicycles, 49, 50
  as evening star, 25, 50
  Galileo's observation of, 103–104, 105
  greenhouse effect on, 319
  as morning star, 25, 50
  motion, 63
  orbits, 31, 66
  phases of, 103–104
  surface temperature, 319
Vernal equinox, 42, 44–45, 46
Vesta, 178
Virgo cluster, 325, 326
Vogel, Herman, 184
Volta, Alessandra, 189
Voltaic pile, 189
Vulpecula, 306

Wave-corpuscle dualism, 219–220, 221, 332
Watch, invention of, 4
Water, as basic element, 24–25
Water clock, 4, 45
Watt, 271
Watzelrode, Lucas, 59
Wave mechanics, 333–334
Wave theory, of light, 180
Weight, distinguished from mass, 188–189
Wëiszacker, C.F., 289
Welser, Mark, 104–105
Whirl Pool Nebula, 168
White dwarfs, 255–256, 270, 279, 300–302
Wien, Wilhelm, 215, 216
Wien displacement law, 215–216
Wilser, Mark, 95
Winter solstice, 42
Wren, Christopher, 137
Wright, Thomas, 137–138, 309, 310

Xenophanes of Kalophon, 23
X-ray tubes, 206

Year
  length, 6–7, 10, 28
  lunar, 7
  sidereal, 6–7, 44, 45–46
  solar, 28
  tropical, 7, 44–46
Young, Thomas, 180

Zeeman effect, 228
Zenodotus of Ephesus, 33
Zodiac, signs of, 5, 10, 18
Zone of avoidance, 316–317
Zurcky, Fritz, 325–326

EAST BATON ROUGE PARISH LIBR
DISCARD
02235 5347